Spiders of Panama

Biogeography, investigation, phenology, check list, key and bibliography of a tropical spider fauna.

by

Wolfgang Nentwig

in collaboration with
Bruce Cutler and Stefan Heimer

Flora & Fauna Handbook No. 12

Flora & Fauna Books

This series of handbooks provides for the publication of book length working tools useful to systematists for the identification of specimens, as a source of ecological and life history information, and for information about the classification of plant and animal taxa. Each book is sequentially numbered, starting with Handbook No. 1, as a continuing series. The books are available on a standing order basis, or singly.

Each book treats a single biological group of organisms (*e.g.*, family, subfamily, single genus, etc.) or the ecology of certain organisms or certain regions. Catalogs and checklists of groups not covered in other series are included in this series.

The books are complete by themselves, not a continuation or supplement to an existing work, or requiring another work in order to use this one. The books are comprehensive, and therefore, of general interest.

The books in this series to date are:

Handbook No. 1, 1985. **THE SEDGE MOTHS,** by John B. Heppner

Handbook No. 2, 1992, 2nd printing, slightly revised. **INSECTS AND PLANTS: Parallel Evolution and Adaptation,** by Pierre Jolivet

Handbook No. 3, 1988. **THE POTATO BEETLES,** by Richard L. Jacques, Jr.

Handbook No. 4, 1988. **THE PREDACEOUS MIDGES OF THE WORLD,** by Willis W. Wirth and William L. Grogan, Jr.

Handbook No. 5, 1990. **A CATALOG OF THE NEOTROPICAL COLLEMBOLA,** by José A. Mari Mutt and Peter F. Bellinger

Handbook No. 6, 1990. **THE ENDANGERED ANIMALS OF THAILAND,** by Stephen R. Humphries and James R. Bain

Handbook No. 7, 1990. **A REVIEW OF THE GENERA OF NEW WORLD MYMARIDAE,** by Carl M. Yoshimoto

Handbook No. 8, 1990. **MAYFLIES OF THE WORLD: A Catalog of the Family and Genus Group Taxa,** by Michael D. Hubbard

Handbook No. 9, 1993. **CATALOG OF THE SOFT SCALE INSECTS OF THE WORLD,** by Yair Ben-Dov

Handbook No. 10, 1993. **NORTH AMERICAN PSOCOPTERA,** by Edward L. Mockford

Handbook No. 11, 2nd Edition, 1993. **INSECT AND SPIDER COLLECTIONS OF THE WORLD,** by Ross H. Arnett, Jr. and G. Allan Samuelson

Handbook No. 12, 1993. **SPIDERS OF PANAMA** by W. Nentwig

Flora & Fauna Handbook No. 12

Spiders of Panama

Biogeography, investigation, phenology, check list, key and bibliography of a tropical spider fauna.

by

Wolfgang Nentwig

in collaboration with
Bruce Cutler and Stefan Heimer

Sandhill Crane Press, Inc.
Gainesville, Florida
Leiden, The Netherlands
1993

Editor: Ross H. Arnett, Jr., Ph.D.
Production: Michael C. Thomas, Ph.D.

ISBN 1877743-18-6

Library of Congress Cataloging-in-Publication Data:

Nentwig, Wolfgang, 1953-
Spiders of Panama : biography, investigation, phenology, check list, key, and bibliography of a tropical spider fauna / by Wolfgang Nentwig, in collaboration with Bruce Cutler and Stefan Heimer.
p. cm. -- (Flora & Fauna handbook ; no. 12)
Includes bibliographical references (p.).
ISBN 1-877743-18-6 : $35.00
1. Spiders--Panama. 2. Spiders--Panama--Classification.
I. Cutler, Bruce, 1943- . II. Heimer, Stefan. III. Title.
IV. Series.
QL453.2.P2N46 1993
595.4'4'097287--dc20 93-2016
CIP

COVER: *Thaumasia* sp., an ambushing pisaurid spider on shrups and trees in Panama, covers a leaf with silken threads so that the spider sits on a silk carpet and is surrounded by a low silk wall.

Manufactured in the United States of America

Sandhill Crane Press, Inc.
MAIL: P.O. BOX 147050, GAINESVILLE, FLORIDA, U.S.A.
OR WARMONDERWEG 80, 2341KZ OESTGEEST, THE NETHERLANDS

Table of Contents

Chapter 1. Introduction

Spiders are certainly not the most important, largest or best-known arthropod group of the tropics and Panama is certainly not the most typical part of the tropics. When I nevertheless decided to compile the available information on the spiders of Panama I did this for many good reasons. The central part of Panama (the Isthmus or the former Canal Zone) has very easily accessible areas of tropical rain forest. A sufficient conservation of this forest, an excellent infrastructure for research (provided by the Smithsonian Tropical Research Institute) and a relative political stability of the country are further important points which today seem to be more an exception than the rule for a tropical or a third world country. These advantages have attracted many biologists in the last 60 - 70 years, among them a small but continuous succession of arachnologists.

The first investigators of the Panamanian spider fauna were primarily collectors, producing fauna lists and taxonomic papers. Although nowadays these fields of research are often considered to be of minor value, it is only due to this circumstance that we know today very much about the taxonomy of the Panamanian spider fauna, i.e. that we are able to identify (most of) these spiders. I even dare say that the Panamanian spider fauna today is the best-known of any tropical country (though there is still a tremendous amount of work to be done). This is certainly one of the reasons why in the last 2 - 3 decades the main focus of arachnological research in Panama has changed and is now dominated by ecological and behavioural questions.

Faunal comparisons, species lists or even a key to the spider fauna of a given area or country are, on the other hand, fraught with snares. Besides containing mistakes and errors, they suggest completeness where only a snapshot exists. My tentative calculations indicate that we know today approx. 60 % of the Panamanian spider fauna (but only 30 % of the South American spider fauna and certainly even less of the African spider fauna!). Every record list or key will therefore be only very preliminary (but this will be the case in 10, 20, or 50 years as well!). So it is not astonishing that a leading spider taxonomist commented on my project on Panamanian spiders with an ex cathedra statement "You cannot identify most of these spiders". He probably thought that other zoologists should, if you please, wait until the taxonomists have completed their pioneer work and that completeness in 50 years time is better than a 60 % result today. But this is a matter of opinion.

This helpful comment highlights two important points: (1) taxonomy was never the fastest of sciences; (2) there is not unlimited time to wait. The tropical countries are today in a very precarious situation: The explosive increase in their populations deteriorates all the circumstances of their life: food, space, health, human dignity and hope (chances) for the future. In their desperate attempts to master their living conditions people exploit their environment in an unreasonable and short-sighted way. It is the tragedy of this situation that the increasing population pressure on natural reserves very probably does not help to ameliorate living conditions but certainly ruins irretrievable nature. It is very difficult to convince these people that their future lies not in the tropical rain forest but only in a very restrictive population control. Unless the population of our already overcrowded planet decreases drastically the future of the tropical rain forest is gloomy. It is of great significance in this situation that the future of the tropical rain forest is not an isolated event. The temperate zone is linked in many ways with the tropical zone and the industrialized world cannot afford to be disinterested in the welfare of the remaining world.

My aim is to compile the available information on one facet of the beauties of the tropical rain forest, the spider fauna of Panama. Despite some shortcomings I hope to show at least a few aspects of this eight-eyed and eight-legged world. I hope as well to encourage especially the young to have less children and to exhibit more personal engagement in research in the tropics and in more effective nature conservation work.

This book would not have been possible without the constant help and kindness of Stefan Heimer who identified most of the spiders, read various drafts of this book and provided useful comments and additions. All salticids have been identified by Bruce Cutler. Further identifications, help, criticisms, comments, additions of records or gifts of specimens were provided by B. Baehr, A. R. Brady, J.E. Carico, J. A. Coddington, F. A. Coyle, A. Decae, C. Deeleman-Reinhold, C. Dondale, W. G. Eberhard, G. B. Edwards, H. Höfer, R. G. Holmberg, B. Jantschke, E. Jutzi, S. Keller, B. Kellerer, Y. D. Lubin, C. Manhart, H.-G. Müller, W.B. Peck, N. I. Platnick, D. Quintero, B. and V. Roth, G. Schmidt, R. Schmidt-Effing, P. Sierwald, D. Ubick and C. Warcup. The team of the Smithsonian Tropical Research Institute provided an infrastructure and logistical help. The Deutsche Forschungsgemeinschaft supported my work for many years and the Stiftung zur Förderung der wissenschaftlichen Forschung an der Universität Bern enabled a field trip in 1990. My thanks to them all. All errors, insufficiencies and over-daring simplifications in the following work are of course my own responsability. Most of the spiders I collected is actually in the Museum of Natural History Genova, the salticids are in the collection Cutler, the thomisids in the collection Ori. (I intend to prepare a list of additional records and corrections. Each contribution is highly welcomed. This list will be distributed on request.)

Chapter 2. The biophysical geography of Panama

2.1 Geography

Relief: The Republic of Panama is the most southerly Central-American state, it lies between 7° and 9° 30' North of the Equator and it links North and South America at the narrowest part of Central America (Figure 1).

The Canal Zone (see comment on page 90), a strip of land circa 15 km wide on both sides of the Panama Canal, divides the country into two parts. A volcanic mountain chain reaching from Costa Rica into Panama forms the western part of the country. The highest elevation is El Baru (or El Volcan) in Chiriqui (3427 m, Nr. 1 in Figure 2), another high peak nearby is Cerro Santiago (2852 m, Nr. 2 in Figure 2). The average peak height of the mountain chain varies from 1300 to 1900 m.

The next prominent part of western Panama is the Azuero Peninsula which is the southernmost point of the Panamanian mainland. Its highest elevation is Cerro Cambutal (1946 m, Nr. 3 in Figure 2). The Isthmus of Panama is a land bridge, today divided by the Panama Canal, where the Atlantic and the Pacific are separated by only 50 km and where the continental divide has an elevation of only 61 m.

The eastern part of the country is formed by the Serrania de San Blas, a low mountain chain on the Atlantic coast (high altitudes (elevations) are to be found in the West Cerro Bruja and Cerro Azul, about 1000 m high, Nr. 5 and 6 in Figure 2, and in the East Cerro Tacarcuna near the Colombian border, 1875 m, Nr. 10 in Figure 2). South of the Serrania de San Blas, several highland complexes extend from the Western Serrania de Majé (highest elevation Cerro Majé, 1440 m, Nr. 7 in Figure 2) to Eastern Serranias (with Cerro

Jaqué, 1585 m, Nr. 8, and Cerro Pirre, 1615 m, Nr. 9 in Figure 2) which continue east to the Colombian Sierra system. (Note: Peak heights differ within the literature sometimes considerably.)

Several islands and archipelagos belong to Panama: La Isla de Coiba between the Golfo de Chiriqui and the Azuero Peninsula, the Archipielago de las Perlas (main island Isla des Rey) in the Golfo de Panama and many small islands close to the coast in the Golfo de Mosquitos and on the San Blas coast (Figure 3).

Hydrography: About 300 rivers and streams flow into the Pacific, about 150 into the Atlantic. Most of them, however, are shorter than 60 - 70 km. The most important river to the Atlantic

Figure 1. The position of Panama between Central and South America.

is the Rio Chagres, the largest rivers to the Pacific are the Rio Bayano and the Rio Chucunaque / Rio Tuira which drain the eastern part of Panama. Apart from a few very small lakes at higher elevations in Chiriqui there are no natural lakes in Panama. During the canal construction, the Rio Chagres was dammed to form the Lago Gatun. Later several rivers in Chiriqui, on the Isthmus (Rio Chagres at Madden Lake) and in the eastern part (Rio Bayano) were dammed to generate electricity.

Size and administrative units: The longest east-west extent of Panama measures about 620 km, the width of the country varies between 180 km (Azuero peninsula) and about 50 km (at the Isthmus). The country covers an area of approx. 77,000 km² (this is approx. the size of Bavaria (Germany), South Carolina (USA) or Sri Lanka and double the size of the Netherlands or Switzerland). Panama is divided into 9 provinces (from West to East): Chiriqui, Bocas del Toro, Veraguas, Herrera, Los Santos, Coclé, Colon, Panama and Darien and the autonomous Comarca de San Blas (Figure 3).

Population: Panama is inhabited by approximately 2 million people (1960 1.08 million; 1970 1.43 million; 1980 1.83 million;

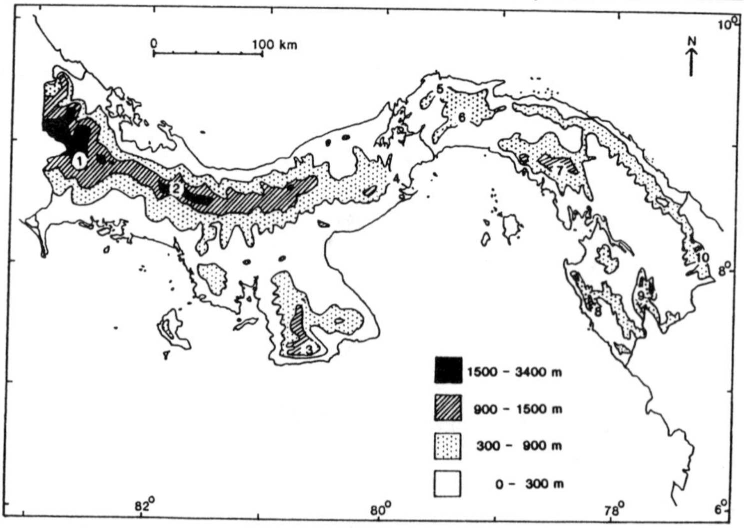

Figure 2. Relief map of Panama with its highest elevations. 1 El Baru / El Volcan 3427 m; 2 Cerro Santiago 2852 m; 3 Cerro Cambutal 1860 m; 4 Cerro Campana 1010 m; 5 Cerro Bruja 990 m; 6 Cerro Azul 1006 m; 7 Cerro Majé 1440 m; 8 Cerro Jaqué 1585 m; 9 Cerro Pirre 1615 m; 10 Cerro Tacarcuna 1875 m (modified according to Bennett 1968).

1985 2.18 million; 1987 2.27 million) with an annual increase of 2.5 % (period from 1970 - 1980). Approx. 400,000 people lived in the capital Panama City (22 %) in 1980, 49 % of the total population lived in cities, 51 % in rural areas. In 1986 the agglomeration of Panama City was estimated to house circa 650,000 people. The population is extremely unevenly distributed over the different parts of the country. The central part of the province of Panama is densely populated (> 100 inhabitants / km²); adjacent areas and the provinces of Chiriqui and Herrera are inhabited by 30 - 40 people / km². The province of Bocas del Toro and Darien and the Comarca de San Blas have the lowest population densities (Figure 4). These population data base on Bennett (1968) and Estadistica Panamena (1980). Further data and more recent and general information on Panama (including economy and agriculture) are available in Statistik des Auslandes (1987).

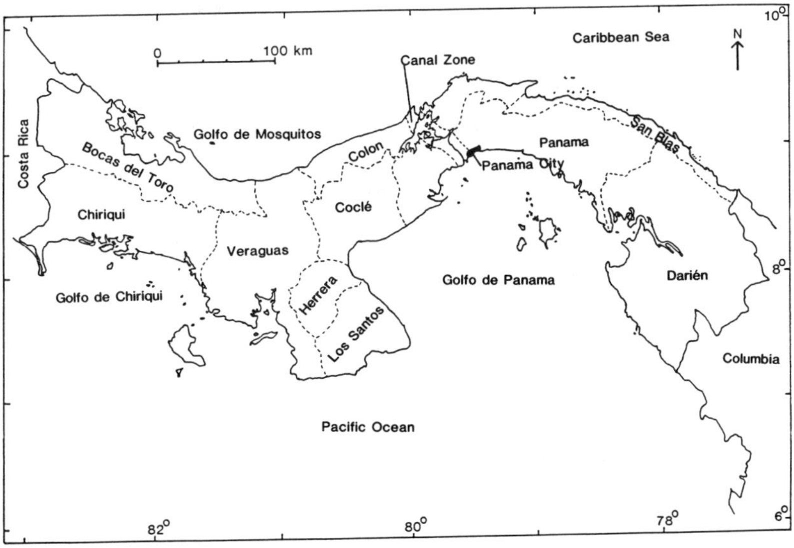

Figure 3. The Republic of Panama, its administrative division into 9 provinces and the Comarca de San Blas and the former Canal Zone which is part of the provinces of Colon and Panama.

2.2 Climate

Annual patterns: The lowlands of Panama have a warm tropical climate; the conditions in the highlands are cooler. This corresponds to the macroclimates A and C of Köppen's system (Figure 5). The climatic heterogeneity is dictated by seasonal changes: the wet season occurs between April and December and the dry season during the remaining period. The intensity of the seasonality varies and is less pronounced in the moist parts of Bocas del Toro (Atlantic-Caribbean influence) and some highlands. This corresponds to the climate f in the Köppen system, while the more seasonal parts are characterized by the climates m and w (Figure 5).

For most parts of the country the annual precipitation is about 2000 - 4000 mm (Figure 6). It increases at the Isthmus from the Pacific coast (1800 mm) to the Atlantic coast (4000 mm), the variation over several years, however, may be ± 30 % (or even more)

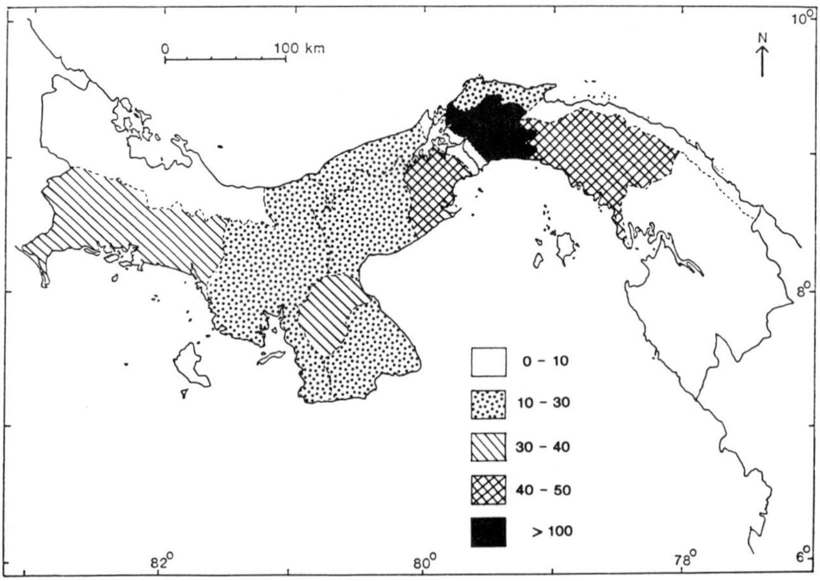

Figure 4. The population density in Panama (inhabitants / km²). No data are given for the former Canal Zone; the province of Panama is divided into 3 parts to show the high population density in the area of Panama City (modified according to Bennett 1968).

of a 10-year-average. Some parts of the provinces of Darien, Herrera, Los Santos and Panama are drier (annual precipitation locally 1000 - 2000 mm); other parts are much wetter with an annual rainfall locally exceeding 5000 mm (Bocas del Toro, Colon, Chiriqui). In the period from 1970 - 1979 the highest precipitations were recorded for Colon (in Coclé del Norte, 1970, 8836 mm) and for Chiriqui (in Macano Arriba, 1975, 8560 mm).

The annual patterns of temperature range between 25 and 30°C for the lowlands with little local variation and little seasonal influence (Figure 7). The relative air humidity is high and averages between 80 and nearly 100 % (about 10 - 20 % lower in the dry season) (Figure 7).

Diurnal patterns: The diurnal patterns for temperature, relative humidity and precipitation are given in Figure 8. There is little variation in air temperature (about 5°C) and relative humidity

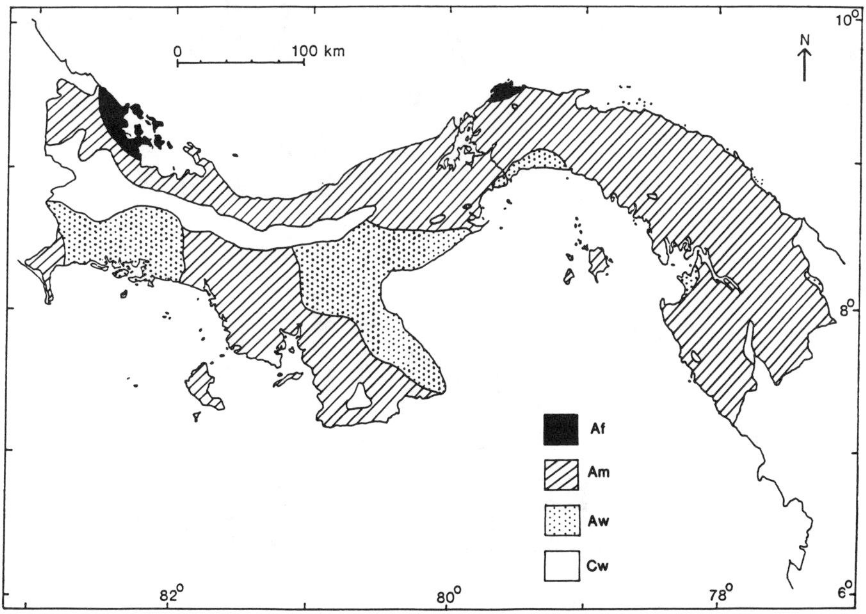

Figure 5. Macroclimatic division of Panama according to the Köppen macroclimates. A: tropical climate (no month colder than 18°C); C: humid mild winter climate (coldest month between -3 and +18°C, warmest month > 10°C); f: wet, all months with precipitation; m: short dry season; w: winter dry season / wet and dry (modified according to Bennett 1968).

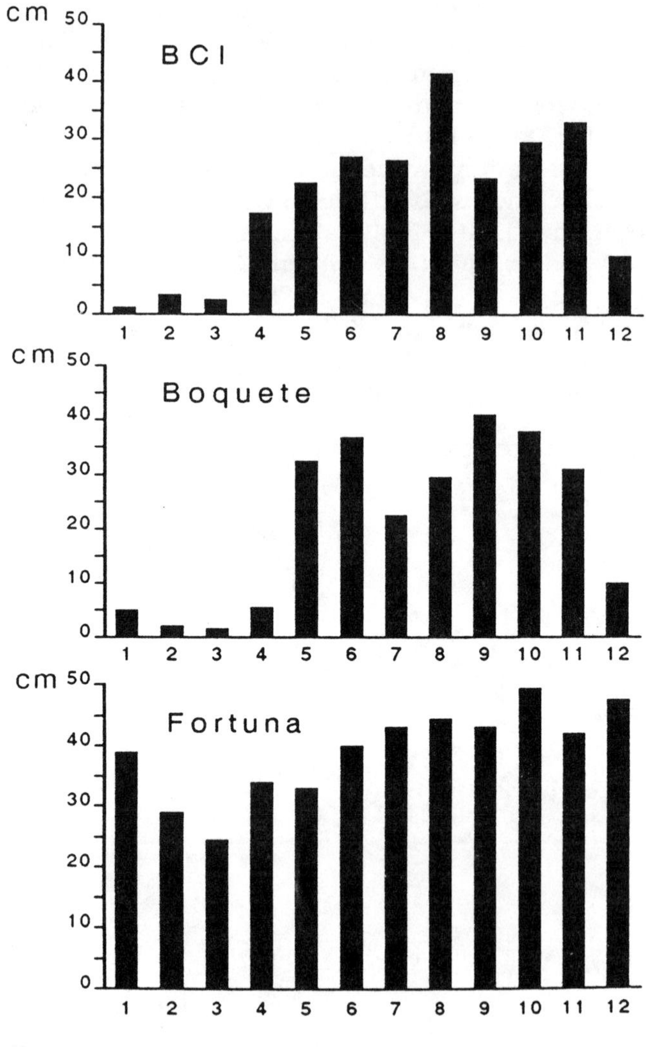

Figure 6. Annual precipitation pattern for BCI (Barro Colorado Island in the Isthmus of Panama) and two locations in Chiriqui. All climatological data were taken from Panama Canal Company (1972), Estadistica Panamena (1980), Dietrich et al. (1982) and Wolda (1982); data are usually given as monthly averages over several years, changed to metric units and modified.

(about 20 %). A marked diurnal change, how-ever, is shown by the precipitation data. Most rainfall occurs in the afternoon and early evening (14 - 21 h). The same patterns can be found in the average hourly wind speed which is usually around 10 - 15 km / h but increases up to 25 km / h between 13 and 17 h (for the sources of meteorological data see Figure 6).

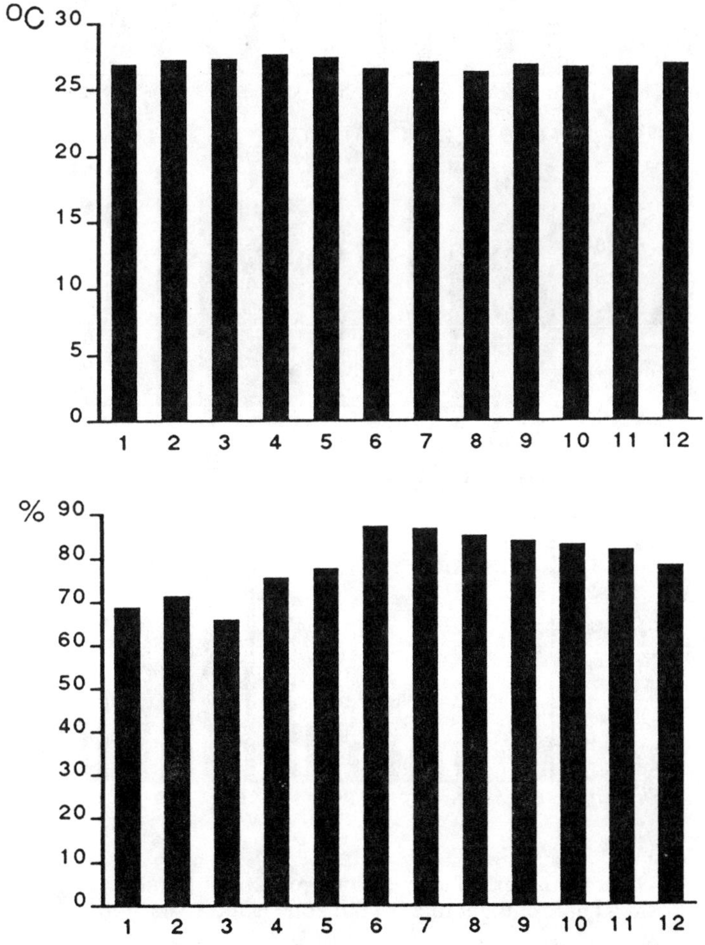

Figure 7. The annual pattern of temperature (°C) and relative air humidity (%) for the central part of the former Canal Zone (source see Figure 6).

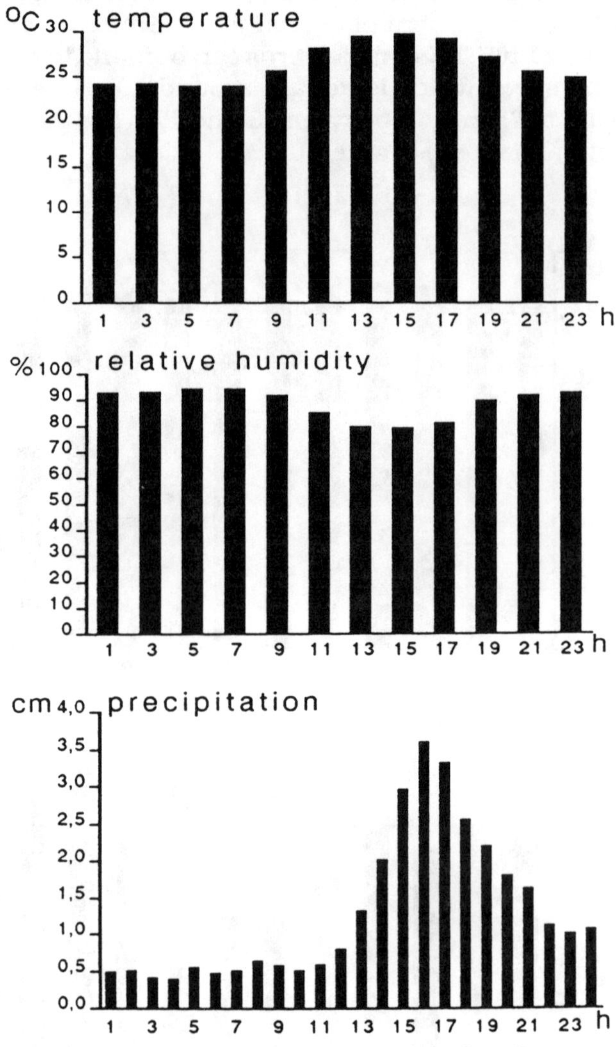

Figure 8. Diurnal pattern of temperature, relative air humidity and precipitation for the central part of the former Canal Zone (source see Figure 6).

2.3 Vegetation

According to Bennett (1968), several main vegetation types can be classified for Panama (compare Figure 9):

Old mature forest: Probably close to a climax vegetation, almost entirely at higher elevations and at places without human

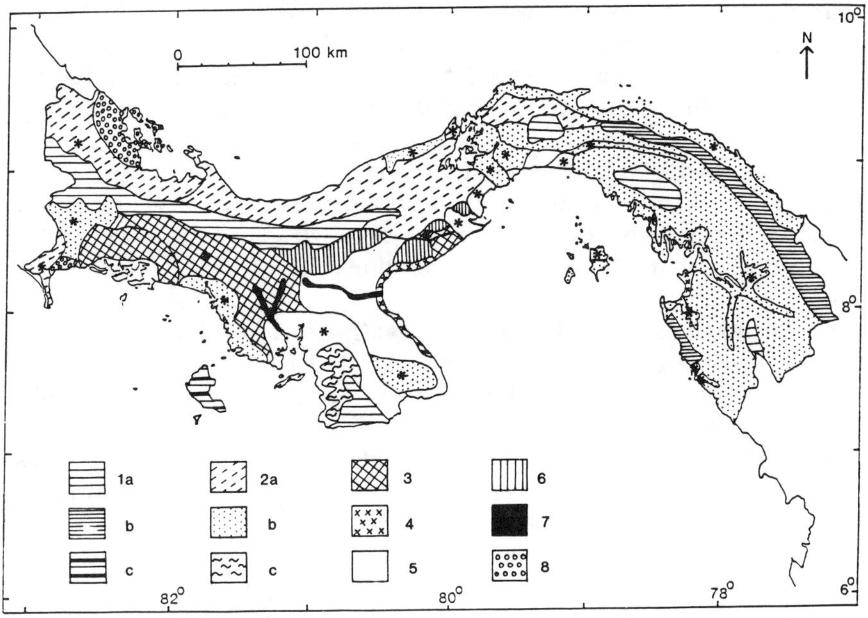

Figure 9. Phytophysiognomic map of Panama (from Bennett 1968, data from 1963, greatly modified). 1: mature forest; probably only minor or very intermittent cultural disturbance; a: montane; b: upland; c: lowland; 2: forest; a: evergreen; b: deciduous; c: mixed; previously culturally disturbed by various influences; * indicates still persisting disturbances; 3: low trees and / or bush, mostly non-deciduous, often fire and drought resistant, grass usually abundant; conditions very culturally disturbed due to fire, overgrazing and associated changes; without pronounced annual period of reduced precipitation; * indicates areas with a moderate seasonality; 4: thorn-bush and cactus, non-culturally induced, climatic and edaphic factors seem to be of chief importance although burning may have contributed; 5: grass with isolated stands of trees and palms, very marked cultural disturbance, area of this designation would normally be forested, with pronounced annual period of reduced precipitation; * indicates areas with moderate seasonality; 6: Grassy plains or steep slopes, a result of edaphic, climatic and cultural factors, would support shrub or tree cover if left unburned; 7: galeria forest, a culturally induced phenomenon resulting from uncleared forest remnants; 8: banana plantations.

influence. Above 1500 m characterized by markedly wet conditions throughout the year.

Other forest types: Similar to the vegetation type above, but different due to major disturbances at present or in the past. These remaining forested regions are consequently reduced to isolated patches and will probably soon disappear.

Low trees, bushes and grass vegetation: Often fire- and drought- resistant, culturally highly affected (disturbed) by burning and overgrazing.

Thorn-bush and cactus: A formation under strong edaphic control (saline and droughty soils), landward of mangroves, on the East and West coast of the Golfo de Panama.

Grassy vegetations: With isolated stands of trees and often palms; culturally induced formation due to a long history of burning, tree removal (felling) and overgrazing. In wetter areas with forest remnants, in drier areas similar to the thorn-bush vegetation.

Grassy plains and steep slopes: Caused by edaphic, climatic and cultural factors, often seriously disturbed by burning and overgrazing, would probably revert to shrub and low tree formations if left undisturbed.

Galeria forest: Widespread in Panama (only 2 major examples are shown in Figure 9), culturally induced stream-side remnants of once extensive woodlands or forests.

More information on the structure of the plant communities is given by Porter (1973) and literature mentioned there. A flora of Barro Colorado Island is given by Croat (1978) and the flora of Panama can be found as a (not yet completed) series of publications in the Annales of the Missouri Botanical Garden (1943-).

2.4 The future of the natural environments

The first people may have arrived in Panama 20,000 years ago, though the earliest archeological date available for Panama is 4850 years B.C. In the course of time the population increased to considerably high levels and when the Spaniards arrived around 1500 A.D., the Indian population was so large that less of Panama was forested than was the case until recently (Bennett 1968). It is likely that Balboa passed through planted fields and not through virgin forest in crossing the Isthmus to reach the Pacific (Ridgely

1976). Later, the Indian population was substantially reduced for many reasons and large parts of the forest could regrow from that period.

In the 20th century forest clearing has occurred much more consequently and thoroughly than in all the years before. After the construction of the Panama Canal the roads which had been built to construct the canal provided an infrastructure which enabled an increasing population to clear forested areas. The usual pattern consists of a road which cuts into a formerly virgin area. The land near the road is soon cleared and within a few years no forest remains except on steep slopes or along streams. Examples of such destruction are provided by the Trans-Isthmian-Highway from Panama City to Colon and the Carti Road. Examples of this progressive destruction can be seen everywhere in the country, especially in Chiriqui and parts of Darien. The largest remaining part of the lowland evergreen rainforest is a possible candidate for the next destruction if the proposals to complete the last link of the Pan-American Highway are realized.

According to Myers (1980), 54 % of Panama's area are officially classified as forests, 95 % of which may be classified as lowland rain forest and 5 % as moist montane forest. At least 25 % of the forest is reported to have been seriously disrupted by slash-and-burn agriculture; the real figure is probably much higher. The destruction of the forests has continued into the nineteen-eighties and the end is not yet in sight. Exact and up to date figures do not exist and accurate prognoses are impossible. It is, however, certain that the tropical rain forest in Panama has little chance to survive the next 50 years if the population growth continues.

Chapter 3. The investigation of the spider fauna

3.1 The increase in the number of species known from Panama

The first record of a spider from Panama is probably one concerning *Mygale Emilia* (= *Euathlus emilia*), published by White in 1856. Other records by Simon, Keyserling and Ausserer increased the number of species known from Panama to 6 in 1888.

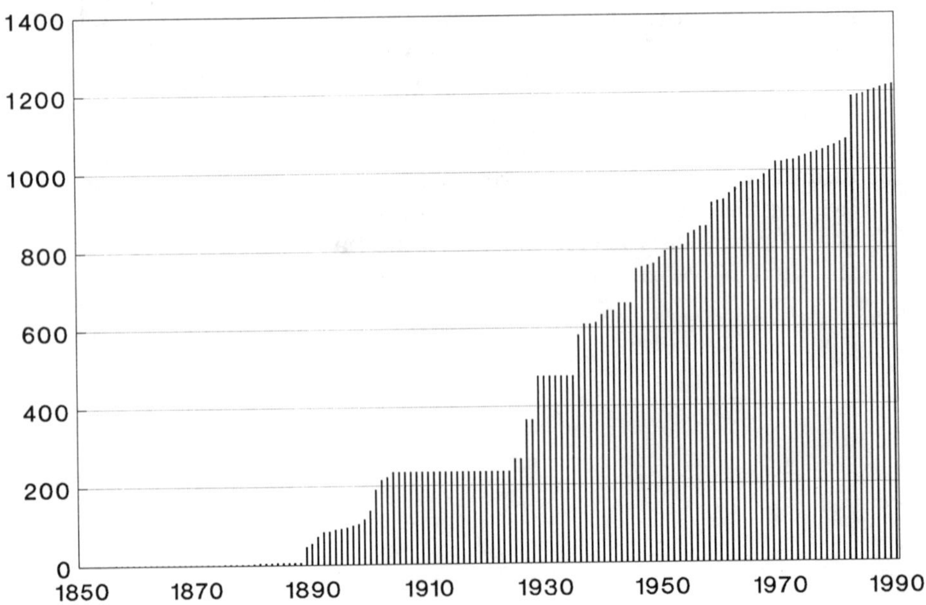

Figure 10. The number of spider species recorded for Panama from the first record in 1856 to 1991.

Table 1: First records of spider species from Panama according to author.

Author	N species	%
Chickering	346	28.4
Levi	138	11.1
F.O. Pickard-Cambridge	120	9.8
Petrunkevitch	119	9.8
Banks	113	9.3
O. Pickard-Cambridge	106	8.7
Nentwig	106	8.7
Chamberlin	60	4.9
Platnick	27	2.2
Gertsch	22	1.8
others	66	5.4
total	1223	100.0

This number may be slightly higher as all 19th century records from the Spanish colony New Granada concern both today's Panama and today's Colombia. F.O. Pickard-Cambridge and O. Pickard-Cambridge described 227 new species for Panama in their enormous work "Biologia Centrali-Americana" (1889 - 1905). From then until 1925 no publication on Panamanian spiders appeared but then the number of known species increased continually (Figure 10). Major milestones ("jumps") in the increasing number of known species were provided by the work of Petrunkevitch (1925), the revision of Salticidae by Chickering (1946) and my collection in 1983/84 which each contributed about 100 new records or species for Panama. In 1991 a total of 1223 species was known for Panama.

Regarding the life work of individual arachnologists, it must be remarked that the continuous collections and revisions of Chickering and Levi, each covering nearly 40 years, represent the largest contribution to the registration of the Panamanian spider fauna. Nearly 40 % of all Panamanian records are based on their work.(including publications with co-authors). The early workers (O. and F.O. Pickard-Cambridge, Chamberlin, Petrunkevitch and Banks) contributed another 40 % (Table 1).

The number of spider species known from Panama increased from 1930 to 1960 almost linearly; the increase from 1960 to 1980 was slightly less. An extrapolation of this trend may suggest an idea of the probable number of spiders which remain to be discovered in Panama with increasing research activities. A total number of 1500 species would then appear to be a reasonable estimate for the year 2030 or 2040. My 1983/84 collections, however, "disturbed" the trend and showed that we are far from reaching the final number of species. In many parts of the country almost no collections have been made (e.g. the Azuero peninsula, Darien or the north-western parts). Many habitats deserve much more intensive investigation; especially the arid areas, swamps, mangrove, mountain forests and microecosystems such as termite or ants' nests have completely been neglected.

At the moment, a definite estimate of the total number of species is not easy but I think that about 2000 species may eventually be recorded for Panama.

3.2 Principal arachnologists involved in the taxonomy of Panamanian spiders

In the last century, field collectors like Champion (in Bugaba, Chiriqui), Boucard (in Veraguas) and Sarg (Guatemala to Panama) collected (among other animals) spiders and sent them to interested taxonomists. Some of these specimens were described by O. and F.O. Pickard-Cambridge.

Octavius Pickard-Cambridge (England) received large collections from Central America and described a considerable number of new species, published in Biologia Centrali-America 1889 - 1902.

Frederic Octavius Pickard-Cambridge (England), nephew of O. Pickard- Cambridge; described spiders from various collections from Central and South America and also contributed (1897 - 1905) to the arachnological part of the Biologia Centrali-Americana which his uncle began.

Alexander Petrunkevitch, American Museum of Natural History, New York; several collections in the Central American region and two major publications on Panamanian spiders (1925, 1939).

Nathan Banks, Curator, Museum of Comparative Zoology, Harvard (1916 - 1944); a comprehensive taxonomic study on Panamanian spiders with many descriptions of new species in 1929.

Arthur Merton Chickering, Associate in Arachnology, Museum of Comparative Zoology, Harvard (1951 - 1971); intensive collections in Panama and adjacent areas, descriptions of many new species and taxonomic revisions of numerous groups, about 50 publications in nearly 40 years.

Herbert W. Levi, Curator, Museum of Comparative Zoology, Harvard (since 1956); systematic revisions of nearly all neotropical genera in the Theridiidae, Metidae, Tetragnathidae and Araneidae, about 40 publications in the last 35 years.

Norman I. Platnick, Curator, American Museum of Natural History, New York; revisions of many families and genera, about 20 publications in the last 15 years (many with coauthors) concerning Panamanian spiders.

Many spider species have been named after these persons: *Scytodes, Araneus, Paratobias, (Myrmarachne), (Toxeus) championi; Caponina sargi; Caraimatta, Chrysso, (Theridion), Teudis, Trachelas, Cobanus cambridgei; Mimetus, Dipoena, Selenops, Carabella, Habronattus, (Lyssomanes), Parafluda banksi; Lygromma chamberlini; Freya, (Dipoena), Stenoonops petrunkevitchi; Gertschosa, Scytodes, Scaphiella, Gelanor, Argyrodes, Ogulnius, Pseudanapis, Tetragnatha, Eustala gertschi; Anopsicus, Modisimus, Anelosimus, Argyrodes, Dipoena, Styposis, Thymoites, Lycosa, Eilica, Neozimiris, Zimiromus, Tentabunda, Myrmarachne chickeringi* (names in brackets refer to synonyms.)

3.3 The taxonomic situation

The first investigators of Panamanian spiders described species in all families, often independently of existing publications concerning a given taxon. Since it is practically impossible to be a specialist for all families, the quality of their descriptions and the validity of their new taxa was very varied and the publications for a genus or family were widely scattered over many journals. Today, the identification of a species is usually only possible with the help of modern revisions of genera or families and descriptions of new species should only be carried out in the light of such a work. Such revisions exist for many families or genera of the Panamanian fauna:

Mygalomorphae (general key by Raven 1985);
Scytodidae (Valerio 1981);
Loxoscelidae (Gertsch 1958, Gertsch & Ennik 1983);
Caponiidae (Chickering 1967);

Pholcidae (several publications of Gertsch);
Oonopidae (many publications of Chickering);
Mimetidae (Chickering 1947, 1956 a);
Uloboridae (Chickering 1968 b, Opell 1979 - 1983);
Nesticidae (Gertsch 1984);
Theridiidae (many publications of Chickering and Levi, Exline & Levi 1962, 1965);
Theridiosomatidae (Coddington 1986);
Symphytognathidae (Forster & Platnick 1977);
Anapidae (Platnick & Shadab 1978 b, 1979);
Mysmenidae (Platnick & Shadab 1978 a);
Tetragnathidae (many publications of Chickering and Levi);
Araneidae (many publications of Chickering and Levi);
Lycosidae (Dondale & Redner 1983 - 1984, Wallace & Exline 1978);
Pisauridae (Carico 1976 - 1981);
Oxyopidae (Brady 1964 - 1975, Griswold 1983);
Senoculidae (Chickering 1940 b);
Anyphaenidae (many publications of Chickering, Platnick & Lau 1975);
Clubionidae (many publications of Chickering);
Corinnidae (many publications of Chickering, Platnick 1974 a,b, Reiskind 1969);
Zodariidae (several publications of Jocqué)
Gnaphosidae (Chickering 1949, many publications of Platnick);
Ctenidae (Lachmuth et al. 1984);
Selenopidae (Muma 1953);
Philodromidae (Dondale & Redner 1969);
Thomisidae (Chickering 1950 b, Dondale & Redner 1975, Lise 1981);
Salticidae (Chickering 1946, Griswold 1987, several publications of Galiano and Cutler).

Among many groups which occur in Panama, however, taxonomic work is urgently necessary. This is especially true in the case of two large families, the Thomisidae and Salticidae, but also for smaller families such as Linyphiidae, Lycosidae, Agelenidae, Dictynidae, Ctenidae, Heteropodidae and nearly all mygalomorph spiders. Even in small families many surprises can occur, for instance in the well-known deinopid genus *Deinopis* where for 90 years only one large species, D. *longipes* F.O.P.C., was known for Panama. Coddington (pers. comm. 1988) believes that there are at least two species in Panama, probably more. Major revisions nearly always lead to changes in the species lists, for example the revision of the araneid genus *Alpaida* by Levi (1988). This revision added to the 40 known species 94 new species for the whole neotropis. For

Panama Levi found 1 new species; a large number of synonyms, however, reduced at the same time the number of valid *Alpaida* species for this country.

The taxonomic nomenclature used here largely conforms with Platnick (1989). This includes the revision of the mygalomorph families by Raven (1985) and the use of the family name Heteropodidae (instead of Sparassidae). In contrast to Platnick, Dolomedidae are not considered to be a valid family and are listed among Trechaleidae (a not yet valid family) and Pisauridae. *Synotaxus* is maintained among the Theridiidae but may be a family Synotaxidae (Forster et al. 1990). Metidae are included as a subfamily within the Tetragnathidae which also includes *Nephila*. Certainly the whole metid-tetragnathid-araneid complex requires much clearer delimitations. The same is true for the complex of Corinnidae, Clubionidae, Miturgidae and Liocranidae (whose limitations are largely unclear). In both cases genera were attributed to families according to Platnick's catalogue (1989).

Chapter 4. The spider fauna in the lowland rainforest

4.1 Description of locations

Nearly all investigators of the Panamanian spider fauna to date have tried to collect as many species as possible. In only very few cases (e.g. Harrison 1968) was the goal of spider collections the sampling of a complete habitat. Thus we have in general very limited knowledge of the overall composition of a tropical spider fauna. During my first stay in Panama I collected spiders at three different locations in 2-week-intervals by means of pitfall traps, colour bowls (trapping liquid 5 - 10 % formaldehyde with 1 % detergent) and sweepnet. These locations are Cerro Galera, a hill near the Pacific Ocean, West of the Panama Canal, at the border between the Canal Zone and the province of Panama; Cerro Pelado, a hill near Gamboa; and Pipeline Road, a road leading from Gamboa to the NE end of Lake Gatun. The position of these locations is indicated in Figure 34.

All locations are covered by a deciduous forest which has previously suffered from some kind of disturbances. These were probably greatest and most recent on some locations on Pipeline Road and less pronounced (or longer ago) at the two other sites (perhaps due to their steeper position). In the last few decades all study sites have remained largely undisturbed.

Cerro Galera: On the access road to the summit (approximately 3 km in length) 5 subsites were chosen, subsite 1 behind the gate to the access road, subsite 5 at the summit. At two subsites a total of 6 pitfall traps and 4 colour traps were set up; at one subsite sweepnetting was carried out. Additional collections per hand were made at all subsites. Both subsites with traps were close to

Table 2. Number of spider species collected in 1983 (months 3 - 12) and 1984 (months 1 - 2) at 3 different locations by sweepnetting, pitfall traps and colour traps.(- no collections)

location	method	3	4	5	6	7	8	9	10	11	12	1	2	average
Cerro Galera	sweepnet	-	26	5	10	21	11	14	15	14	13	20	5	14.0
	pitfall trap	8	24	11	7	15	9	10	9	17	11	4	7	11.0
	colour trap	8	28	5	7	18	20	11	13	11	15	8	11	12.9
Cerro Pelado	sweepnet	-	8	12	5	5	8	4	-	-	8	7	-	7.1
	pitfall trap	-	7	11	7	6	3	3	-	-	2	5	8	5.8
	colour trap	-	13	18	12	14	13	9	-	-	8	10	7	11.6
Pipeline Road	pitfall trap	-	2	2	6	27	37	20	15	22	21	26	16	17.6

Table 3: Composition of the spider communities at 3 locations according to family (addition of Table 4).

family	individuals (%)	species (%)
Salticidae	22.7	26.7
Lycosidae	17.4	0.6
Linyphiidae	15.9	2.4
Araneidae	5.1	9.7
Theridiidae	4.8	12.1
Oonopidae	4.7	10.3
Mysmenidae	4.2	0.6
Tetragnathidae	3.3	2.4
Philodromidae	3.0	1.2
Corinnidae	2.4	5.5
Theridiosomatidae	2.4	0.6
Ochyroceratidae	2.3	1.2
23 other families	11.8	26.7
total N	1913	165

Table 4. Composition of the spider communities at 3 locations (Cerro Galera, Cerro Pelado, Pipeline Road) according to 3 different collecting methods (s = sweepnet, p = pitfall trap, c = colour trap).

	C. Galera			C. Pelado			P. R.	
	s	p	c	s	p	c·	p	total
Dipluridae								
Ischnothele guyanensis		3						3
Theraphosidae								
Mygalarachne rubronitens		1						1
Scytodidae								
Scytodes championi						1	1	2
Scytodes gertschi		1	1				2	4
Loxoscelidae								
Loxosceles panama		1	1				1	3
Caponiidae								
Caponina sargi		1						1
Nops coccineus					1			1
Nops guanabacoae			6		1	1		8
Nops largus	1							1
Nops simla			1.					1
Pholcidae								
Anopsicus zeteki			1				24	25
Ochyroceratidae								
Speocera sp.			1		2		23	26
sp.					1		17	18
Oonopidae								
Dysderina dura		1	1				1	3
Dysderina recondida		1						1
Ischnothyreus browni		1						1
Ischnothyreus peltifer		3	1		3	3	1	11
Oonopoides bryantae		6	1					7
Oonops amacus			2					2
Oonops balanus			1				2	3
Oonops castellus							1	1
Oonops ornatus						1		1
Oonops pallidulus							1	1
Oonops persitus							1	1
Oonops tenebus		1					1	2
Scaphiella kalunda		4	3				19	26
Scaphiella scutata		1						1

steep parts of the hill at the roadside and dominated by shrubs and larger herbs such as *Heliconia* or ferns.

Cerro Pelado: On the access roads to the watertank of Gamboa a total of 4 pitfall traps and 4 yellow traps was set up; additional sweepnetting was performed in the area of the traps. This area was

Table 4: continued.

	C. Galera			C. Pelado			P.R.	
	s	p	c	s	p	c	p	total
Scaphiella simla		6						6
Scaphiella williamsi		8	1				13	22
Triaeris stenaspis							1	1
Palpimanidae								
sp.			1					1
Mimetidae								
sp.	3							3
Oecobiidae								
Oecobius concinnus		1				1		2
Deinopidae								
Deinopis longipes					1	1		2
Uloboridae								
Philoponella tingens						1		1
Uloborus trilineatus	1			1				2
Theridiidae								
Achaearanea hermosillo	1							1
Argyrodes projiciens	1							1
Chrysso vexabilis	1		1					2
Dipoena copiosa	1							1
Dipoena hortoni						1		1
Dipoena insulana	1				1			2
Episinus erythrophthalmus			1					1
Euryopis spinigera		1	1					2
Phoroncidia sp.						1		1
Steatoda moesta							31	31
Stemmops mellus		2	1					3
Stemmops orsus		4	7				12	23
Stemmops questa			3				3	6
Stemmops servus			1					1
Theridion dilucidum				1		1		2
Theridion lathropi			1					1
Theridion metabolum	1							1
Thymoites chickeringi		1						1
Thymoites notabilis	1	1	6					8
Thymoites sp.			1			1		2
Theridiosomatidae								
sp.		3	12	1		5	25	46
Symphytognathidae								
Anapistula secreta						1	16	17

situated on a steep slope and was dominated by tall ferns with grasses and sedges on the ground. This fern clearing was surrounded by tall forest.

Pipeline Road: On the first 8 km of Pipeline Road 3 subsites were selected: subsite 1 in a wet place within dense forest and

Table 4: continued.

	C. Galera			C. Pelado			P. R.	
	s	p	c	s	p	c	p	total
Mysmenidae								
Mysmena incredula		11	38		1	2	28	80
Linyphiidae								
Ceraticelus similis						1		1
Grammonota tabuna		3	2					5
Pocobletus coroniger				1				1
sp.	8	24	42		7	12	205	298
Tetragnathidae								
Glenognatha foxi						1	43	44
Leucauge globosa	3		2		1	5		11
Leucauge venusta	7							7
Tetragnatha pallescens				1				1
Araneidae								
Bertrana striolata	1							1
Cyclosa caroli							1	1
Eriophora fuliginea	2							2
Eustala devia						1		1
Eustala sp.	5							5
Mangora pia	2							2
Mangora picta	3			2				5
Mangora spinula	36	2	3	8		7		56
Micrathena horrida				1				1
Micrathena sexspinosa	2							2
Neosconella styligera	1							1
Pronous tuberculifer						4	2	6
Scoloderus cordatus				1				1
Verrucosa mexicana				1		2		3
Wagneriana tauricornis	6							6
Wittica crassicauda						4		4
Lycosidae								
sp.	13	63	33	2	24	8	189	332
Pisauridae								
Thaumasia argenteonotata					4	5		9
sp.			2				1	3
Agelenidae								
sp.							5	5
Dictynidae								
Dictyna sp.	2	2	16	1	1	4	2	28
Thallumetes pusillus				1				1

covered with tall *Heliconia* and *Carludovica*; subsites 2 and 3 in a less thickly forested area with numerous vines. A total of 21 pitfall traps was set up at the 3 subsites. For the presented calculations here all records from the subsites of one location were combined.

Table 4: continued.

	C. Galera			C. Pelado			P. R.	
	s	p	c	s	p	c	p	total
Titanoecidae								
Goeldia mexicana							1	1
Oxyopidae								
Hamataliwa puta				1		1		2
Oxyopes acleistus	5							5
Oxyopes salticus						1	1	2
Peucetia sp.				20		1		21
Anyphaenidae								
Teudis juradoi			1					1
Wulfila mandibulata	1							1
sp.	2						7	9
Clubionidae								
Clubiona sp.	1							1
Corinnidae								
Castianeira alba	6	12	6		1		8	33
Castianeira memnonia						1		1
Corinna robusta			1					1
Corinna sylvatica							1	1
Corinna variegata			1					1
Mazax ajax		1	1			1		3
Myrmecotypus sp.		1						1
Trachelas barroanus			1					1
Trachelas bispinosus			3					3
Zodariidae								
Naibena barroana	1	18	4				2	25
Selenopidae								
Selenops sp.	3		2	1				6
Heteropodidae								
Olios stylifer							1	1
sp.	4			12		3	2	21
Philodromidae								
Apollophanes punctipes	4	2	9	14		27		56
Thanatus sp.						1		1
Thomisidae								
Epicadinus trifidus	1							1

4.2 Species composition

The number of spider species which were collected monthly at the different locations was generally low. The highest value was 28

Table 4: continued.

	C. Galera			C. Pelado			P. R.	
	s	p	c	s	p	c	p	total
Epicadinus trispinosus				1				1
Misumenoides rugosa	1							1
Thomisus odiosus	1			1				2
Tmarus aculeatus				1				1
Tmarus jocosus	1							1
Tmarus probus	1							1
Tmarus studiosus	1							1
Salticidae								
Acragas fallax			1					1
Albionella propria							1	1
Atelurius incertus	3							3
Breda notata	1							1
Chapoda panamana	1							1
Chira trivittata			1					1
Chira sp.	1							1
Corythalia bicincta	5		2					7
Corythalia obsoleta	1	1	21			3	17	43
Corythalia opima						1		1
Corythalia pulchra	3		2			2		7
Corythalia serrapophysis						1		1
Corythalia spiralis			*1*					*1*
Corythalia sulphurea	19	11	33		1		21	85
Corythalia sp.	1							1
Fluda principes			7					7
Freya albosignata	1	3	3		2	10	1	20
Freya bifurcata	1	1		6	3	14	3	28
Freya demarcata	13	1	6		1	29	1	51
Freya infuscata	1							1
Freya simplicicava	2			6	2	7	4	21
Hapaeus benignus	1							1
Hentzia parallela	1							1
Jollas minutus	3	6	1		1	1	1	13
Lyssomanes unicolor	1			1				2
Neonella sp.							1	1
Nycerella delecta	6							6
Nycerella donaldi	4							4
Nycerella neglecta	10							10
Nycerella sp.	1							1
Parafluda banksi	1							1
Parnaenus recurvus	1							1
Pellenes paratus	7			2	2	4	6	21
Phiale crocea	1					1		2
Phiale guttata			—	2				2
Sarinda hentzi	6	1			1		2	10
Scopocira panamena	28	1	2	1		7		39
Synemosyna edwardsi				1				1

species (Cerro Galera, colour trap), the lowest value was zero. The monthly average number of species was highest at Pipeline Road (17.6 species), medium at Cerro Galera (11.0 - 14.0 species depending on the trapping method) and lowest at Cerro Pelado (5.8

Table 4: continued.

	C. Galera			C. Pelado			P. R.	
	s	p	c	s	p	c	p	total
Thiodina sp.	7		1	2		2	1	13
Zuniga laeta			1					1
Zuniga magna			1					1
Zygoballus electus	1							1
Zygoballus optatus	3		1	1				5
Zygoballus sp.	13						1	14
total	284	216	308	95	63	192	755	1913

- 11.6 species). There was no clear influence of the collecting method on the number of species collected: at Cerro Galera and at Cerro Pelado pitfall traps yielded the lowest species number, on Pipeline Road, however, this collecting method yielded the highest species number of all sites or methods. At Cerro Galera sweepnetting yielded the highest species number whereas at Cerro Pelado colour traps represented the most successful method (Table 2). The number of traps set up per site is perhaps only a partial explanation of these discrepancies. Differences in the vegetation structure or the degree of disturbance are other possible explanations.

A total of 165 species was collected at the 3 study sites during one year. This represents only 13.6 % of the known spider fauna of Panama. Among my 1-year-collection, Salticidae were the most common family with 44 species (27 % of all species) and 22.7 % of all individuals. Lycosidae (17.4 % of all individuals) and Linyphiidae (15.9 %) were the next most common families. Unfortunately, the taxonomic situation for both families with respect to neotropic species is very poor, so we were unable to determine the exact number of species involved in both groups. Lycosids are probably represented by 2 - 3 species, linyphiids by 4 (or 5 - 6) species. The families of Araneidae, Theridiidae and Oonopidae comprise about 5 % of individuals and 16 to 20 species respectively. Twelve families occur more frequently (more than 2.0 % of individuals); 23 families are represented by less than 1.6 % of individuals (Table 3).

A complete list of species collected at 3 locations using 3 methods is given in Table 4. The influence of the collecting method is obvious. Araneids, thomisids and salticids are caught much more frequently by the sweepnet whereas no clear difference exists between the returns of pitfall traps and colour traps. The methods seem to complement each other by collecting certain species (usually in low individual numbers) which the other method did not collect.

Only a few species occurred in high densities and comprise more that 2 % of the total catch. These are lycosid (N = 332) and linyphiid (N = 298) species (although they were not further identified), *Corythalia sulphurea* (N = 85), *Mysmena incredula* (N = 80), *Mangora spinula* and *Apollophanes punctipes* (N = 56 each), *Freya demarcata* (N = 51), a theridiosomatid species (N = 46), *Glenognatha foxi* (N = 44), *Corythalia obsoleta* (N = 43) and *Scopocira panamena* (N = 39). Only a few of these very common species were found to be restricted to two investigation areas: *Glenognatha foxi* was not found at Cerro Galera; *Scopocira panamena*, *Mangora spinula* and *Apollophanes punctipes* were not collected at Pipeline Road (during other collections these species were, however, collected elsewhere in the vicinity of the trap sites) (see chapter 9).

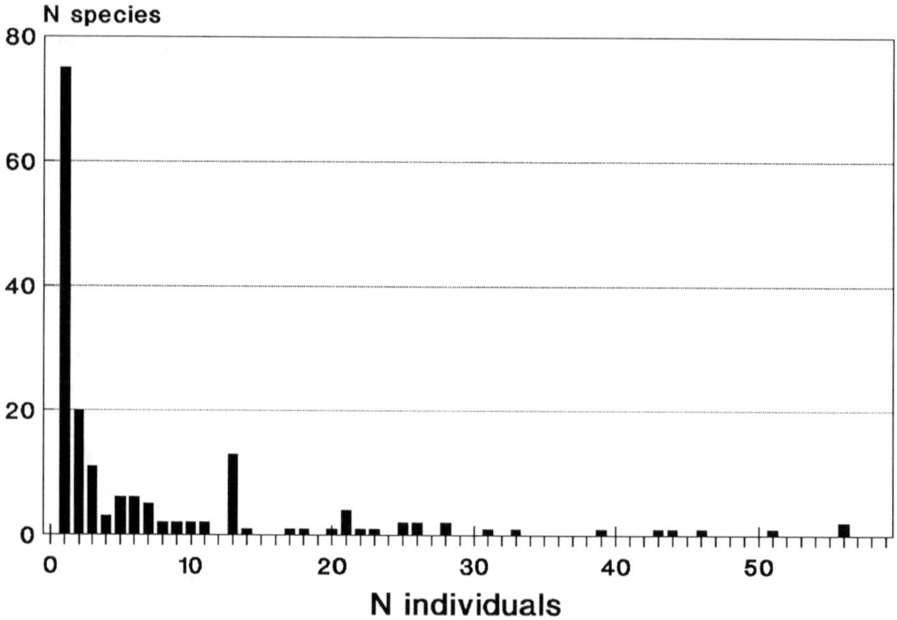

Figure 11. Number of individuals per spider species; combined data from 3 locations on the Isthmus of Panama (see Table 4). Four species with 80, 85, 298 and 332 individuals are not included in this graph.

Table 5. Monthly values of diversity H_S (after Shannon-Weaver) and evenness E for the spider communities at 3 different locations (months 3 - 12 in 1983 and months 1 - 2 in 1984). The data of Cerro Pelado and Cerro Galera base on pitfall traps, colour traps and sweep netting, the data of Pipeline Road base only on pitfall trapping. (- not investigated)

month	3	4	5	6	7	8	9	10	11	12	1	2
diversity H_S												
Cerro Galera	2.53	3.59	2.62	2.94	3.39	3.0	3.04	3.08	3.21	3.0	2.93	2.66
Cerro Pelado	-	2.52	3.01	2.86	2.83	2.87	2.08	-	-	2.67	2.47	2.6
Pipeline Road	-	0.69	0.64	1.46	2.47	2.96	2.90	2.31	2.17	2.11	2.43	1.99
evenness E												
Cerro Galera	0.96	0.86	0.89	0.93	0.89	0.84	0.88	0.89	0.92	0.90	0.87	0.94
Cerro Pelado	-	0.8	0.88	0.94	0.92	0.92	0.84	-	-	0.94	0.87	0.99
Pipeline. Road	-	0.99	0.92	0.81	0.75	0.82	0.97	0.85	0.70	0.69	0.75	0.72

The dominant species in the vegetation (sweepnet catches) were *Mangora spinula, Scopocira panamena* and *Corythalia sulphurea* at Cerro Galera; *Peucetia* sp. (juvenile), *Apollophanes punctipes* and *Olios stylifer* at Cerro Pelado. On the ground (pitfall traps and colour traps) other species dominated: at Cerro Galera a lycosid, a linyphiid and *Mysmena incredula*; at Cerro Pelado a lycosid, *Freya demarcata* and *Apollophanes punctipes*; at Pipeline Road a linyphiid, a lycosid and *Apollophanes punctipes*.

Many species were only found in low numbers, among them 75 species (= 45 %) with only one specimen. Of most species only a few individuals were collected and few species are common (Figure 11).

Therefore it is difficult to discuss species distributions or exclusions since the impression remains that the spider communities were not exhaustively investigated. This impression is reinforced by additional collections of spiders near the trap sites which yielded further species records. The general structure of the spider community with high species numbers and low individual numbers per species is well documented and is characteristic for tropical communities (see p. 43).

The necessarily limited collecting effort of a given study will, therefore, register a tropical spider community much less completely than one in the temperate zone.

To calculate the diversity index H_s of the spider community the Shannon-Weaver index was used (compare remark on p. 45). However, this index does not show if a high value is caused by a high species number with an irregular distribution or by a low species number with a regular distribution. This problem is solved by the measure of the evenness E. E ranges from 0 to 1 and the higher the value the more regular is the distribution of the respective species (Pielou 1966). The monthly diversity values for Cerro Galera range between 2.53 and 3.59. Most values for Cerro Pelado range between 2.5 and 3.0 and are generally somewhat lower than the Cerro Galera values. Pipeline Road nearly always produced the lowest values. This is probably due to the collecting methods which consisted only of pitfall traps at this site. Diversity values vary considerably during the year but no clear trend can be determined (seen). Most high values, however, were obtained in the late dry season and early wet season; many low values occurred in the late wet season and early dry season (Table 5).In contrast to this, the evenness values do not fluctuate very much. For Cerro Galera and Cerro Pelado E varies between 0.8 and nearly 1.0; only some Pipeline Road values are lower (s. above). Evenness values around 0.9 must be considered as being extremely high, i.e. the (small) number of individuals is extremely regularly distributed among the (large) number of species (Table 5). Very abundant species do not occur and even "quite common" species are rare. This may be illustrated by the example of the extremely high E value of 0.99 (February, Cerro Pelado) which bases on 14 species. Twelve of them were represented by only 1 specimen, 2 species were represented by 2 individuals.

Calculated over the complete year of investigation, the diversity H_s is highest for Cerro Galera (4.09) and very similar for Cerro Pelado (3.82). The data of Pipeline Road (only pitfall traps) yield a H_s-value of 2.8. The evenness values for Cerro Pelado (0.84) and Cerro Galera (0.80) are similar; Pipeline Road again yields a lower value (0.64).

4.3 Phenology

4.3.1 Annual occurrence of major groups

When does the maximum number of spider species occur and how strong are seasonal influences? In Figure 12 the number of species found per month is plotted as a percentage of the total number of species found during the complete investigation period, i.e. in one year. A clear species maximum can be seen from July to September (the beginning of the wet season). During this period 11 - 13 % of the total number of species found in the complete investigation period were found within one month. The number of species declined from this peak in August continually until the end of the wet season in December. The beginning of the dry season is characterized by a smaller species peak (maximum) in January,

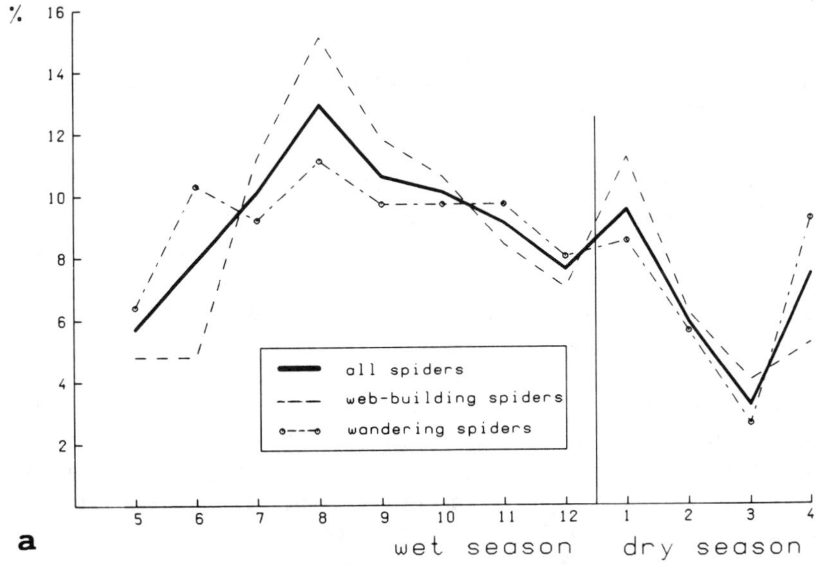

a

Figure 12. The number of spider species collected monthly (combination of 3 collection sites and 3 methods, see Table 4) as a percentage of the total number of species of the complete investigation period of one year (compare text). The vertical bar divides the two years of investigation and the wet and dry season. **a:** all spiders (solid line) and two sub-groups, web-building spiders and wandering spiders;

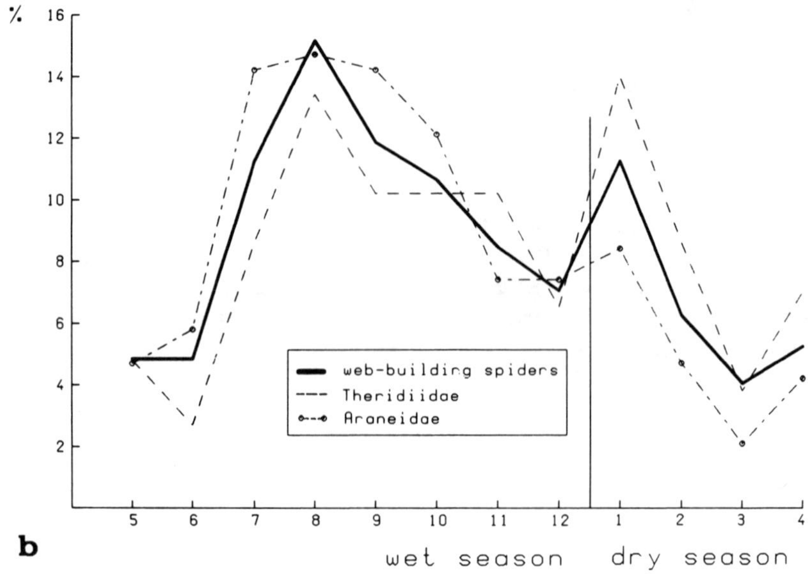

b

Figure 12. continued. **b:** web-building spiders (solid line) and the two most abundant web-building families, Theridiidae and Araneidae;

the end of the dry season by an absolute species minimum in March (less than 4 % of the total number of species) and an increase towards the beginning of the wet season. The periods of richest and poorest spider communities (with respect to species numbers) are separated by approximately 6 months and differ from each other in species number by a factor of 3.

Do the most abundant spider families react in the same manner to seasonal changes? Figure 12a differentiates between two major groups, the web-building spiders (comprising approximately 45 % of all species) and the wandering (or non web-building) spiders (comprising approximately 55 % of all species). Since the biology of many species is inadequately researched and since many families comprise species of both types, this approach has only a tentative character. Seasonal differences are more pronounced among web-building spiders than among wandering spiders. Figure 12b shows that among the web-building spiders Araneidae (38 % of all web-building species) have a clear species maximum from July to October (first part of the wet season) and

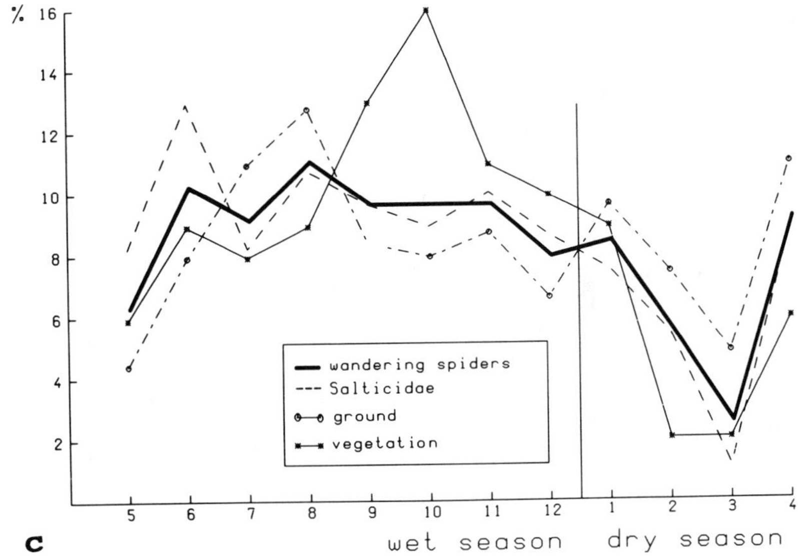

Figure 12. continued. **c:** wandering spiders (solid line), the most abundant wandering family Salticidae and the remaining groups, divided into families living mainly on the ground and families living mainly in the vegetation.

that the wettest part of the wet season and the entire dry season is rather unfavourable for them. Theridiidae (37 % of all web-building species), in contrast, show a seasonality with two clear peaks, one in the early wet season (August) and another, even higher peak in the early dry season. I am not sure if there are araneids which are more or less restricted to the wet season or theridiids which are restricted to the dry season but these data may suggest this to some degree.

Among wandering spiders (Figure 12c) Salticidae (46 % of all wandering species) show an early wet season peak, more or less constant species numbers during the following wet season and an absolute species minimum in the latter half of the dry season. The remaining families of wandering spiders, which can be separated into those which live primarily on the ground and those in the vegetation, deviate from the general pattern of wandering spiders. Families which live primarily in the vegetation (17 % of all wandering species, among them Oxyopidae, Philodromidae, Heteropodidae, Ctenidae, Thomisidae and Hersiliidae) show a very

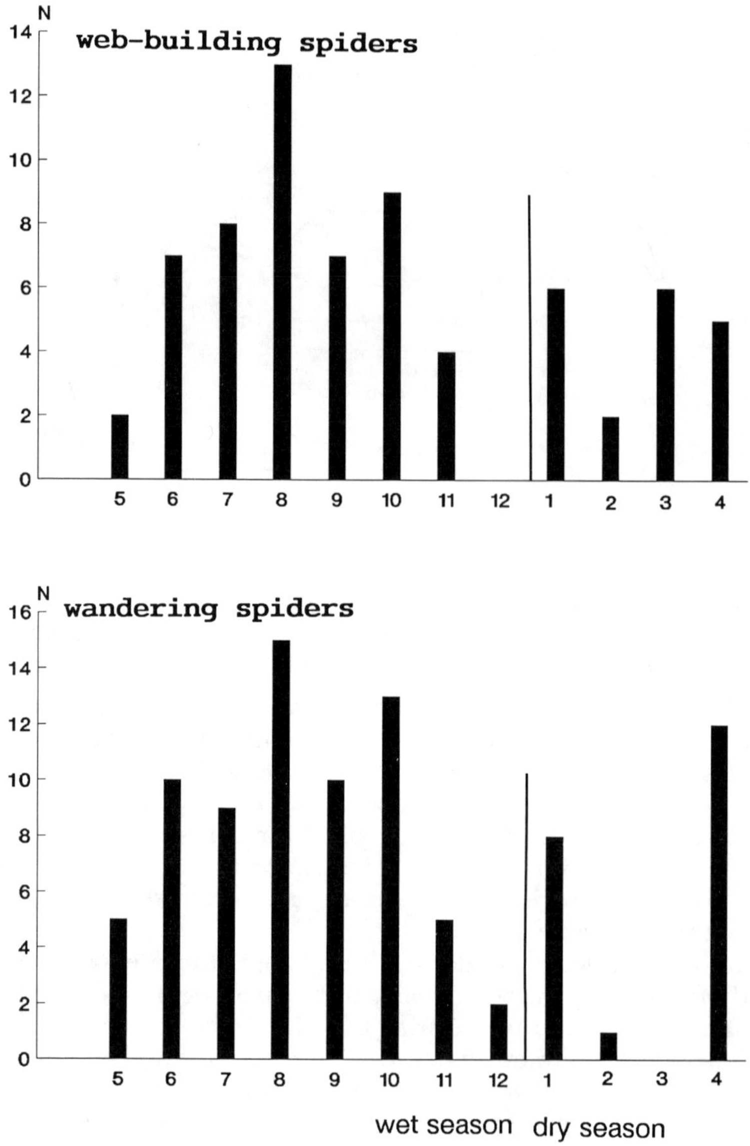

Figure 13. The occurrence of highly seasonal spider species, i.e. the number of species which were found during only 1 month of the 12 months of investigation (same data as Figure 12).

pronounced species peak in the mid-wet season and a strong decline towards the dry season, where their species number is nearly as low as that of the Salticidae. Wandering spiders living primarily on the ground (37 % of all wandering species, among them Caponiidae, Oonopidae, Ochyroceratidae and Theraphosidae) show the least clearly pronounced seasonality of all the wandering spiders.

4.3.2 Seasonality of species

How seasonal are spiders and are extremely seasonal species specialized to certain parts of the year? I attempted to find the most seasonal species by considering those species which were found

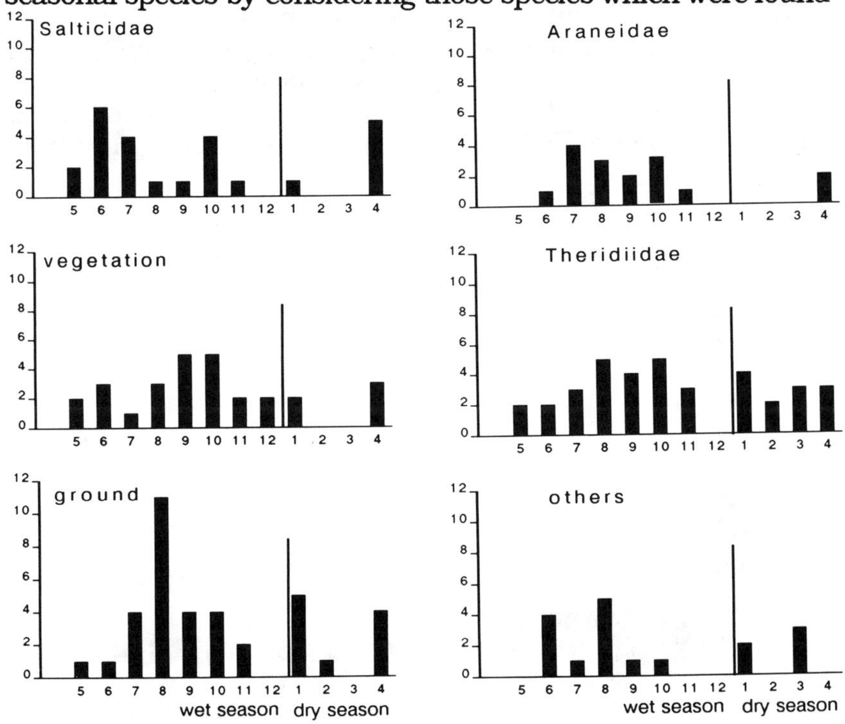

Figure 14. The occurrence of highly seasonal spider species, i.e. the number of species which were found only during 1 month of the 12 months of investigation. Left: The most important families of the web-building spiders, Araneidae, Theridiidae and others; right: the most important groups of wandering spiders, Salticidae, families living mainly on the vegetation and families living mainly on the ground (compare Figures 12 and 13).

only during one month of the 12 months of collection. Figure 13 shows that these spiders occur primarily in the middle part of the wet season (June to October). The wettest part of the wet season (December) has the smallest number of such highly seasonal species. Each of the two halves of the dry season which were investigated shows a small species peak. This suggests that the dry seasons differ in some way, but it suggests as well the existence of specialized dry season species.

A separate plot of web-building and non web-building species (Figure 14) yields no remarkable differences compared with Figure 13. This may indicate that the selection pressure of the seasons is similar for both groups.

The seasonal width of a given spider species or group (i.e. the number of months in which it was collected) is similar for all species (Figure 15) and for selected groups e.g. web-building and wandering spiders (Figure 16). Most species were found only during 1 month as adults (46 %), 25 % of the species occurred during 2 or 3 months. Fifteen percent were found during more than

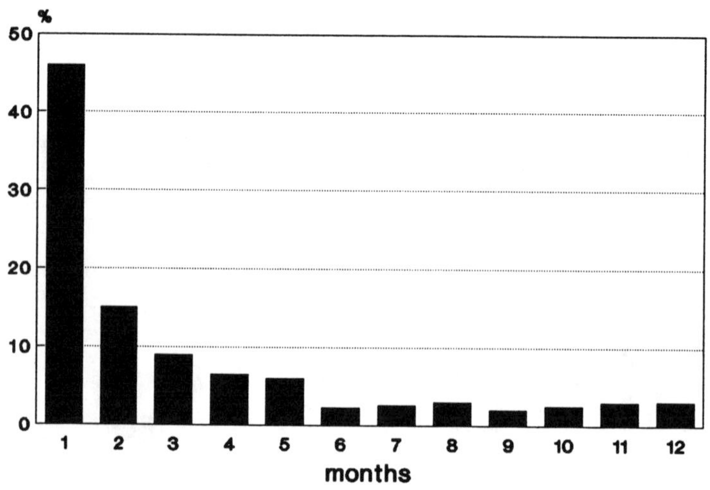

Figure 15. The seasonal width of spider species, i.e. the number of months in which a given species could be collected (combination of 3 collection sites and 3 methods, see Table 4).

6 months but only a few species were found throughout the year. These are long-lived species (e.g. Theraphosidae or the filistatid *(Kukulcania hibernalis)* or species which live in permanent colonies *(Anelosimus eximius, Ischnothele guyanensis)* or synanthropic species such as *Scytodes longipes, Oecobius concinnus* and *Theridion rufipes*.

Although no principal differences between web-building and wandering spiders could be found, some families may deviate from the general scheme. Among wandering species few salticids occur for 1 month; more species were found for 2 or 3 months, i.e. the average season of a salticid is longer than the average season of other wandering species. Among web-building species, Araneidae

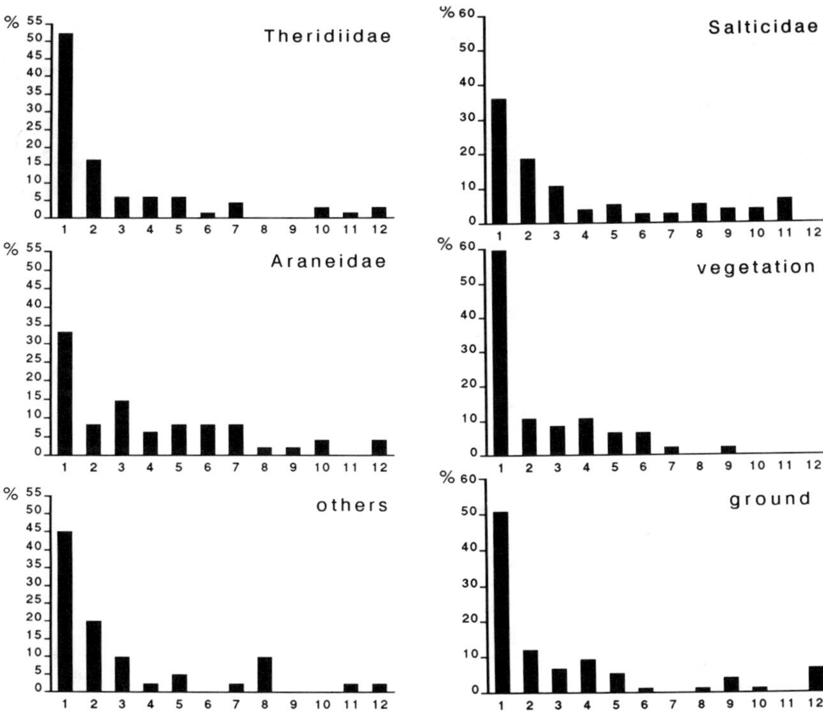

Figure 16. The seasonal width of spider species (data from Figure 15). Left: The most important families of the web-building spiders, Theridiidae, Araneidae and others; right: the most important groups of wandering spiders, Salticidae, families living mainly in the vegetation and families living mainly on the ground.

contain less and Theridiidae contain more species with a short phenology (Figure 16).

These data, however, are somewhat problematical. They refer only to adult spiders since juveniles cannot be identified and "rare" species are probably underrepresented (see p. 31). The general trend of an extremely short seasonal appearance (70 % of all species were found during less than 3 months within the year), however, is certainly correct.

In some cases, species occurred for more than 3 months per year and showed a scattered seasonal distribution. This may indicate that these spiders have two generations per year. This phenomenon is clear in the case of some large orb-weaving spiders, e.g. *Eriophora fuliginea, Argiope argentata* or *Nephila clavipes*. They show two distinct seasonal peaks from July to October and from January to March (Figure 17). In other cases, however, it is very difficult to describe two distinct generations since the phenological data are often very inadequate.

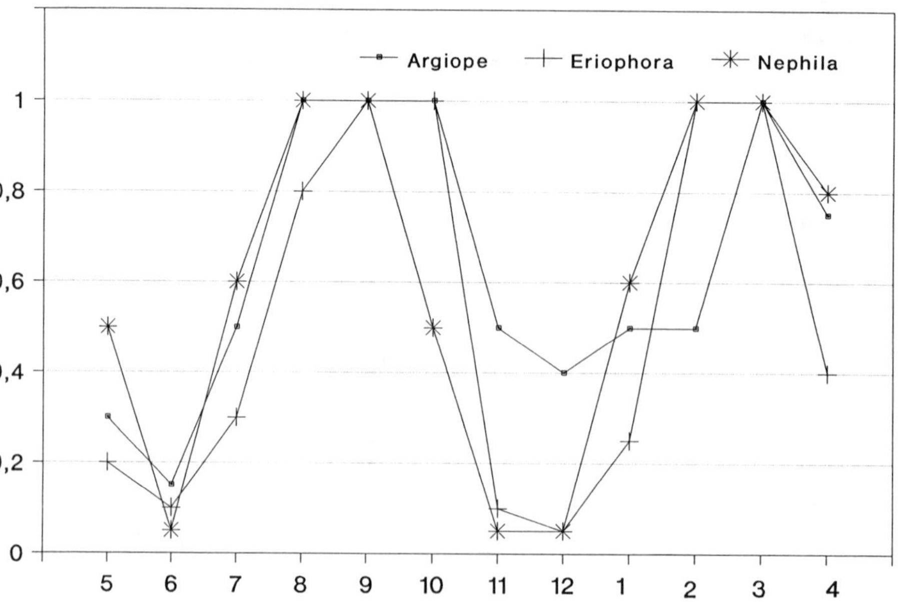

Figure 17. The relative abundance of the araneid spiders *Argiope argentata, Eriophora fuliginea* and *Nephila clavipes* (combined data from Cerro Galera, Cerro Pelado and Pipeline Road) in the wet season (months 5 - 12) and the dry season (months 1 - 4).

In some other cases it is possible to characterize phenological types or seasonal maxima. *Castianeira alba* represents a species which most often occurs in the dry season. Some species have their population maximum in the early wet season (*Speocera* sp., *Scaphiella kalunda*, *Corythalia sulphurea*, *Freya albosignata* or *Stemmops orsus*), some have their maximum in the middle of the wet season (*Cyclosa caroli*, *Micrathena sexspinosa*, *Thaumasia uncata* or some *Argyrodes* species), others undoubtedly have a population maximum, but it is difficult to define (e.g. *Leucauge venusta*, *Anopsicus zeteki*, the *Corythalia* and *Freya* species or *Glenognatha foxi*). In some of the latter cases two overlapping generations are possible; for further conclusions, however, more data are needed.

Chapter 5. A comparison of the spider communities in the tropical and temperate zones

For a comparison with the tropical spider communities at Cerro Galera, Cerro Pelado and Pipeline Road, 68 published spider faunas of different habitats in Central Europe were analysed. These European spider communities cover a wide range of habitats and were performed with different methods (e.g. pitfall trapping, sweep netting, handsorting of litter samples etc.), but usually only one method per site. All studies cover at least 8 months of the year (usually spring to autumn, i.e. the vegetation period in the temperate zone), many of them also several years.

For the comparison of the spider fauna of Panama with North America and Europe all records from Panama (pages 93 - 175) are compared with check lists from Costa Rica (Vega 1980, Ubick 1988), North America (Roth 1985) and British Columbia (West et al. 1984); for Central Europe data provided by Heimer & Nentwig (1991) were used. A sufficient comparison of tropical and temperate zone spider communities would require a comparison with Mexico, at least some parts of South America and / or Africa. Unfortunately, check lists do not exist for most of these countries and their degree of arachnological investigation is still completely inadequate.

5.1 Habitat comparison

5.1.1 Number of species

A total of 68 analyses of Central European spider faunas yielded an average number of 65.6 species per habitat. The variation between habitats, however, is high and the numbers range from 16 to 209 species (standard deviation ± 40.2). The distribution of species (Figure 18) is uneven (skewed to the left with a skewness of 1.56 and a kurtosis of 2.68) and most values are between 40 and 65 species per habitat.

In Panama spider communities of 31 to 83 species were found per habitat and trapping method, most values being between 60 and 70 species. This covers the same range as in Figure 18 in which most values of European spider communities are to be found. By combining the 3 trapping methods at Cerro Galera and Cerro Pelado, the species number increases to 96 for Cerro Pelado and 163 for Cerro Galera. Taking into consideration an additional number of not yet recorded species, the total number may increase to 200 or even more species for Cerro Galera. Additionally, there are guilds of spiders which are difficult to collect such as burrowing species or spiders which live on bark. This, however, is the case for

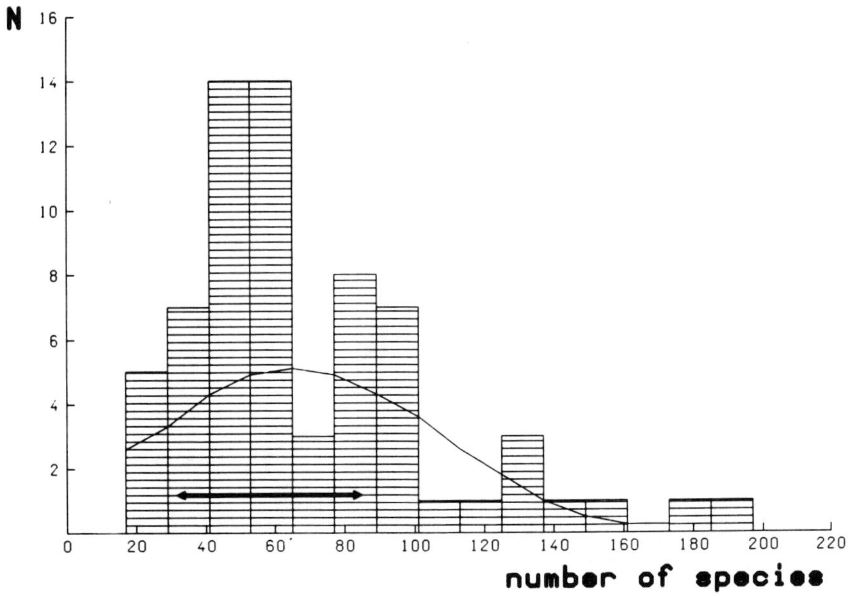

Figure 18. A comparison of the species numbers of 68 spider communities in Central Europe (histogram and computer-fitted curve) with the spider communities at Cerro Galera, Cerro Pelado and Pipeline Road in Panama (range of values indicated by a solid arrow) (compare text).

all investigations. Two hundred (or more) species at Cerro Galera appears to be an enormous number but it includes several subsites (with respect to plant communities) such as forest edges, dense forest, plots with more grasses, ferns or herbs. So a good comparison with Central Europe should include differently structured subsites and several trapping methods as well and should probably only refer to the studies which yielded higher numbers of species (circa 180 - 200). Bearing this in mind (and referring to Figure 18) the species number of spiders in the temperate and the tropical zone is astonishingly similar.

5.1.2 Individuals per species

In the 68 Central European spider communities the average number of spiders per species was 42.6 ± 90.5. The distribution is considerably skewed to the left (skewness 6.98, kurtosis 53.0) and

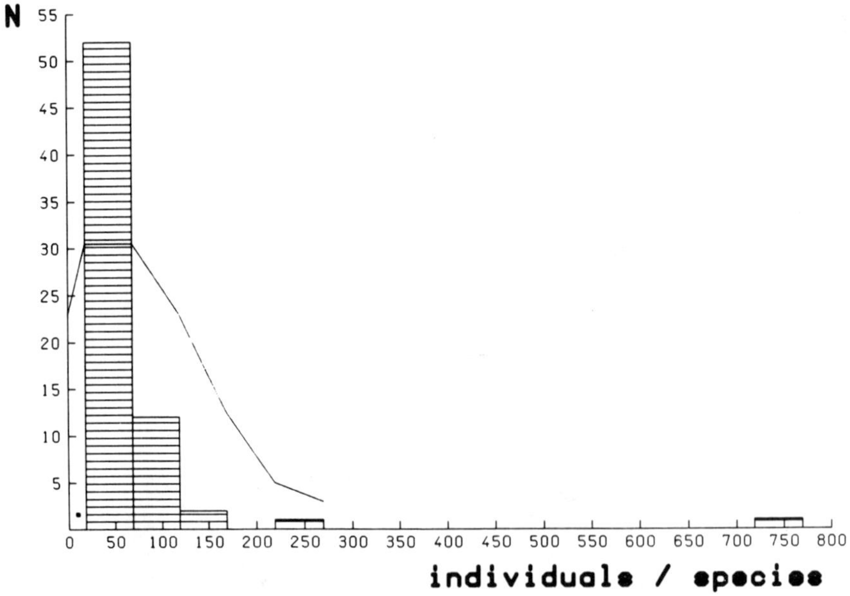

Figure 19. A comparison of the number of individuals per species in 68 spider communities in Central Europe (histogram and computer-fitted curve) with the spider communities at Cerro Galera, Cerro Pelado and Pipeline Road in Panama (range of values indicated by a dot on the left) (compare text).

the values range from 3.6 to 734.7 individuals per species. In the Panamanian investigations the 3 trapping methods at Cerro Galera yielded 3.7, 4.1 and 4.1 individuals per species; at Cerro Pelado the corresponding figures are 3.1, 3.1 and 3.7. On Pipeline Road 10.5 spiders per species were collected by pitfall traps; on exclusion of the two most common species (a linyphiid and a lycosid) this value decreases to 5.2. Thus the abundance of tropical spiders is considerably lower than in the temperate zone and averages about one tenth of the temperate zone value (Figure 19). The extremely low number of individuals per species is the most pronounced and the most impressive difference between the temperate and the tropical spider communities.

5.1.3 Diversity and evenness

Some authors (e.g. Pielou 1966, Lloyd et al. 1968) argue that the diversity index of Shannon-Weaver is not the best measure of diversity for trap data and recommend the diversity index of

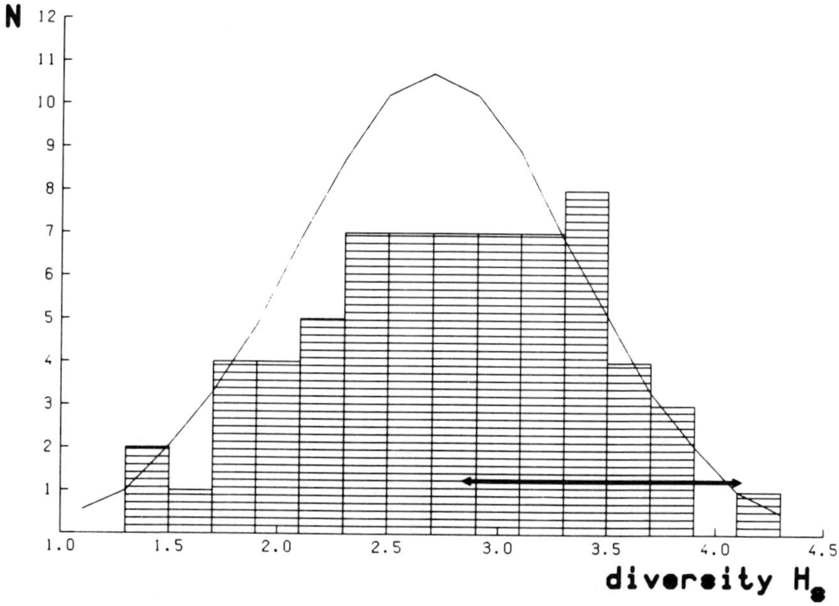

Figure 20. A comparison of the diversity H_s in 68 spider communities in Central Europe (histogram and computer-fitted curve) with the spider communities at Cerro Galera, Cerro Pelado and Pipeline Road in Panama (range of values indicated by a solid arrow) (compare text).

Brillouin (1956). Since nearly all diversity calculations up to now, however, have been made using the Shannon-Weaver formula, it seems more appropriate to use this index for the purpose of comparison.

The diversity as calculated by the Shannon-Weaver index varies in 68 temperate zone spider communities from 1.24 to 4.2. The average value is 2.71 ± 0.66 and the distribution of values is relatively even (skewness -0.11, kurtosis -0.35). The diversity of the tropical spider communities is 2.8 for Pipeline Road (pitfall traps), 3.8 for Cerro Pelado (3 combined trapping methods) and 4.1 for Cerro Galera (3 combined trapping methods). This range covers the right half of the distribution of the temperate zone diversity (Figure 20). Although only 3 values are compared here to the 68 values of the temperate zone, the figures suggest that the diversity of the tropical spider community is generally much higher than the diversity of the temperate zone communities.

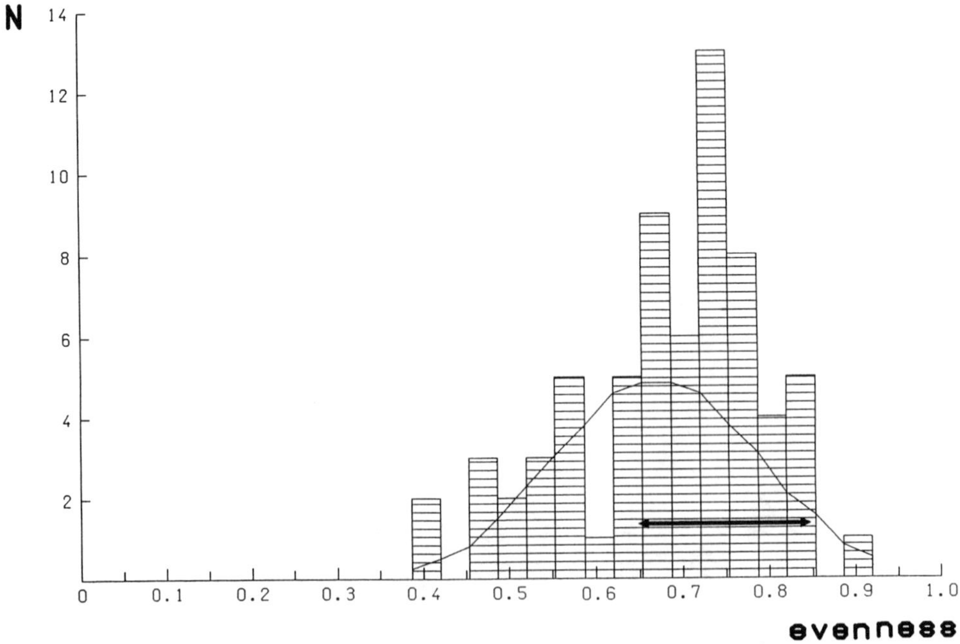

Figure 21. A comparison of the evenness of 68 spider communities in Central Europe (histogram and computer-fitted curve) with the spider communities at Cerro Galera, Cerro Pelado and Pipeline Road in Panama (range of values indicated by a solid arrow) (compare text).

Table 6: List of spider species found in British Columbia (Canada) (West et al. 1984, 1988), USA (Roth 1985), Costa Rica (Vega 1980, Ubick 1988), Panama (this study) and Central Europe (Heimer & Nentwig 1991).

family	British Columbia	North America	Costa Rica	Panama	Central Europe
Atypidae		8			3
Antrodiaetidae		2	25		
Cyrtaucheniidae		39		1	
Idiopidae			1	2	
Ctenizidae		16		2	
Actinopodidae				1	
Mecicobothriidae		6			
Microstigmatidae				1	
Dipluridae		4	4	5	
Nemesiidae		2			
Barychelidae			2	1	
Theraphosidae		40	16	17	
Paratropididae			2		
Hypochilidae		6			
Filistatidae		13	2	1	1
Sicariidae			1		

Since the diversity index of Shannon-Weaver does not show if its value is caused by a high number of species with irregular distribution or by a low number of species with regular distribution, an additional measure, the evenness (Pielou 1966), is calculated. The higher the value, the greater the evenness and the more regular the distribution. Evenness values for the temperate spider communities range from 0.37 to 0.94 with an average of 0.67 ± 0.12. The distribution is skewed to the right (skewness -0.55, kurtosis 0.07) (Figure 21).

The evenness in tropical spider communities is 0.64 at Pipeline Road, 0.80 at Cerro Galero and 0.84 at Cerro Pelado. These values do not exceed the temperate comparison but lie in the right half of its distribution range (Figure 21).

The difference in evenness between the temperate and tropical samples is not very pronounced but may indicate that the tropical values are distributed more evenly than the temperate values. The higher tropical diversity is therefore partly caused both by a moderately higher species number and a more regular distribution with low numbers of individuals per species.

Table 6. Continued.

family	British Columbia	North America	Costa Rica	Panama	Central Europe
Scytodidae		9	6	5	1
Loxoscelidae	1	13	1	2	
Diguetidae		6			
Plectreuridae		15			
Caponiidae		6	2	7	
Tetrablemmidae			1	2	
Pholcidae	3	42	12	21	5
Ochyroceratidae		1	2	2	
Leptonetidae		34			
Telemidae		10			
Segestriidae	1	6	1	1	3
Dysderidae	1	1			14
Oonopidae		25	18	42	6
Palpimanidae				1	
Dimetidae	2	13	5	16	4
Eresidae					1
Oecobiidae		7	1	1	2
Hersiliidae		2	1	2	
Deinopidae		1	1	1	
Uloboridae	1	15	12	21	2
Nesticidae	1	31	2	3	4
Theridiidae	41	231	116	227	69
Theridiosomatidae		2	6	8	1
Symphytognathidae		1	1	2	
Anapidae		1	1	7	
Mysmenidae		6	6	5	3
Linyphiidae	193	845	40	29	407
Tetragnathidae	15	38	40	56	21
Araneidae	25	133	102	171	46
Lycosidae	46	234	25	17	80
Pisauridae	1	14	11	21	3
Agelenidae	31	372	7	4	29
Argyronetidae					1
Desidae		1			
Hahniidae	6	19	1		10
Dictynidae	16	159	3	10	21
Amaurobiidae	10	78	2	1	8
Titanoecidae	2	4		1	4
Tengellidae		6	2	2	
Oxyopidae	1	20	17	25	4
Senoculidae			2	9	
Anyphaenidae	2	37	27	43	2
Miturgidae		9	5	4	
Liocranidae	5	122	3	2	17
Clubionidae	16	62	13	13	38
Corinnidae	2	37	36	48	1
Zodariidae		5	2	3	3

Table 6. Continued.

family	British Columbia	North America	Costa Rica	Panama	Central Europe
Homalonychidae		2			
Gnaphosidae	47	248	9	18	114
Ctenidae		6	25	18	
Zoridae		1	1		9
Selenopidae		5	5	7	
Heteropodidae		8	10	12	1
Philodromidae	29	95	2	3	24
Thomisidae	31	128	20	59	42
Aphantochilidae		1		3	
Salticidae	34	288	149	237	75
N families	28	62	52	55	38
N species	565	3614	779	1223	1079

5.2 Continental comparison

In Table 6 the number of species in 73 families which are known from North America, Central America and Central Europe is given. The richest spider fauna is reported for North America (more than 3600 species). From Central Europe and Panama 1100 to 1200 species are known; the corresponding figures for Canada and Costa Rica are much lower. However, two points should be borne in mind when comparing these data. Firstly: North America and Europe are relatively thoroughly investigated areas and the number of species discovered will certainly not increase much more. The Panamanian fauna is perhaps the best-investigated spider fauna in the tropics, but the total number of species is certainly much higher than the number reported here (compare p. 16). Costa Rica is far less well investigated and will certainly prove to have at least as many species as Panama. Secondly: British Columbia, Costa Rica and Panama are small states; North America and Central Europe are continents. This hundredfold difference in size and correspondingly more contrasting landscapes and habitat types cannot be compensated by the general richness of the tropics.

Table 6 shows that the fauna of British Columbia is a small subset of the North American fauna with only 1/6 of the species and roughly 1/2 of the families. Both Central American faunas are

Table 7: Families endemic or restricted to North America, Central America and Central Europe. Only the families which occur in these 3 areas are considered here for these areas, i.e. some families may also occur in other parts of the world which are not compared here (compare Table 6).

North America	Central America	Central Europe
Atypidae	Idiopidae	Atypidae
Antrodiaetidae	Actinopodidae	Dysderidae (1)
Mecicobothriidae (2)	Microstigmatidae	Eresidae
Nemesiidae	Barychelidae	Argyronetidae
Hypochilidae	Paratropididae	
Diguetidae (2)	Sicariidae	
Plectreuridae	Tetrablemmidae	
Leptonetidae	Palpimanidae	
Telemidae	Senoculidae	
Desidae		
Homalonychidae		

(1) One European species introduced to North America.
(2) Also known from Argentina.

fairly similar and differ from the North American fauna. The differences are exemplified by a lack of families, a few additional families and different species compositions. Nearly as many families are known from Panama as from North America; the number of species is approximately one third. Taking into consideration an enormous number of yet undiscovered species (compare p. 18), the Panamanian spider fauna will certainly comprise at least half the number of species recorded from the North American continent. A comparison with Central Europe shows that this continent has far fewer families and species than Panama and probably also than Costa Rica.

A number of the spider families in Table 6 is endemic to the areas of comparison or is restricted to these areas but may occur as well in other parts of the world not taken into consideration here. These families are listed in Table 7. Ten families from North America are not reported for Central America and Europe; 5 of them are endemic to North America and Argentina (Antrodiaetidae, Mecicobothriidae, Diguetidae, Plectreuridae, Homalonychidae). Atypidae occur in North America and Europe. Three European families are restricted to the Old World. Of the nine families mentioned for Central America, only 3 are neotropical endemites (Paratropididae, Sicariidae, Senoculidae); the other families are restricted to the tropics or to the Southern hemisphere.

Table 8: Species percentage of the 10 most common spider families in 5 different areas (compare Table 6).

family	British Columbia	North America	Costa Rica	Panama	Central Europe
Theridiidae	7.3	6.4	14.9	18.6	6.4
Linyphiidae	34.2	23.4	5.1	2.4	37.7
Tetragnathidae	2.7	1.1	5.1	4.6	1.9
Araneidae	4.4	3.7	13.1	13.9	4.3
Lycosidae	8.1	6.5	3.2	1.4	7.4
Agelenidae	5.5	10.3	0.9	0.3	2.7
Gnaphosidae	8.3	6.9	1.2	1.5	10.6
Philodromidae	5.1	2.6	0.3	0.2	2.2
Thomisidae	5.5	3.5	2.6	4.8	3.8
Salticidae	6.0	8.0	19.1	19.3	7.0
others	12.9	27.6	34.5	33.0	16.0

Table 8 gives a list of the above mentioned spider faunas with respect to those families which exceed in at least one area 5 % of the total number of species (these are the 10 most common families). While the percentages of the most common families in British Columbia and Central Europe are very similar, North America as a whole is different. The differences between North America and the Central American countries are even more distinct (i.l. larger). The dominant family in the Northern hemisphere is that of the Linyphiidae (more than one third of the European species). Their importance decreases towards the South and in Panama they comprise only 2.4 % of all species (8 other families have more species). Agelenidae are relatively important in North America, Gnaphosidae in Europe (about 10 % each). Theridiidae, Salticidae and Lycosidae comprise 6 - 8 % in North America and Europe. Towards the South these families increase or decrease in species number considerably and Panama always has more extreme values than Costa Rica. In Panama the most important families of web-building species are Theridiidae and Araneidae; the most important family of wandering species are Salticidae (with one fifth of all species the most common family of all) and Thomisidae. Further, the number of species from other families increases towards the South from 13 % in British Columbia to 33 % in Panama (Figure 22).

Among the other families some 10 families comprise only a few species and do not show clear distribution maxima or trends. These are Dipluridae, Caponiidae, Ochyroceratidae, Hersiliidae,

Deinopidae, Symphytognathidae, Mysmenidae, Selenopidae and Aphantochilidae. Most of them have tropical distribution maxima but the spider fauna of these countries deserves further investigation. The Zoridae are a family with a primarily European distribution.

The remaining families show more or less clear maxima or trends in species numbers (Table 9). Amaurobiidae, Liocranidae, Agelenidae, Cyrtaucheniidae, Philodromidae and Linyphiidae have an extremely high species number in the North and a low species number in the South (species proportions ranging from 1:78 to 1:29). The families with the most species in the tropics are Anapidae, Ctenidae and Theridiosomatidae (proportions from 1:7

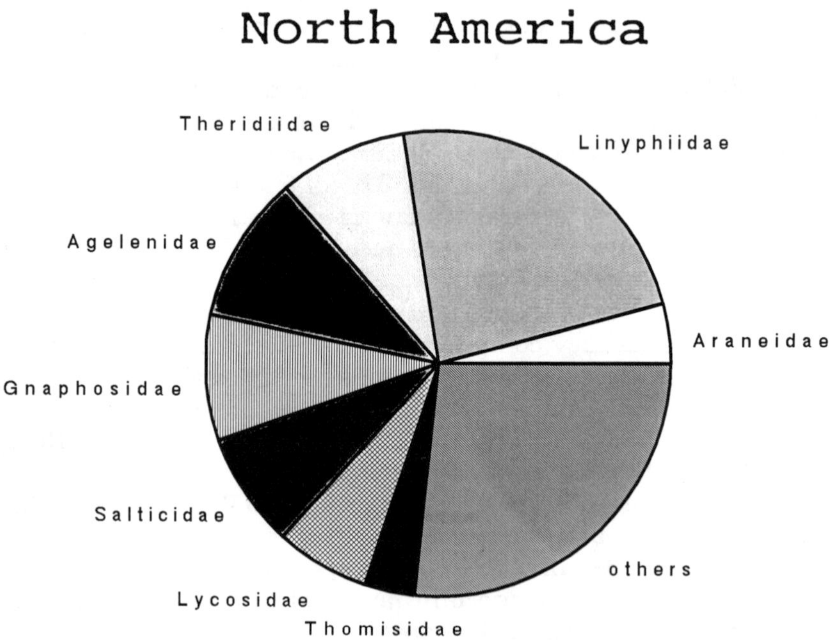

Figure 22a. Composition of the spider fauna in North America (data from Roth 1985).

to 1:4). Many species seem to be evenly distributed in both climate regions but the different degrees of intensity of the investigations and the areas under comparison should be taken into account as well. Theraphosidae, Thomisidae, Pholcidae, Salticidae and Theridiidae, which have according to Table 9 in North America higher species numbers than in Central America, are certainly families with more species in the tropical regions. Further investigations here will probably confirm this hypothesis. Correspondingly, the small majority of species in the South for Oxyopidae, Mimetidae, Anyphaenidae, Araneidae and Uloboridae will presumably increase at the same time (Table 9).

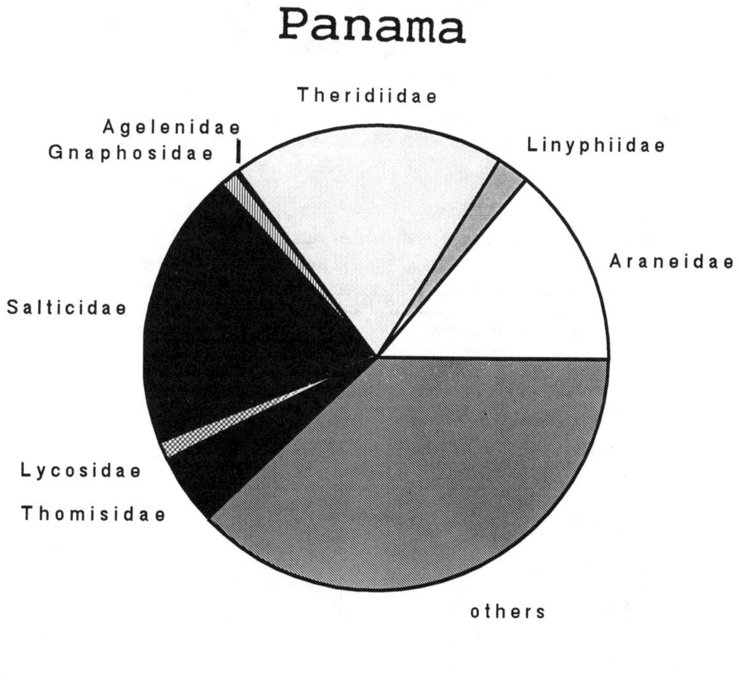

Figure 22b. Composition of spider fauna in Panama (data from Heimer & Nentwig 1991) with respect to the most common families.

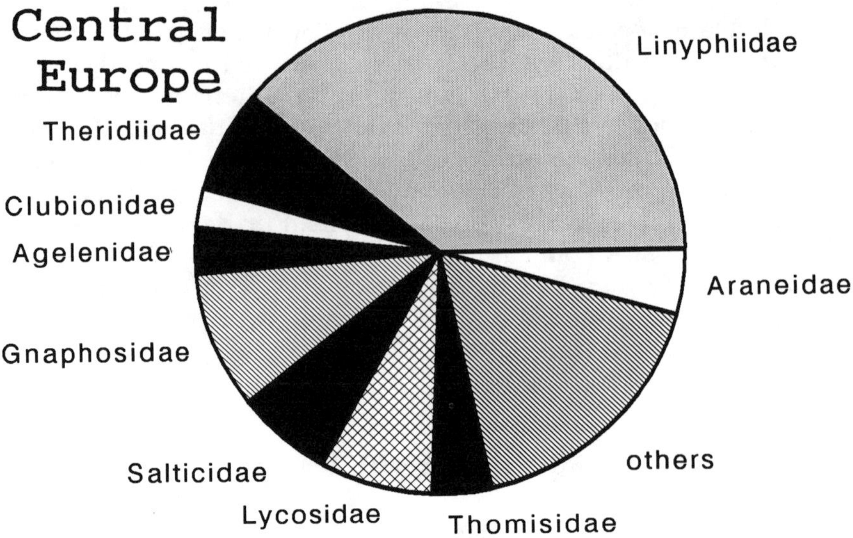

Figure 22c. Composition of spider fauna in Central Europe (data from Heimer & Nentwig 1991) with respect to the most common families.

Table 9: Relative species numbers of spider families in North America or Panama (compare Table 6). Numbers indicate the factor by which the species number in one area exceeds the species number in the other area (in decreasing order).

most species in North America		most species in Panama	
Amaurobiidae	78	Anapidae	7.0
Liocranidae	61	Theridiosomatidae	4.0
Agelenidae	53	Ctenidae	3.0
Cyrtaucheniidae	39	Heteropodidae	1.5
Philodromidae	32	Pisauridae	1.5
Linyphiidae	29	Tetragnathidae	1.5
Hahniidae	19	Uloboridae	1.4
Dictynidae	16	Araneidae	1.3
Gnaphosidae	14	Clubionidae	1.3
Lycosidae	14	Anyphaenidae	1.2
Filistatidae	13	Mimetidae	1.2
Nesticidae	10	Oxyopidae	1.2
Ctenizidae	8		
Loxoscelidae	7		
Oecobiidae	7		
Segestriidae	6		
Clubionidae	4.8		
Titanoecidae	4.0		
engellidae	3.0		
Zodariidae	2.5		
Theraphosidae	2.4		
Thomisidae	2.2		
Pholcidae	2.0		
Miturgidae	1.8		
Scytodidae	1.8		
Salticidae	1.2		
Theridiidae	1.02		

Chapter 6. Origin of the Panamanian spider fauna

6.1 The evolution of spiders and the continental drift

How old are spiders? The oldest known fossils (from the Devonian and Carboniferous periods (380 and 300 million years old)) have been described by Petrunkevitch (1955) and Shear et al. (1984) from the Devonian and Lower Carboniferous (380 and 300 million years old). These descriptions concern the most plesiomorphic group of spiders, the suborder Mesothelae. Another fossil from the Upper Carboniferous, *Megarachne servinei*, is possibly a mygalomorph spider or belongs to an extinct group from the ancestral stock of spiders (Hünicken 1981). For the subsequent period of nearly 200 million years, the Mesozoic, no spider fossils have been found.

Many fossil spiders have been found from the tertiary period, most of them in amber. Lebanese amber is 100 - 120 million years old, Baltic amber 40 million years and Dominican amber 15 - 40 million years. Hundreds of species are known and their degree of preservation is so excellent that most of them can be attributed to recent families, many even to existing genera (Wunderlich 1986, 1988). The evolution of the majority of recent spider families and genera, therefore, probably occurred between the Devonian and the early Mesozoic (200 - 400 million years ago).

The geographical position of Central America and especially of Panama changed several times during this time. At the end of the Paleozoic (500 million years ago) the supercontinent Pangea comprised North America, South America, Europe and Africa (and other continents). In the Permian (250 million years ago) Pangea divided into a northern part (Laurasia) and a southern part (Gondwana) (Figure 23). The rift in Central America changed its form several times and increased in size. The Caribbean Sea, the Gulf of Mexico and the North Atlantic developed and in the Cretaceous (100 million years ago), North and South America were widely separated (though a row of islands connected both continents). South America and Africa were still connected until the South Atlantic developed and separated both continents. During

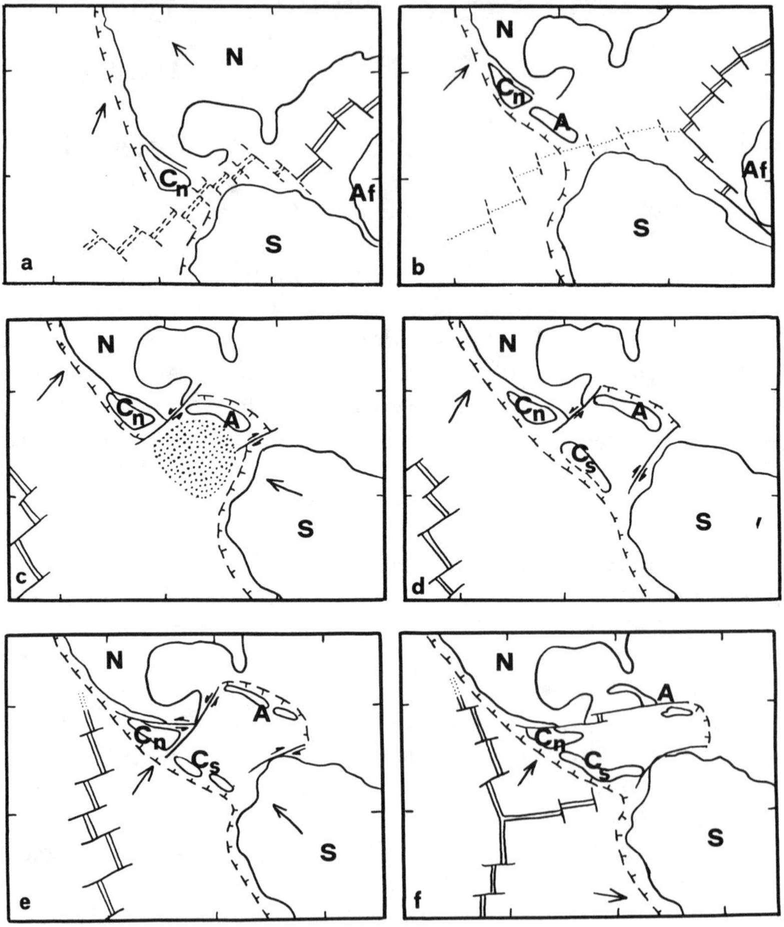

Figure 23. The Central American land bridge from the Cretaceous to the present (from Gursky 1987, 1989, modified). **a.** lower Cretacious; **b.** middle Cretacious; **c**; upper Cretacious; d. end of Cretacious; **e.** Eocene; **f.** lower Miocene. A Antilles, Af Africa, Cn Central America (north), Cs Central America (south) (south = Costa Rica and Panama), N North America, S South America.

the Mesozoic both Americas were separated by a row of islands and parts of the Central American land mass which sometimes certainly functioned as a land bridge between both continents. In the Cretaceous this land bridge was pushed eastward (Cuban and Puerto Rican Trench) and the Americas were separated until the Pliocene (6 million years ago) (Figure 23).

With respect to spider evolution, both Americas, Europe and Africa were connected for the long time in which most recent taxa (families and genera) probably evolved. South America and Africa were connected for a longer period than North and South America, so the zoogeographical similarities between the southern continents should be greater than those between both American continents. For at least 70 million years South America was completely isolated from other continents but for the last 3 million years it has been reunited with the North.

Thus the southern part of Central America which is today a part of Panama was for most of its existence a zone of transition (Schmidt 1954, Halffter 1987), or a filter area (Simpson 1950, Udvardy 1969) between the Nearctic and the Neotropical spheres. According to Udvardy (1975) it has the status of a biogeographical province which he called Panamanian.

6.2 The origin of Panamanian spiders

Banks (1929) discussed the origin of the Panamanian spider fauna and (on the basis of 240 species) suggested that there is a marked affinity between the fauna of Panama and that of South America. Ten years later Petrunkevitch (1939) analysed the Panamanian spider fauna and (on the basis of 503 species) came to the opposite conclusion. He stated that

- the Central American fauna is quite distinct from the South American one;
- the transition from the one to the other is quite abrupt;
- Central America was separated at one time from both North and South America.

His more detailed analysis showed that 37 % of all species known from Panama were known only from that country, 41 % of all species were known from Panama and Central America, 7 %

were known in Panama and South America, and about 20 % were cosmopolitan, cosmotropical or neotropical. According to Petrunkevitch (1939) an analysis of the older faunal data of Banks (1929) yielded quite similar results and the corresponding figures were about 30, 40, 10 and 20 %.

Several and more general aspects, however, were not taken into account by Petrunkevitch.

1. The spider fauna of Panama was not sufficiently well known (Petrunkevitch knew of only 41 % of the species which we know today); the fauna of Central America as a whole is probably less well known and the South American fauna is even today inadequately investigated. The low percentage of known species occurring in both Panama and South America, therefore, does not indicate a great discrepancy between these areas but rather demonstrates our lack of knowledge of the southern fauna.

2. Taxonomical problems in many groups excluded them for detailed calculations or led to wrong conclusions.

3. Insufficient knowledge of the dispersal capabilities of spiders. Petrunkevitch assumed that spiders migrate by land, through transport by man or by floating but greatly underestimated the importance of ballooning. Today we know that probably all spiders (including mygalomorphs) can (at least as hatchlings) balloon and that this is an important means of dispersal (Decae 1987).

4. Data on spider evolution and on the geological history of the continents were in Petrunkevitch's time far less precise than today.

The area of today's Panama consists of parts of the old Central American landmass, parts of South America and new connections resulting from recent tectonic shifts (Figure 23). In view of the enormous dispersal capabilities of spiders, this area was certainly invaded from the North and from the South. Regarding the origin or composition of the Panamanian spider fauna, therefore, one may postulate that

- the Panamanian spider fauna consists of both Central American and South American species;
- the percentage of Central American species should decrease towards the South and the percentage of South American species should decrease towards the North;

- the number of endemic species should be low;
- general factors such as similarities in vegetation or climate will even out ancient faunal differences and no sharp boundaries or abrupt transitions should exist.

Today, such a North - South transition (which is within the area of Panama a West - East - transition) is difficult to demonstrate because the most thoroughly investigated area is the central Isthmus (the former Canal Zone) and the eastern part of Panama (Darien) in particular has been only sketchily investigated. In view of our yet insufficient knowledge of the South American fauna continent-wide comparisons of complete faunas are even more difficult and will not be attempted here. These "white spots" also render discussions about discussions endemic species or genera less convincing because in my opinion most "endemic species of Panama" will be found in Costa Rica or Colombia sooner or later (see also p. 61). An additional point of interest would be a comparison with the African spider fauna. The long common history of South America and Africa should result in many faunal similarities but the African spider fauna (especially the tropical) is even less well investigated than the South American fauna.

6.3 The distribution of selected taxa

The following, more detailed considerations refer only to a few genera of the Panamanian spider fauna which have been intensively investigated in modern revisions. I chose (my own subjective selection) *Achaearanea* (Levi 1959 b, 1963 d), *Dipoena* (Levi 1963 b), *Theridion* (Levi 1963), *Micrathena* (Levi 1985), *Chrysometa* (Levi 1986 a), *Anapis*, *Mysmenopsis*, *Anapisona* and *Pseudanapis* (Platnick & Shadab 1978 a, b, 1979), *Cesonia* (Platnick & Shadab 1980), *Apollophanes* (Dondale & Redner 1975) and (for North and Central America only) *Oxyopes* and *Hamataliwa* (Brady 1964, 1970, 1975) and *Selenops* (Muma 1953). These papers cite 282 Panamanian species (about 23 % of the known spider fauna of Panama), but in contrast to earlier treatments (see above) this quarter of the Panamanian fauna enjoys a clearly defined taxonomic status and its distribution in America is fairly well known (but see below).

Where else do these species occur in the Americas? The greatest affinities were established with Costa Rica and Mexico where 30 - 32 % of the selected Panamanian species have been found as well. In the other Central American countries these

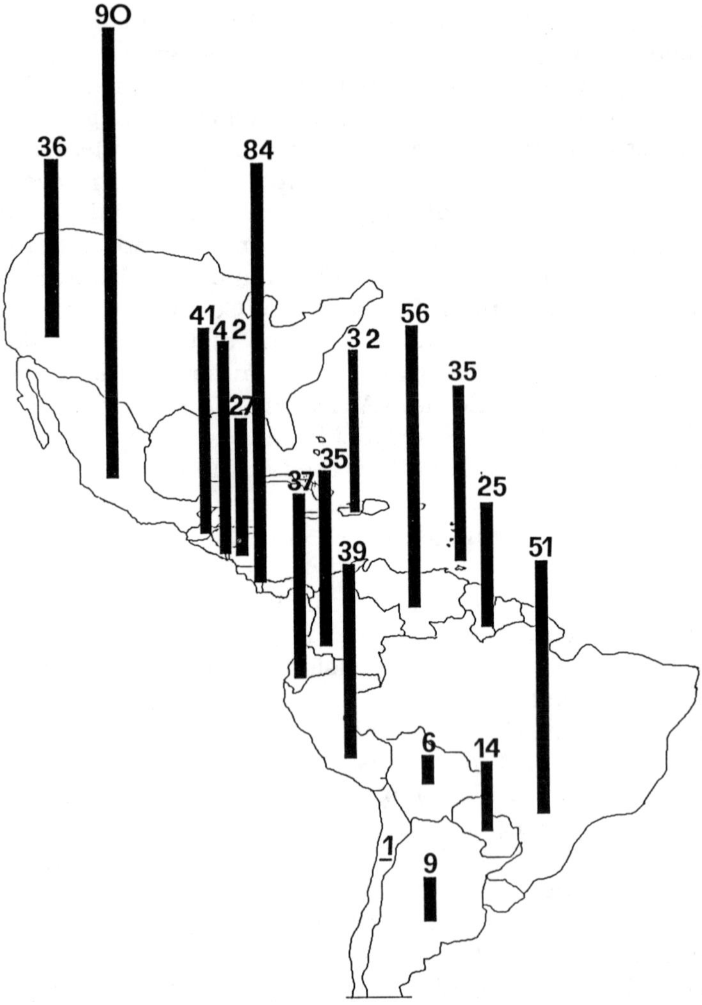

Figure 24. The distribution of 282 selected Panamanian species in the Americas; figures indicate the number of species per country (for composition and references compare text).

figures range between 10 and 15 %; probably mainly a problem of neglected field research. In the USA, 13 % of the selected Panamanian spiders have been found as well. This figure is nearly as low as the corresponding one for the Caribbean area (11 % for the Greater Antilles, 12 % for the Lesser Antilles). In South America, Brazil and Venezuela show a 18 and 20 % correspondance, other tropical countries 9 - 14 %. In Uruguay, Argentina and Chile the correspondance drops from 3% to almost zero (Figure 24).

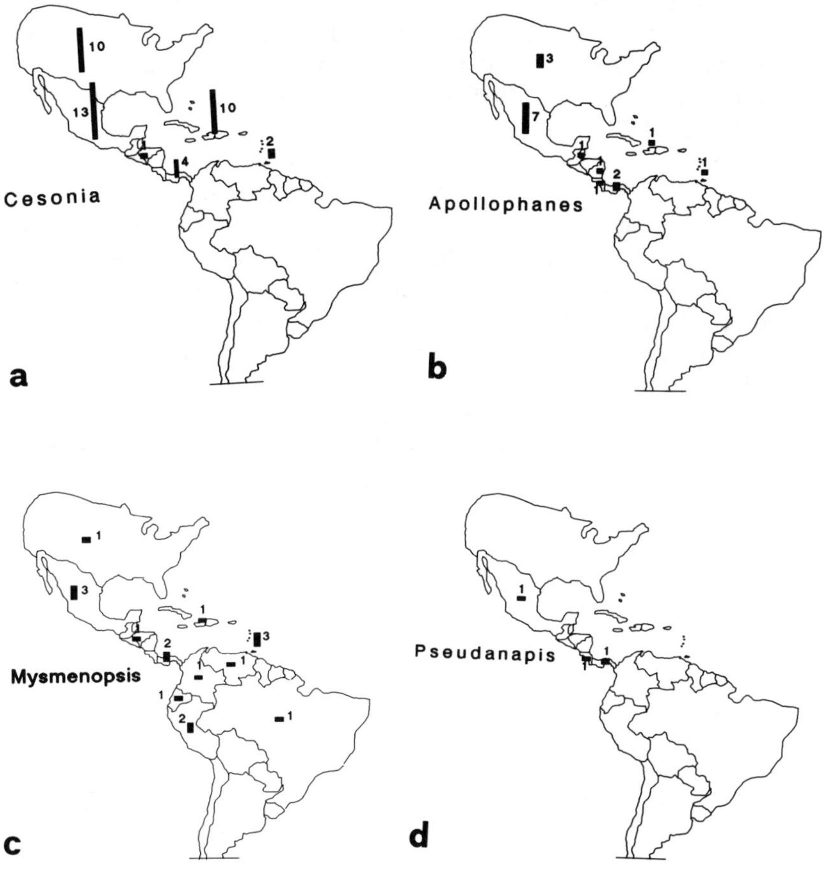

Figure 25. The number of species in the Americas of **a**. the gnaphosid genus *Cesonia*, **b**. the philodromid genus *Apollophanus*, **c**. the mysmenid genus *Mysmenopsis* and **d**. the anapid genus *Pseudanapis* (data from Dondale & Redner 1975, Platnick & Shadab 1978a, 1979, 1980; compare text).

Species which were found in Panama and in more than one other country are rare, with figures ranging from 4% to 15%. The combination of Panama and any northern country excluding the Caribbean was found in 49 % of all records, the combination of Panama and any southern country excluding the Caribbean was found in 38 %; Panama and any Caribbean country in 20 % of all records. The last three figures suggest that there must still be an enormous deficiency of field records. If 49% of the selected species were found in any northern country, and if the highest individual record (Panama and Mexico 32%) is much lower, hundreds of species remain to be discovered in the other countries. The discrepancy towards the South is even greater (the corresponding figures are 38% and 20%). A rough estimate basing on these figures and for these selected taxa (which are assumed to be thoroughly investigated, see above) could suggest that at least one third of the Central American spider fauna remains yet to be found. Taking into account my earlier estimate of the possible total number of Panamanian spider species (about 2000, see page 18), this would reduce the percentage of the known fauna to approximately 60% for Panama, to 40% for Central America and to 30% for tropical South America.

These calculations also relativize the value of giving percentages for endemic species. A total of 90 from 282 species (32%) is only reported from Panama, in most cases only from one locality. I believe that it is much more probable that these species have not

Figure 26. The number of species in North and Central America of the oxyopid genera *Hamataliwa* and *Oxyopes* and the selenopid genus *Selenops* (data from Muma 1953 and Brady 1964, 1970, 1975; compare text).

yet been collected in adjacent countries where they occur than that they are really restricted to small areas within Panama.

Do there exist centres of origin? On the generic level, many taxa clearly have distinct centres of origin.

Central American genera: The genus *Cesonia* has a high number of species in North America with Mexico and the Greater Antilles, low numbers in Central American countries and does not occur in South America (Figure 25).

Apollophanes has its highest species numbers in Mexico, is represented in other Central American countries by only 1 - 2 species and has not been found in South America (Figure 25). *Pseudanapis* and *Mysmenopsis* are genera with low species numbers and occur mainly in Central America and adjacent countries (Figure 26). Considering the distribution in North and Central America alone, most species of *Hamataliwa*, *Oxyopes* and *Selenops* are mainly to be found in Mexico (Figure 26).

South American genera: *Anapisona* and *Anapis* are represented in all American regions by at least one species, but most species are known from Colombia. For *Anapisona*, the Panamanian species number is about half that of Colombia (Figure 27); for

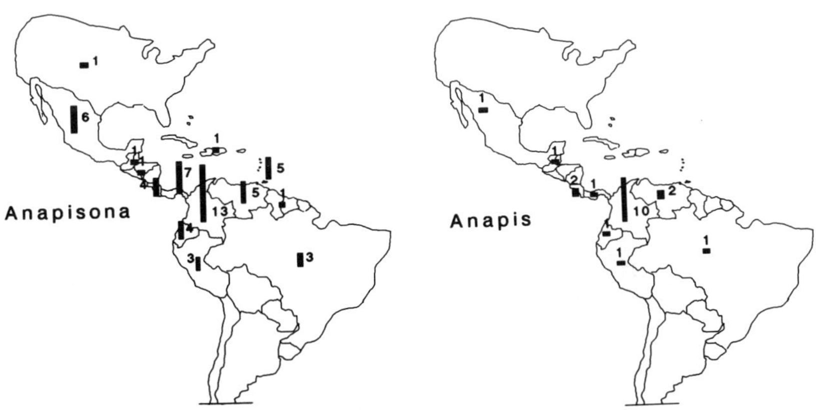

Figure 27. The number of species in the Americas of the anapid genera *Anapisona* and *Anapis* (data from Platnick & Shadab 1978 b, 1979; compare text).

Anapis there is only 1 species known from Panama but 10 from Colombia (Figure 27). The following five genera considered here all have a very high species number. *Dipoena* has the most species in Panama but is also well represented in Brazil (Figure 28). *Chrysometa* has most species in Colombia and Ecuador; *Theridion*, *Micrathena* and *Achaearanea* have their species maxima in Brazil. *Theridion* also has high species numbers in Peru, Ecuador and Venezuela; *Micrathena* has many species in several other countries ranging from Argentina to the United States. High species numbers of *Achaearanea* occur between Peru, Panama and Mexico (Figure 28).

On the species level these distribution patterns are similar and they show even more clearly the transitional position of Panama between the North and the South. *Philoponella semiplumosa*

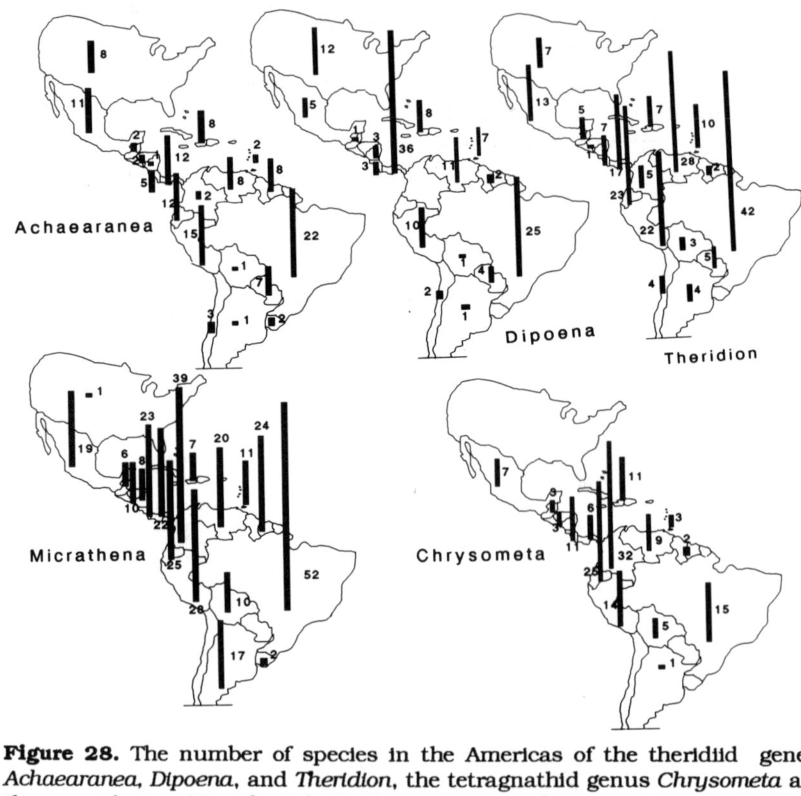

Figure 28. The number of species in the Americas of the theridiid genera *Achaearanea*, *Dipoena*, and *Theridion*, the tetragnathid genus *Chrysometa* and the araneid genus *Micrathena* (note: there are 4 *Micrathena* species in the US) (data from Levi 1959 a, b, c, 1963 d, 1985, 1986 a; compare text).

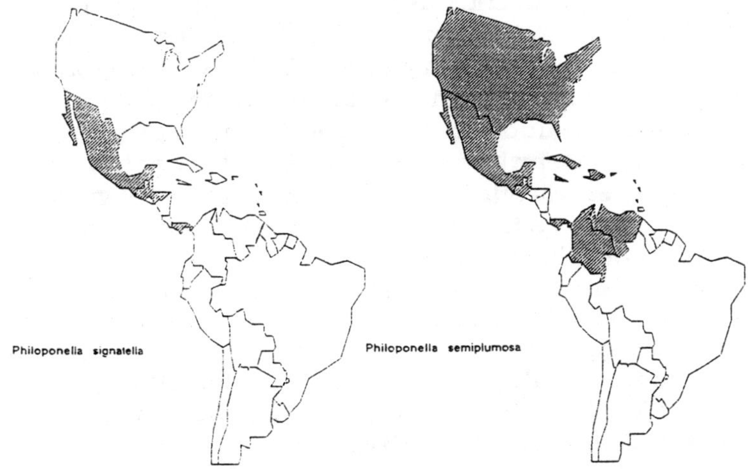

Figure 29. The distribution of selected North and Central American species which reach Panama and the most Northern South American countries: the uloborids *Philoponella semiplumosa* and *P. signatella*, the philodromid *Apollophanes punctipes*, the theridiid *Thymoites delicatulus* and the araneids *Micrathena sagittata* (additional records in the eastern United States) and *M. quadriserrata* (data from Levi 1964 c, 1985, Dondale & Redner 1975, Opell 1979).

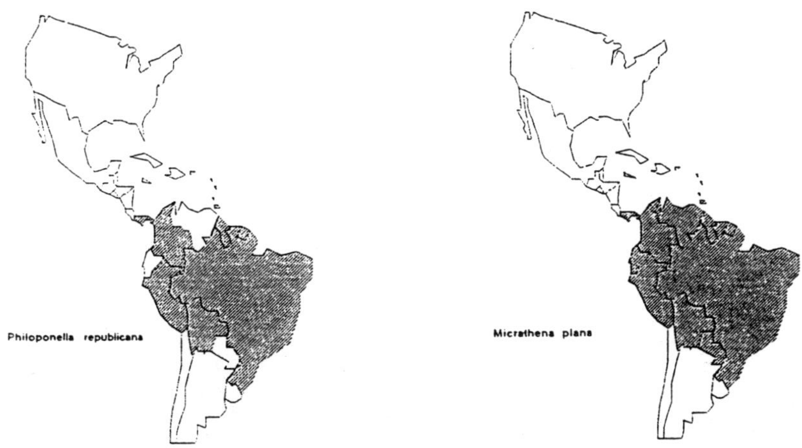

Figure 30. The distribution of selected South American species which reach Panama from the South: the uloborid *Philoponella republicana* and the araneid *Micrathena plana* (data from Levi 1985, Opell 1979).

occurs from North America to the northern countries of South America (Figure 29). The Central American species *Apollophanes punctipes*, *Philoponella signatella* and *Micrathena sagittata* occur from Mexico (or eastern North America) to Panama; no records are known for Caribbean or South American countries (Figure 29). *Micrathena quadriserrata* is widely distributed in Central America but also in Venezuela (Figure 30) (and certainly in Colombia too). *Thymoites delicatulus* is known from Mexico to the northern part of South America (Figure 30).

The South American *Philoponella republicana* and *Micrathena plana* (Figure 30) are widely distributed in the tropical countries of South America but do not occur north of Panama. Species with a more general tropical distribution are *Zosis geniculatus*, *Theridion hispidum*, *Micrathena sexspinosa* and *Eriophora ravilla* which even occurs in the USA (Figure 31).

Figure 31. The distribution of selected species which are widespread from North to South America: the uloborid *Zosis geniculatus*, the theridiid *Theridion hispidum*, and the araneids *Micrathena sexspinosa* and *Eriophora ravilla* (data from Levi 1963 c, 1970, 1985, Opell 1979).

Chapter 7. Spider associations

Although I do not wish to claim that associations between spiders and spider webs or specific spider-plant communities are exclusively tropical phenomena, we become aware of them more often in the rich tropical environment where these characteristics appear more pronounced than in the temperate zone. Therefore I present some data here on a limited number of subjects, most of them only preliminarily studied. The selection itself is my own arbitrary choice, reflecting more my areas of interest than the importance or degree of research done in the respective field. Thus, my main interest is not to give a serious and complete overview but to increase the personal curiosity of the interested reader.

7.1 Spiders and plants

Characteristic plant communities bear a specific arthropod fauna. In many cases some details are known; in respect of spiders, however, almost nothing has been investigated. There are spider species or species combinations which obviously occur anywhere while other species are restricted, for instance, to large and rough leaves, bark types, special flowers or specific ground or litter structures. A few data are given in the following sections.

7.1.1 Flowers

For spiders, inflorescences are ideal places in which to live. This habitat is often highly structured (for protection, camouflage etc.) and the potential prey density is high (pollinators, herbivorous insects, plant suckers etc.). An example of a typical floral spider community is given in 10 for the weed shrub *Lantana camara* (Verbenaceae). More than 70 % of all spiders which I collected from the inflorescences were thomisid spiders and more

Table 10: Spiders at the inflorescences of *Lantana camara* L. (Verbenaceae), Pipeline Road, June to September 1983, 350 flowers. Since many spiders were juvenile identification could not always be made to genus or species level.

family	genus / species	N
Thomisidae	*Misumenops dubius* *Runcinia* sp., *Thomisus* sp.	67
Clubionidae		9
Araneidae		8
Salticidae		4
Anyphaenidae	*Teudis mordax*	2
Oxyopidae	*Oxyopes salticus*	1
Pisauridae	*Thaumasia* sp.	1
Theridiosomatidae		1
total		93

than 90 % of all the spiders collected were active searchers or sit and wait predators. The most common web-building family (Araneidae) probably uses the *Lantana* flower heads to build small webs within a flower or (more probably) between the flower and adjacent green parts of the plant.

As typical pollinators of *Lantana*, Lepidoptera, small bees, and some flies were observed. Insects other than true pollinators on the flower heads include large numbers of ants and Tingidae (Heteroptera). Further records mention Aphidina, Cicadina, Pentatomidae, Thysanoptera, Psocoptera, Ensifera, Coleoptera (Staphylinidae, Chrysomelidae, Curculionidae), parasitic hymenopterans and Collembola.

Table 11: Spiders at the inflorescences of *Palicourea guianensis* (Rubiaceae), Cerro Galera and Pipeline Road, June to September 1983, 240 flowers. Since many spiders were juvenile identification could not always be made to genus or species level.

family	genus / species	N
Salticidae		41
Thomisidae	*Synaema haenschi*, *Thomisus* sp.	14
Araneidae	*Verrucosa* sp.	14
Philodromidae	*Apollophanes* sp.	13
Clubionidae		5
Oxyopidae	*Hamataliwa triangularis*	1
Theridiidae	*Dipoena* sp.	1
Pisauridae		1
total		90

Table 12: Spiders at the inflorescences of *Rhynchospora cephalotes* (L.) and *R. nervosa* (Vahl) (Cyperaceae) (several locations in the Canal Zone, April - November 1983). Since many spiders were juvenile identification could not always be made to genus or species level.

family	genus / species	N
web building spiders		
Araneidae	*Argiope* sp., *Bertrana striolata*, *Eustala* sp., *Mangora picta*, *Nephila clavipes*, *Wagneriana tauricornis*	54
Theridiidae	*Dipoena* sp., *Coleosoma acutiventer*, *Euryopis* sp., *Theridion armouri*, *T. dilucidum*, *T. galerum*	36
Dictynidae		7
Linyphiidae		5
Symphytognathidae	*Anapistula secreta*	1
sit and wait spiders		
Salticidae	*Hentzia parallela*	22
Clubionidae		20
Thomisidae	*Epicadinus* sp., *Mysmenoides rugosa*, *Thomisus sp.*	14
Oxyopidae	*Oxyopes aglossus*	5
Pisauridae		3
Oonopidae	*Oonops persitus*	1
Scytodidae	*Scytodes* sp.	1
Philodromidae		1
Lycosidae		1
total		171

In contrast to the *Lantana* flowers, the inflorescences of another shrub, *Palicourea guianensis* (Rubiaceae), are inhabited predominantly by salticids (46 % of individuals); further common ambushing spider families are Thomisidae, Philodromidae and Clubionidae. Web-building families comprise about 17 % of the individuals (Table 11). *Palicourea* flowers house astonishingly few insects. Hummingbirds are important pollinators but I also observed pollinating insects, especially large numbers of butterflies, some bees, bumblebees and larger flies. Ants dominate on all flower parts; in addition I found in low numbers Heteroptera, Cicadina, Aphidina, Psyllina, several Diptera groups, Coleoptera (Curculionidae, Lamellicornia, Chrysomelidae, Staphylinidae), Blattodea, Thysanoptera and Collembola.

The varying degrees of attractivity of the flowers of both shrubs for insects is reflected by the density of spiders (0.27 spiders per *Lantana* flower head; 0.38 spiders per *Palicourea* flower head). The flat flower heads of *Lantana* measure 1.5 - 2.0 cm in diameter, whereas the large inflorescences of *Palicourea* are 12 - 18 cm long and 10 - 15 cm wide. Taking into consideration the much larger surface and volume of the *Palicourea* flowers, their spider density is approx. 1/50 to 1/100 of the *Lantana* spider density.

There are also flowers which look quite attractive but do not house spiders. Such an example is *Psychotria marginata* Sw. (Rubiaceae), a common shrub, 2 - 3 m tall, with large composed stands of white flowers. The flowers and most parts of the plant are covered with ants. Only a few insects (Brachycera, parasitic hymenopterans, Thysanoptera, Collembola) could be detected. I observed *Trigona* bees as occasional pollinators. From a total of 69 inflorescences only 1 spider (a juvenile araneid) was collected.

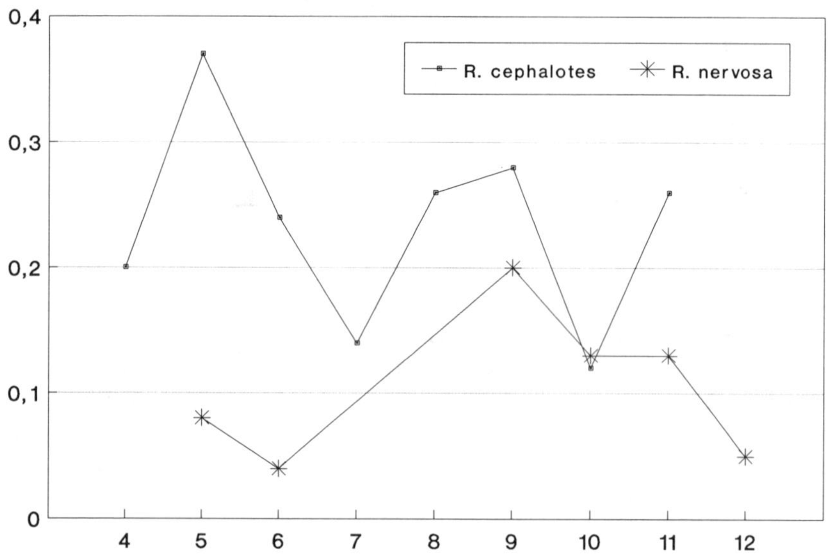

Figure 32. The density of spiders per inflorescence of *Rhynchospora nervosa* Vahl and *Rhynchospora cephalotes* (L.) (Cyperaceae), two entomophilous sedge species from various locations on Pipeline Road, April - November 1983; 50 flower heads for *R. cephalotes* and 200 - 600 flower heads for *R. nervosa* per month.

Grass is normally pollinated by the wind and its inflorescences are not especially attractive to insects or spiders. I found, however, remarkable exceptions. The large inflorescences of tall grasses, widespread along roadsides in Panama, (several species, among them the so-called "elephant grass") offer a good hiding place for araneids which build their webs between the grass stalks. In 80 inflorescences I found 88 spiders, 66 % Araneidae, 19 % Salticidae and among the other spiders of several families 3 % large Heteropodidae (gravid females which had their retreat in the grass inflorescences). *Trigona* bees and Collembola were observed in high numbers and fed on the pollen; further insect records include Cicadina, Aphidina, Thysanoptera, Ensifera, Coleoptera (Chrysomelidae, Cantharidae, Staphylinidae, Curculionidae) and Lepidoptera larvae.

In contrast to the majority of wind-pollinated grasses, a few Cyperaceae have readapted to insect pollination and attract a large number of pollinators (Leppik 1955). One may assume that the spider fauna of these inflorescences has also adapted to these changes and now resembles a typical flower fauna. *Rhynchospora* is a sedge genus which contains several species displaying spikelets which are gathered into a terminal head and involucrated by 4 - 8 upper leaves. These leaves are white at the base and resemble the white petals of an ordinary flower. *Rhynchospora nervosa* (Vahl) (= *Dichromena ciliata*) is a close imitation of a snow-white flower of a simple type; it is widespread at roadsides in the Canal Zone and is very conspicuous. The sedge 'flowers' attract a large number of pollinators, especially *Trigona* bees, honey bees, and smaller bees, parasitic hymenopterans and several dipterans (Syrphidae, Muscidae etc.). Other insects were regularly found on the flowers, among them Cicadina, Aphidina, several Heteroptera families, Caelifera, many Ensifera, Coleoptera (Chrysomelidae, Curculionidae), Thysanoptera and many Collembola. *Rhynchospora cephalotes* (L.) has less white leaves but more compact and larger flowers and is also highly attractive to a wide range of pollinators and other insects.

The spider fauna collected from the inflorescences of the two *Rhynchospora* species is listed in Table 12. Ambushing spiders are not the major group (as one would assume for insect pollinated flowers) but rather the web-building Araneidae and Theridiidae (52 % of all spiders). Sit and wait predators, however, represent a further large group of flower spiders and constitute about 37 % of the total. Among them the most common families are Salticidae, Clubionidae, Thomisidae and Oxyopidae.

The density of spiders per sedge inflorescence varied within a range of 0.05 to 0.45 spiders per flower (Figure 32). The mean densities were 0.11 spiders per *R. nervosa* and 0.24 spiders per *R. cephalotes* inflorescence. The difference between these figures can be explained by the much larger size of the flower heads of the latter species. No clear seasonal influence and no distinct difference between the spider or insect associations of either species could be detected (Figure 32).

7.1.2 Leaves

In July 1990 we collected all the spiders found on the underside of 8500 leaves from a variety of herb and shrub species with leaves at a height ranging from 0.1 - 2.5 m above ground along Pipeline Road near Gamboa. The result is shown in Table 13. The 158 spiders (one every 54 leaves) belong to 11 families and represent about 30 species. Most salticids (except *Chinoscopus flavus)* crawl all over the vegetation and are obviously not restricted to the underside of the leaves, for which reason we did not record them. Orb weavers (Araneidae and Tetragnathidae) are certainly users of the free space between the leaves and should not be counted as true "leaf spiders". The remaining 89 % can be divided into three groups of spiders with varying degrees of specialisation on this microhabitat.

First group (stenecous species): Four species (a theridiid, the salticid *Chinoscopus flavus*, an anyphaenid and a heteropodid are the most common species under leaves. The salticid *Chinoscopus flavus* and the anyphaenid were found with a frequency of once every 370 leaves, the theridiid once every 470 leaves. This specialized group is characterized by an elongated body and very long legs. The colour is bright green or yellow and additional black dots or stripes camouflage the spider. The salticid *Chinoscopus flavus* and the anyphaenid sit in a large, silken retreat which also functions as a catching web. Dust and dirt particles (most of them of plant origin) are placed on the silken construction by the spider and provide an additional optical camouflage for the web resident. The theridiid sits in a fold of the leaf or between major veins but spins

Table 13: Spiders found on the underside of leaves from a variety of shrubs (N leaves = 8500, 0.1 - 2.5 m above ground, July 1990, Pipeline Road / Gamboa).

family	genus / species	N
Theridiidae	species 1	18
	species 2	5
	species 3	3
	species 4	2
	others	5
Salticidae 1)	*Chinopus flavus*	23
Anyphaenidae	1 species	23
	others	23
Araneidae 2)		16
Pisauridae	*Architis*	6
	Thaumasia	2
	Staberius	1
	juvenile	1
Heteropodidae	1 species	9
Scytodidae	*Scytodes* sp.	8
Thomisidae	various species	6
Pholcidae		3
Mimetidae		3
Tetragnathidae	*Tetragnatha* sp.	1
total		158

1) Without a variety of salticid species running all over the vegetation and obviously not restricted to the leaf underside;
2) *Eriophora*, *Verrucosa*, *Witica*, and other genera.

a distinct catching web. The heteropodid folds the margin of the leaf to build a silken retreat.

Second group: Spiders with some adaptations to the leaf underside but not very often found and perhaps more common in slightly different microhabitats or areas. This group comprises 3 families (Theridiidae, Anyphaenidae and Pholcidae) with 6 - 8 species and 21 % of all spiders. All theridiids (probably 3 species) are bright yellow or green; some of them are skilful builders of camouflages, silken retreats or catching webs (as described above), other are difficult to see due to a partially silver pigmentation or black dots. The pholcids belong to *Metagonia panama* and the anyphaenids are probably several species; most specimens were juvenile and the biology of many species is not well known.

A third group consisting of pisaurids, mimetids, scytodids, other theridiids and thomisids comprises about 20 % of the total and consists of approximately 15 species. Many species show some kind of adaptation to living under leaves but they were not commonly found there and occurred frequently in other habitats.

Are the most common spiders specialized to certain kinds of leaves? We distinguished between hairy and smooth leaves, weak and strong veins and leaves of tender or xeromorphic appearence. In general, xeromorphic leaves with strong veins had more spiders than tender leaves with weakly developed veins. Our analysis showed that the anyphaenid species had a strong preference for hairy leaves with a pronounced veinage while the salticids *Chinoscopus flavus*, one theridiid species and the pisaurids were more frequently found under smooth, xeromorphic leaves. The other spider species showed no significant preferences.

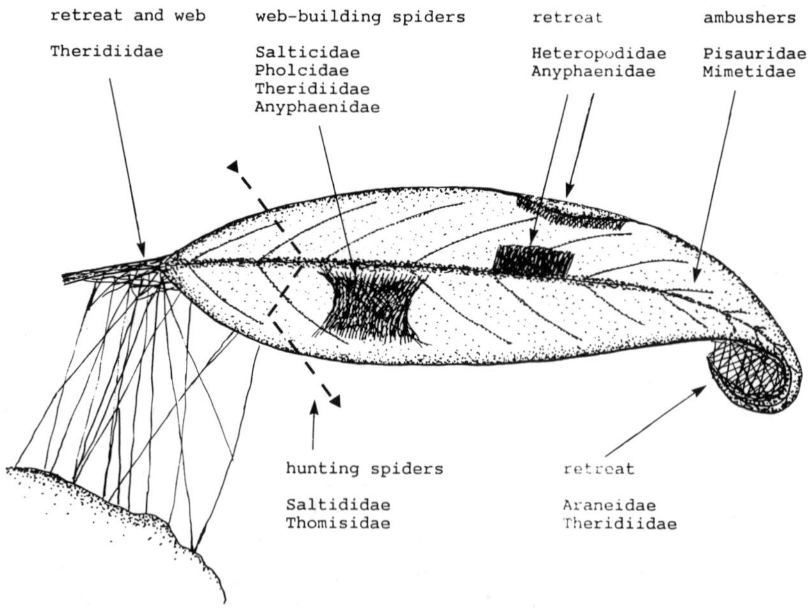

Figure 33. Niche separation among spiders on the underside of a leaf (schematical, compare text).

Table 14: Spiders associated with banana plants at Changuinola (summer 1965, N spiders per stratum) (data from Harrison 1968). *Wittica crassicauda, Nephila clavipes* and *Leucauge argyra* were only collected from banana mats, the clones of a mother and its suckers, and a *Lycosa* sp. and a *Pirata* sp. were taken from bare ground around the bases of banana plants.

species	extended leaves	hanging leaves	pseudostem	ground-cover vegetation	total N
Theridula gonygaster	122	61		12	195
Chrysso clementinae	112	60		5	177
Metaphidippus cf. *centralis*	24	14			38
Chrysso albomaculata	18	12			30
Metaphidippus octonotatus	15				15
Verrucosa mexicana	8				8
Eustala sp.	4				4
Wagneriana undecimtuberculata	3		1		4
Thiodina sylvana	2				2
Corythalia spiralis		11			11
Eriophora edax	2	4	1	4	11
Dictyna sp.		2			2
Mimetus sp.		2			2
Metaphidippus aff. *longipalpus*		2			2
Thiodina sp.	1	2			3
Theridion sp.		1			1

With respect to these preferences and special predatory techniques (such as web building, ambushing or free hunting) it should be possible to characterize different ecological niches on the underside of a leaf. Although our knowledge of the biology of the spiders mentioned is lacking in many details, one may postulate a niche separation schema such as the one in Figure 33 where I have tried to describe some of the ways in which spiders can live under leaves.

The preference of the lower leaf surface has also been documented by Greenberg & Gradwohl (1980) who counted in Panama 334 spiders under the leaf but only 16 spiders on the leaf. In nearly all cases the question of why (or by what) the spider species are linked to a special plant species cannot be answered. Much more general research is needed to throw light on these interesting relationships.

Table 14. Continued.

species	extended leaves	hanging leaves	pseudostem	ground-cover vegetation	total N
Tidarren fordum		1			1
Asthenoctenus sp.			7		7
Ctenus aff. *peregrinus*			2		2
Ctenus cf. *rubripes*			1		1
Olios stylifer			1		1
Mangora spinula		2		35	37
Argiope argentata		1		31	32
Thaumasia argenteonotata				24	24
Metaphidippus aff. *laxus*				15	15
Larinia directa				14	14
Olios sp.			1	13	14
Misumenops dubius				10	10
Oxyopes gracilis				9	9
Gasteracantha cancriformis		2		6	8
Lycosa sp.				6	6
Pellenes paratus				4	4
Cheiracanthium inclusum		1		3	4
Phiale sp.				3	3
Eustala fusco-vittata				2	2
Tmarus aff. *rugosus*				2	2
Corythalia sp.			1	1	2
Misumenoides rugosa				1	1
Metazygia wittfeldae				1	1
Coleosoma acutiventer				1	1
Heteropoda sp.				1	1
Chrysometa alboguttata				1	1
total	311	178	15	204	708

7.1.3 Banana plants and Bromeliaceae

Excellent data from an investigation into the spider community associated with one plant species is given by Harrison (1968) for banana plants (*Musa*, Musaceae) in Changuinola, Bocas del Toro). In Table 14 I have rearranged his data into groups of spider species which are more common on specific plant parts. Assuming that Harrison collected all plant parts with the same thoroughness, one sees clearly that the extended, dry leaves are the most attractive part of the banana plant (most species, highest individual numbers) and that the pseudostem is the least attractive plant part (fewest species and individual numbers). The leaves are dominated by 2 theridiid species; the next most common family is salticids. *Theridula gonygaster* and *Chrysso clementinae* are the

Table 15: Spiders from epiphytic Bromeliaceae and moss (March 1983, Pipeline Road). Since many spiders were juvenile identification could not always be made to genus or species level.

family	genus / species	N
Oonopidae	*Oonops* sp.	7
Scytodidae	*Scytodes longipes*	5
Salticidae		5
Ochyroceratidae		2
Pholcidae	*Modisimus* sp.	2
Prodidomidae		1
Hersiliidae	*Tama mexicana*	1
Loxoscelidae	*Loxosceles* sp.	1
Clubionidae		1
Araneidae		1
Theridiidae		1
Linyphiidae		1
total		28

most common species on both extended and hanging leaves. There are, however, 5 species which were only found on extended leaves and 10 species which were only found on hanging leaves. Only five species are common to both leaf types, indicating that they represent separate ecological niches. The pseudostem is inhabited by a typical ctenid and heteropodid community which does not occur elsewhere.

The ground-cover vegetation which Harrison encountered was probably a diverse herb layer. It contains 24 species (in contrast to 27 - 30 species on the banana plant), 15 of which do not occur on banana plants. The dominant species (4 araneid species and 5 species of wandering spiders) comprise 77 % and do not occur elsewhere.

The large tank bromeliads in the moist mountain forest of Chiriqui offer a special habitat for many arthropods. The spider fauna seems to be very specialized, consisting mainly of web-building spiders (Theridiidae, Araneidae, Linyphiidae and Agelenidae). The latter are of especial interest because the open end of the funnel of their webs lead into the waterbody of the bromeliads. When threatened many spiders flee into the water and stay motionless on the bottom of the tank for a long time.

A spider collection from smaller epiphytic Bromeliaceae and moss from the lowland forest is given in Table 15. It comprises 12 families and most taxa are represented by only 1 or 2 specimens.

Web-building spiders are not obviously dominant but rather Oonopidae, Scytodidae and Salticidae. This spider composition shows some similarities to the results of an investigation of Young & Lockley (1989) who found in Spanish moss (*Tillandsia usneoides* L., Bromeliaceae) in the Mississippi delta (USA) only 3 % web-building spiders, but 45 % salticids, 31 % anyphaenids and 9 % philodromids.

Further special habitats or microhabitats are moss on tree trunks or litter layers on the ground, on trees, in tree ferns, bromeliads etc. Special characteristics of particular plant species, however, often enable spiders to build their web or retreat in typical communities. Large rolled-up leaves form retreats for certain heteropodid and ctenid species. The gaps and holes in the leaves of the palm *Carludovica* offer web-sites for juvenile *Thaumasia* and *Architis*.

7.1.4 Bark

In July 1990 we collected a total of 224 spiders (Table 16) from 45 trees (from the ground up to a height of 2 m, including buttresses, trunk diameter 20 - 80 cm at breast height, average diameter 45 cm) in the area around Pipeline Road and Gamboa. This represents approx. 5 spiders per tree. Nearly half of the spiders were salticids and 4 other families contribute a further 9 - 13 % each (Pholcidae, Scytodidae, Gnaphosidae, Theridiidae). Seven additional families were found, but were represented only by a few specimens. Some of the spiders must be regarded as bark specialists, e.g. some salticids (*Corythalia bicincta, Albionella propria,* perhaps also *Corythalia obsoleta, Cobanus electus* and *Balmaceda picta)* and the hersiliids. Many other species are not primarily restricted to tree bark but occur there regularly as well as in other habitats.

For some species (e.g. gnaphosids), a tree represents an upward extension of the ground (especially if the tree has well-pronounced buttresses); some species (e.g. theridiids, araneids and uloborids) regard it as a downward extension of the tree top. Pholcidae (most of them were *Modisimus pulchellus)* and Scytodidae can be found in a wide range of habitats including tree trunks, which provide very suitable conditions. Pisauridae (*Syntrechalea* amd *Trechalea* species), Philodromidae, Senoculidae and

Table 16:. Spiders (%) collected from the bark of 45 trees in the forest near Pipeline Road / Gamboa (0 - 2 m height, stem diameter 20 - 80 cm) and in Panguana, Peru (data from Manhart, in prep.).

family	Panama	Peru
Salticidae	41	6
Pholcidae	13	62
Scytodidae	10	
Theridiidae	9	
Araneidae	4	1.5
Pisauridae	2	13.5
Hersiliidae	2	2
Philodromidae	1	
Senoculidae	1	3
Uloboridae	1	9.5
Heteropodidae	0.5	
Dipluridae		1.5
Tetragnathidae		1
Cyrtaucheniidae		1
total N	224	136

Heteropodidae prefer all kinds of large, smooth surfaces such as tree bark, rocks, walls etc. Thus, many species of this group are preadapted for living in and on buildings.

We distinguished between smooth and rough bark, bark with or without moss, and trunks with or without buttresses. Our analysis showed some preferences for special bark or tree types: The few hersiliids and pisaurids which we collected were found exclusively on smooth bark. The few philodromids and heteropodids were collected only from trees which were not covered with moss and senoculids were found only on trees with an intermediate degree of moss cover. Salticids were found on rough and smooth bark types without pronounced preference but were more common on trees without moss. Scytodids were more common on trees with smooth bark without moss. Moss cover and bark type did not appear to influence pholcids or gnaphosids but they showed a preference for the spaces between buttresses of large trees, which were also highly attractive for scytodids, senoculids, theridiids and araneids. From these results and our experiences while collecting the spiders we can present a schema which shows how the different spider groups may divide the habitat "tree bark" into different microhabitats (Figure 34).

A comparative collection of bark spiders has been performed by Manhart (in prep.). His results are listed in Table 16 and differ considerably from the Panamanian data. In the Peruvian forest more Pholcidae, Pisauridae and Uloboridae were found, the percentage of Salticidae was much lower and Scytodidae and Theridiidae were lacking completely.

An interesting microhabitat is provided by the surfaces of termite nests (*Nasutitermes* sp.), which resemble rough bark and which are also often attached to tree trunks. We collected 50 spiders from 12 nests. This amounts to an average density of 4.2 spiders per termite nest but three nests were without any spider. This figure is 10 - 100 times greater than the corresponding spider density on tree bark. Scytodidae were the most common family (46 %), but only because one nest was "occupied" by a large group of juveniles. Leaving this group out of consideration salticids are the

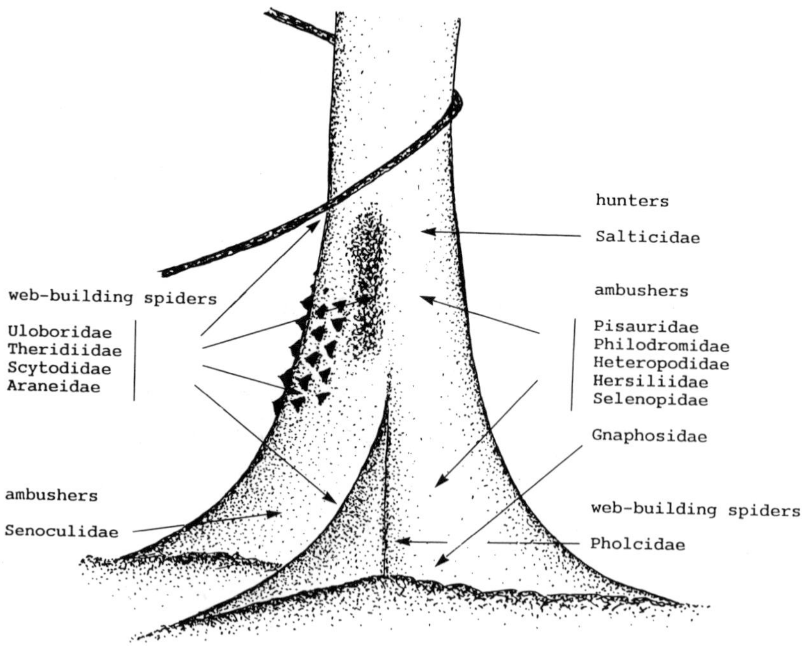

Figure 34. Niche separation of spiders on tree trunks up to a height of 2 m (schematic, compare text). The graph shows a liana, thorns, moss (dark area beside the thorns) and buttresses.

Table 17: Spiders (%) collected from the surface of 12 *Nasutitermes* termite nests (Pipeline Road / Gamboa, July 1990).

family	%
Scytodidae	46
Salticidae	32
Senoculidae	8
Theridiidae	2
Thomisidae	2
Pholcidae	2
Gnaphosidae	2
Linyphiidae	2
juveniles	4
total N	50

predominant family (Table 17). This corresponds to the results obtained from bark (Table 16). Salticids from termite nests were *Albionella propria*, *Corythalia bicincta*, *Freya* (juveniles) and *Nycerella* sp.. With the exception of *Freya*, all of them could also be found on bark. The total number of spiders found during this survey of termite nests may be too low to allow a detailed comparison, but the high percentage of Senoculidae and the low percentage of Pholcidae (*Modisimus dilutus*) are striking. We observed only two spiders feeding on termites (a salticid and a theridiid), so it is difficult to say whether the high spider density on termite nests can be explained by the high abundance of this potential prey. In feeding tests, however, most spiders refused termites (Nentwig unpublished). Other arthropod taxa which were regularly observed on termite nests include Heteroptera (3 reduviid species, 1 nabid species, the reduviids at least feeding on termites), ants (feeding on termites) and opilionids. Small opilionids were also the only non-termite arthropods which we found inside termite nests. We expected to find spiders in the termite nests and analysed several nests thoroughly, but did not find any spiders.

7.2 Spiders on spider webs

A web-building spider is of course closely associated with its web. Other interactions involve insects which may be prey or predator of the web resident spider. This microcosmos of web-building spider and various insects is attractive for other spiders as well. They may steal prey from the web (kleptoparasites) or prey on the resident spider. For a recent review on the biology of kleptoparasitic spiders, see Vollrath (1987).

The numbers of 'tenant' spiders per larger spider web are impressive. Most orb webs are occupied by kleptoparasitic *Argyrodes* (Theridiidae), often represented by 2 - 3 different species. The frame threads of orb webs or the peripheral web parts of larger space webs offer a suitable habitat for other orb weavers which build their own small webs there. This is true for Uloboridae (mainly *Philoponella* species) and several araneid and tetragnathid species. Besides some other web building spiders (e.g. Pholcidae and Linyphiidae) wandering spiders such as Lycosidae, Thomisidae and Salticidae can be found. They probably invade the webs and prey on the resident spider. In the case of salticids and thomisids this could be observed several times on the webs of *Argiope argentata*

In the course of 1 year of collecting I analysed the spider fauna associated with several spider species at several locations and I give these data here.

The spider fauna of *Nephila clavipes* webs consists mainly (93 %) of kleptoparasitic *Argyrodes* species, but among the 7 species found at 3 locations only 3 species are more common (*A. caudatus, dracus* and *elevatus*). The other spiders are uloborids, araneids and tetragnathids (6 %) and a few spiders from other families (< 1 %) (Table 18). Among adult *Argyrodes*, *A. caudatus* dominates with 32 % (Pipeline Road) to 51 % (Cerro Pelado), *A. elevatus* is dominant at Cerro Galero (54 %) and contributes 22 - 35 % at the other locations. The percentage of *A. dracus* ranges from 3 % (Cerro Galero) to 22 % (Pipeline Road).

The spider fauna of *Eriophora fuliginea* webs resembles that of the *Nephila* webs. *Argyrodes* comprise a high percentage of all

Table 18: Spiders collected at the webs of *Nephila clavipes* at 3 locations in the Canal Zone during 1 year.

	Cerro Galero	Cerro Pelado	Pipeline Road	total N
Theridiidae				
Argyrodes caudatus	89	159	88	336
A. dracus	7	42	37	86
A. elevatus	120	108	38	266
A. fictilium			1	1
A. projiciens	1			1
A. spinosus			5	5
A. ululans	3		1	4
A. juvenile	216	320	107	643
Theridion sp.		1	1	2
Uloboridae	17	20	5	42
Araneidae/Tetragnathidae				
Cyclosa caroli	1	1		2
Gasteracantha cancriformis	1	1		2
Leucauge argyra		1		1
Leucauge lugens		1		1
Leucauge venusta	3			3
Leucauge sp.	16	8	1	25
Mangora picta	1	1		2
Micrathena mitrata		1		1
Micrathena sp.		1		1
Tetragnatha sp.			1	1
Wagneriana undecimtuberculata	1			1
sp.	10	1	1	12
other families				
Pholcidae	2			2
Linyphiidae		1		1
Lycosidae	2			2
Thomisidae	1	2	2	5
total				1450

spiders (98 %), but hardly any uloborids (0.1 %) or other spiders (about 1 %) were found. One explanation of this contrast is probably that *Nephila* webs are fairly permanent orb webs with a pronounced barrier web while the *Eriophora* webs are built at night and removed in the early morning hours and, thus, do not offer a permanent web support. The *Eriophora* webs were studied at the same locations as the *Nephila* webs and I found 5 *Argyrodes* species. Two of the rare species (*A. fictilium* and *A. spinosus*) were not detected on *Eriophora* webs. *A. caudatus* in *Eriophora* webs comprises a much higher percentage than in *Nephila* webs (56 - 75 %); the number of *A. elevatus* is much less in *Eriophora* webs (2 -

Table 19: Spiders collected at the webs of *Eriophora fuliginea* at 3 locations in the Canal Zone during 1 year.

	Cerro Galero	Cerro Pelado	Pipeline Road	total N
Theridiidae				
Argyrodes caudatus	91	55	65	211
A. dracus	31	16	20	67
A. elevatus	35	8	2	45
A. projiciens	1	3		4
A. ululans		1		1
A. sp.	83	60	56	199
Uloboridae			1	1
Araneidae/Tetragnathidae				
Argiope savignyi		1		1
Eustala sp.	1			1
Leucauge venusta		1		1
Leucauge sp.			1	1
Mangora spinula	1			1
Wagneriana sp.			1	1
sp.			1	1
other families				
Anyphaenidae	1			1
total				536

22 %) while *A. dracus* is represented on both webs with about 20 % of the total (Table 19).

The smaller orb webs of *Argiope argentata* and *A. savignyi* are as ephemeral as the *Eriophora* webs but are built during the day and their spider fauna resembles the *Eriophora* fauna. Nearly all spiders are *Argyrodes* (98 %) and the most common species are the same: *A. caudatus* (55 %) and *A. elevatus* (37 %) for the webs of *A. argentata* (Table 20).

The spider faunas in the large and permanent webs of the social spiders *Anelosimus jucundus* and *A. eximius* (Theridiidae) differ in many ways from the araneid spider faunas. *Argyrodes* constitute only 67 % (5 species, with *A. caudatus* as the most common species among the adult *Argyrodes* (74 %)) and uloborids comprise about 20 %. Differences among species and locations are, however, considerable. The *Anelosimus eximius* webs at El Valle support 3 *Argyrodes* species but almost no uloborids, whereas at Cerro Galera the *A. eximius* webs housed only one *Argyrodes* (*A. caudatus*)

Table 20: Spiders collected at the webs of *Argiope argentata* and *A. savignyi* at 2 locations in the Canal Zone during 1 year.

	Argiope savignyi	*Argiope argentata*		
	all locations	Cerro Pelado	Pipeline Road	total N
Theridiidae				
Argyrodes caudatus	18	31	35	84
A. dracus	3	5	5	13
A. elevatus	35	18	4	57
A. sp.	28	176	73	277
Uloboridae		2	1	3
Araneidae/Tetragnathidae				
Gasteracantha cancriformis		1		1
Leucauge sp.			3	3
Micrathena sp.		1		1
Tetragnatha sp.			1	1
sp.		1	1	2
other families				
Pholcidae		1		1
Thomisidae		1	1	2
Salticidae	1			1
total	85	237	124	446

but at least 3 Uloboridae species (*Philoponella diversa*, *P. signatella* and *P. tingens*). The other groups comprise a rich orb weaving fauna (10 %) and 5 additional families, among them Gnaphosidae, Palpimanidae, the scytodid *Scytodes longipes* and the caponiid *Nops guanabacoae* (Table 21).

Collections of spider associations from other web-building spiders yield a heterogenous picture. A sample from webs of the araneid *Mecynogea lemniscata* was populated by 11 *Argyrodes*, 2 uloborids and 1 salticid. Webs of 4 theridiid species (*Achaearanea tesselata*, *Tidarren fordum*, *T. sisyphoides* and *Theridion rufipes*) yielded a collection of 23 *Argyrodes*, 9 Uloboridae and 1 juvenile *Witica*. In the webs of several diplurid spiders (*Euagrus* and 1 *Ischnothele*) I found only 1 *Argyrodes* but 14 Uloboridae, as well as *Scytodes longipes* and 2 *Mysmenopsis dipluramigo* (Symphytognathidae). Uloborids were also the only inhabitants of pholcid (*Modisimus inornatus*) and pisaurid (*Architis nitidopilosa*)

Table 21: Spiders collected at the webs of *Anelosimus jucundus* and *A. eximius* at Cerro Galera and El Valle during 1 year.

	Anelosimus jucundus	*Anelosimus eximius*	
	El Valle	El Valle	Cerro Galera
Theridiidae			
Argyrodes atopus		3	
A. caudatus	19	6	29
A. dracus		10	
A. elevatus	5		
A. spinosus	1		
A. sp.	38	18	48
Uloboridae		2	50
Araneidae/Tetragnathidae			
Argiope sp.			1
Cyclosa sp.	1		
Leucauge moerens		4	
L. venusta		6	
L. sp.	1	8	3
Mangora spinula		1	
Micrathena horrida			1
M. sexspinosa		1	
other families			
Caponiidae			1
Scytodidae			3
Palpimanidae			1
Gnaphosidae			1
Salticidae			1
total	65	59	139

webs. A small colony of *Kukulcania* webs (Filistatidae) was also inhabited by 1 *Scytodes* and 1 Zodariidae.

On the basis of approximately 400 webs per species (collected over 1 year) the average kleptoparasitic load for *Argiope argentata* is 0.6 *Argyrodes* per web; the maximum value observed was 6 spiders per web. The much larger webs of *Eriophora fuliginea* and *Nephila clavipes* contain about 3 - 4 times as many: *Eriophora* 2.75 *Argyrodes* per web (maximum values about 15 - 20); *Nephila* 2.02 *Argyrodes* per web (maximum value 27 spiders per web).

The number of free-living *Argyrodes* which permanently search for webs must be high. In a simple experiment I collected all *Argyrodes* from a sample of 10 *Nephila* webs and made a daily collection of the newly arrived kleptoparasites for a total period of 24 days. On the first day, I found 1.63 *Argyrodes elevatus*, 0.25 *A. caudatus* and 2.8 juvenile *Argyrodes* per *Nephila* web. The mean number of newly arriving *A. elevatus* on days 2 - 24 was only 0.15/day, while 0.26 *A. caudatus* and 1.55 juveniles arrived daily per web (Figure 35). These data indicate that the initial density of *A. elevatus* is 11 times higher than the daily arrival rate, while arrival rate and initial density is more or less balanced in *A. caudatus* and varies for juvenile *Argyrodes* only by a factor of 2. Apart from seasonal and local differences, there are obviously different population strategies, migration intensities and interspecific hierachies which can explain these results.

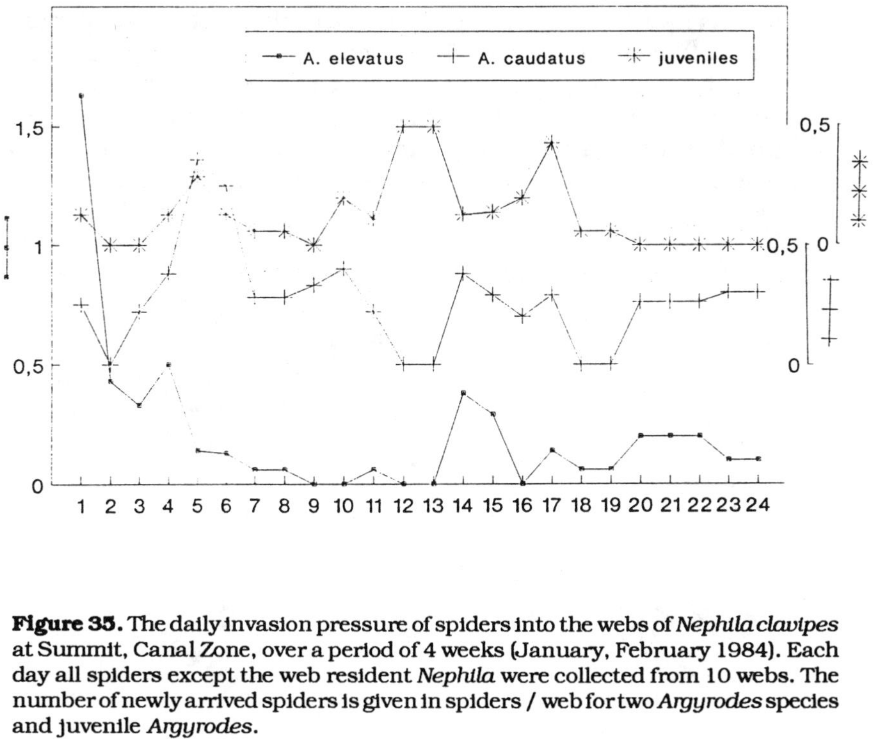

Figure 35. The daily invasion pressure of spiders into the webs of *Nephila clavipes* at Summit, Canal Zone, over a period of 4 weeks (January, February 1984). Each day all spiders except the web resident *Nephila* were collected from 10 webs. The number of newly arrived spiders is given in spiders / web for two *Argyrodes* species and juvenile *Argyrodes*.

Chapter 8. Sites of spider collections

Collecting spiders in Panama is greatly restricted by the infrastructure of the country, especially the road network. So it is no wonder that most collectors only worked at a few sites and new collections have often been repeated at the old locations. Not more than 80 spider collection sites have been reported for Panama and most of them are grouped in certain areas. Figure 36 gives the position of locations in the Republic of Panama and Figure 37 gives the position of locations in the former Canal Zone. I could not localize the published collection site "Bella Vista". This name is

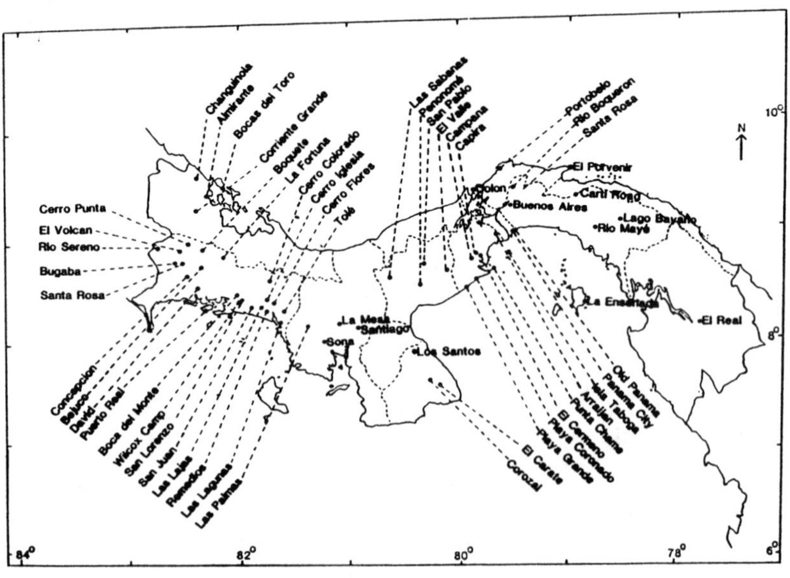

Figure 36. Position of places in the Republic of Panama from where spider collections have been reported. Additional locations are: Gatuncillo 10 km NNW of Buenos Aires; La Yeguada 25 km SW of Las Sabanas; Bambito is close to Cerro Punta; Aguadulce is 30 km SW of Penonomé.

cited several times for 5 provinces in "Censos Nationales de 1960" (1961), which gives a complete list of all localities in the Republic of Panama. Three sites could only be localized on province level: (1) Finca Lerida in Chiriqui, (2) Desierto de Sarigua in Herrera and (3) Villa Darien in Panama (or in Darien, compare Levi (1991b)). There are 5 El Salto in Veraguas.

In several cases the orthography of collection sites as given by the collector differed from the official orthography (as given in "Censos Nationales de 1960"). Examples are (incorrect citations in brackets) El Cermeno (El Cermano), Bocas del Toro (Boca del Toro), Boca del Monte (Boca de Monte) or Portobelo (Porto Bello, Porto Belo). Alto Limon Lo Yegueda (Veraguas) probably refers to a location near La Yeguada. Articles and accents are often omitted. Spanish parts of names are often translated: Rio (River), Isla (Island), Cerro (Peak), Lago (Lake).

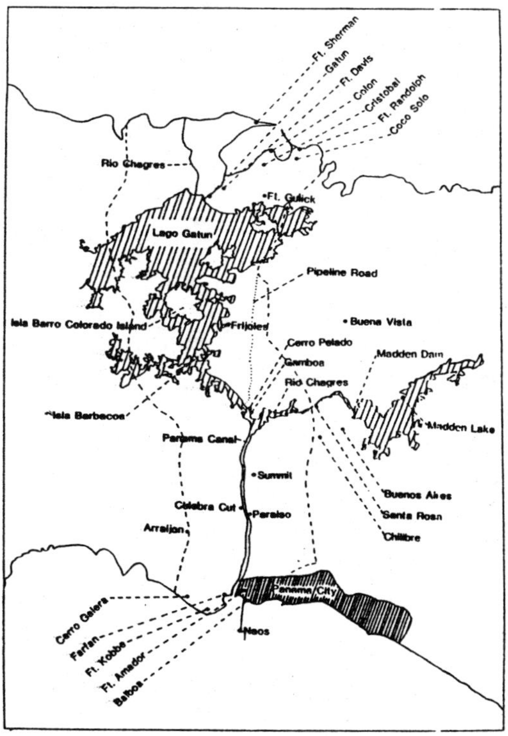

Figure 37. Position of places in the former Canal Zone from where spider collections have been reported.

Many spider species have been named after the land, the name of a province or another location. Among the species listed in the subsequent check list, the species name *panamensis / panamanus* occurs very often (29 x), followed by *barroanus* (10 x). The names *balboanus* and *chiriqui* occur 3 times, the names *isthmicus, volcanensis, zonatus / zonensis, canaliculatus / canalis, boquete, vallensis* and *arraijanicus* twice and the name *gamboa, taboga, galero* and *changuinola* once.

It is evident that nearly half of all the collection sites are in the former Canal Zone and that most other sites lie near the Pan American Highway (towards Costa Rica; eastwards there is no land connection with Colombia). Thus, there are enormous gaps where almost no collections have been made: the part of the country which lies east of the Isthmus, the Azuero peninsula, the mountainous parts of Chiriqui and the northern coastal areas.

A small strip of land across the Isthmus of Panama was rendered to the USA in 1903 and named the Canal Zone (width circa 15 km, including both Canal sides, the total area of Lake Gatun and the Madden Lake, excluding the cities of Panama and Colon). In 1977 the USA agreed to give this area back to the Republic of Panama until the end of 1999. As a consequence of these political events the names of the area and of places within it have changed several times (and probably will change in the foreseeable future even more). In this book I prefer to use the by far most often cited name Canal Zone (or former Canal Zone) but there is no doubt that this area belongs to the Republic of Panama and is part of the province of Panama (Southern half of the Canal Zone) and of the province of Colon (Northern half of the Canal Zone). Parts of the Canal Zone in the area around Barro Colorado Island (also called Isla Barro Colorado) were sometimes called the Canal Zone Forest Reservation (or simply the Forest Reservation); the Barro Colorado Island area was also called the Barro Colorado Natural Monument. Today this area forms a National Park called Parque National Soberania.

Chapter 9. Species list

The following list of species reported from Panama gives:

- the valid genus and species name (underlined);
- the author who described the species;
- synonymies which are important for understanding the nomenclature of Panamanian species (in brackets);
- references;
- localities in Panama (see Figure 35, 36).

Species List

Cyrtaucheniidae

(Ancylotrypa panamana Petrunkevitch = Bolostromus p.)

Bolostromus panamanus (Petrunkevitch)
(= Ancylotrypa p.)
Petrunkevitch (1925), Decae (pers. comm. 1988)
Wilcox Camp, Canal Zone
(This species occurs sympatrically with a second as yet undescribed species, Decae, pers. comm. 1988)

Idiopidae

Idiops sp.
Chickering (1936), Decae (pers. comm. 1988)
Barro Colorado Island, El Valle

Neoctenlza occulta Platnick & Shadab
Platnick & Shadab (1981 b), Decae (pers. comm. 1988)
Barro Colorado Island, Gamboa

Ctenizidae

(Pachylomerus = Ummidia)

Ummidia eremus (Chamberlin)
(= Pachylomerus e.)
Chamberlin (1925)
Barro Colorado Island
Ummidia rugosus (Karsch)
(= Pachylomerus r.)
Reimoser (1939)
Panama

Actinopodidae

Actinopus robustus (O. Pickard-Cambridge)
(= Pachyloscelis r.)
O. Pickard-Cambridge (1892), F.O. Pickard-Cambridge (1897), Petrunkevitch (1925), Decae (pers. comm. 1988)
Veraguas, Coclé, Province of Panama, Herrera, Los Santos, Chiriqui, Canal Zone, Gamboa

(Pachyloscelis robustus O. Pickard-Cambridge = Actinopus r.)

Microstigmatidae

Micromygale diblemma Platnick & Forster
Platnick & Forster (1982)
Cerro Colorado (Chiriqui)

Dipluridae

(Accola downeyi Chickering = Masteria d.)
(Accola spinosa Petrunkevitch = Masteria spinosa)

Diplura guatemalensis (F.O. Pickard-Cambridge)
(= Euagrus g.)
Petrunkevitch (1925), Nentwig & Heimer (1983), Decae (pers. comm. 1988)
Wilcox Camp, Canal Zone, Gamboa

(Diplura sp. see Linothele sp.)
(Euagrus guatemalensis F.O. Pickard-Cambridge = Diplura g.)

Ischnothele guyanensis Walckenaer
Petrunkevitch (1925, 1939 b), Banks (1929), Vollrath (1978), Nentwig & Heimer (1983), Jantschke & Nentwig (1987), Decae (pers. comm. 1988)
Barro Colorado Island, Canal Zone, Santiago, Azuero Peninsula

Linothele sp.
(= Diplura sp.)
Vollrath (1978), Decae (pers. comm. 1988), Paz (1988), Nentwig (1990)
Gamboa, Bayano, Carti Road

Masteria downeyi (Chickering)
(= Accola d.)
Chickering (1966)
Boquete, El Volcan

Masteria spinosa (Petrunkevitch)
(= Accola spinosa)
Petrunkevitch (1925), Chickering (1936, 1945 a, 1966), Decae (pers. comm. 1988)
Barro Colorado Island, Canal Zone, Wilcox Camp, Cerro Campana

BARYCHELIDAE

(Merothele zebra Petrunkevitch = Trichopelma z.)

Trichopelma zebra (Petrunkevitch)
(= Merothele z.)
Petrunkevitch (1925), Decae (pers. comm. 1988)
Cerro Flores, San Blas

(At least 2 other, undescribed species in the Panama Canal Area, Decae, pers. comm. 1988)

Theraphosidae

(Aphonopelma balboanum Chamberlin = Mygalarachne rubronitens (Ausserer))

Aphonopelma sabulosa (F.O. Pickard-Cambridge)
(= Delopelma s.)
(= Eurypelma s.)
(= Rhechosticta s.)
Petrunkevitch (1925)
Canal Zone

Avicularia embrithes (Chamberlin & Ivie)
(= Eurypelma e.)
Chamberlin & Ivie (1936)
Barro Colorado Island

Avicularia glauca Simon
Simon (1891 a, b), F.O. Pickard-Cambridge (1899), Petrunkevitch (1925)
Panama

Avicularia panamense (Simon)
(= Eurypelma p.)
Simon (1891)
Panama

(Barropelma parvius (Chamberlin & Ivie) = Crypsidromus p.)
(Brachypelma emilia (White) = Euathlus e.)
(Brachypelma vagans (Ausserer) = Euathlus v.)

Crypsidromus colorata (Valerio)
(= Metriopelma c.)
Valerio (1982 a)
Panama

Crypsidromus panamanus (Petrunkevitch)

(= Neischnocolus p.)
Petrunkevitch (1925)
Cerro Iglesia

Crypsidromus parvius (Chamberlin & Ivie)
(= Eurypelma p.)
Chamberlin & Ivie (1936)
Barro Colorado Island

Cyclosternum stylipum Valerio
Valerio (1982 a), Nentwig & Heimer (1983)
Barro Colorado Island, Gamboa

(Delopelma sabulosum (F.O. Pickard-Cambridge) = Aphonopelma s.)

Euathlus emilia (White)
(= Brachypelma e.)
(= Eurypelma e.)
White (1856), F.O. Pickard-Cambridge (1897), Petrunkevitch (1925)
Panama

Euathlus vagans (Ausserer)
(Brachypelma v.)
(= Eurypelma v.)
Petrunkevitch (1925)
Canal Zone

(Eurypelma embrithes Chamberlin & Ivie = Avicularia e.)
(Eurypelma emilia (White) = Euathlus e.)
(Eurypelma panamense Simon = Avicularia p.)
(Eurypelma parvius Chamberlin & Ivie = Crypsidromus p.)
(Eurypelma sabulosum F.O. Pickard-Cambridge = Aphonopelma s.)
(Eurypelma vagans Ausserer = Euathlus v.)
(Metriopelma colorata Valerio = Crypsidromus c.)

Mygalarachne communis (F.O. Pickard-Cambridge)
(= Sericopelma c.)
F.O. Pickard-Cambridge (1897), Petrunkevitch (1925), Schmidt (1990)
Canal Zone, Chiriqui

Mygalarachne rubronitens (Ausserer)
(= Aphonopelma balboanum Chamberlin)
(= Sericopelma r.)
(= Sericopelma panamanum (Karsch))
(= Theraphosa panamana Karsch)
Ausserer (1875), Karsch (1880), Simon (1891), Petrunkevitch (1925), Banks (1929), Nentwig & Heimer (1983), Schmidt (1990)
Canal Zone, Barro Colorado Island, Bocas del Toro, Gamboa, Chiriqui

(Neischnocolus panamanus Petrunkevitch = Crypsidromus p.)

Psalmopeus intermedius Chamberlin
Chamberlin (1940), Roth (pers. comm. 1987)
Panama

Psalmopeus rufus Petrunkevitch
Petrunkevitch (1925)
Panama

Psalmopeus pulcher Petrunkevitch
Petrunkevitch (1925), Nentwig & Heimer (1983)
Gamboa, Canal Zone

(Rhechostica sabulosa (F.O. Pickard-Cambridge) = Aphonopelma s.)
(Sericopelma communis F.O. Pickard-Cambridge = Mygalarachne c.)
(Sericopelma panamanum (Karsch) = Mygalarachne rubronitens (Ausserer))
(Sericopelma rubronitens Ausserer = Mygalarachne r.)
(Theraphosa panamana Karsch = Mygalarachne rubronitens (Ausserer))

Xenesthis colombiana Simon
Simon (1891 a)
Panama

Xenesthis immanis (Ausserer)
(= X. colombiana Simon)
F.O. Pickard-Cambridge (1897), Petrunkevitch (1925)
Panama

Filistatidae

(Filistata hibernalis Hentz = Kukulcania h.)

Kukulcania hibernalis (Hentz)
(= Filistata h.)
Petrunkevitch (1925, 1939 b), Banks (1929), Chickering (1936), Nentwig & Heimer (1983)
Barro Colorado Island, Canal Zone

Scytodidae

Scytodes championi F.O. Pickard-Cambridge
F.O. Pickard-Cambridge (1899), Petrunkevitch (1925), Banks (1929), Chamberlin & Ivie (1936), Valerio (1981), Nentwig & Heimer (1983)
Barro Colorado Island, Cerro Flores, El Volcan, Gamboa, Carti Road

Scytodes fusca Walckenaer
(= S. nannipes Chamberlin & Ivie)
Petrunkevitch (1925), Chamberlin & Ivie (1936), Valerio (1981), Nentwig & Heimer (1983)
Puerto Royal, Barro Colorado Island, Gamboa

Scytodes gertschi Valerio
Valerio (1981), Nentwig & Heimer (1983)
Barro Colorado Island, Gamboa, Canal Zone

(Scytodes guttipes Simon = S. championi F.O. Pickard-Cambridge, misidentification of Panamanian specimens by Banks (1929), (Valerio 1981))

Scytodes intricata Banks
Valerio (1981), Eberhard (1986)
El Volcan, Canal Zone, Gamboa

Scytodes longipes Lucas
F.O. Pickard-Cambridge (1899), Petrunkevitch (1925), Banks (1929), Valerio (1981), Nentwig & Heimer (1983), Nentwig (1985 b)
Gamboa, Barro Colorado Island, Canal Zone, Santiago, Remedios, Wilcox Camp, Cerro Flores, David

(Scytodes nannipes Chamberlin & Ivie = S. fusca Walckenaer)

Loxoscelidae

Loxosceles rufipes (Lucas)
Petrunkevitch (1925), Banks (1929), Gertsch (1958), Gertsch & Ennik (1983)
Santa Rosa, Panama City, Barro Colorado Island, Wilcox Camp

Loxosceles panama Gertsch
Gertsch (1958), Gertsch & Ennik (1983), Nentwig & Heimer (1983)
Barro Colorado Island, Canal Zone, Taboga Island, Bejuco, Barbacoas Island, El Valle, Bayano / Rio Mayo, Penonome, Panama City

Caponiidae

Caponina sargi F.O. Pickard-Cambridge
Nentwig & Heimer (1983)
Gamboa

Nops coccineus Simon
Nentwig & Heimer (1983)
Canal Zone

Nops guanabacoae Mac Leay
Nentwig & Heimer (1983)
Canal Zone, Gamboa

Nops largus Chickering
Chickering (1967), Nentwig & Heimer (1983)
Canal Zone, Barro Colorado Island, El Valle, El Volcan, Arraijan

Nops maculata Simon
Banks (1929), Petrunkevitch (1939 b)
Barro Colorado Island

Nops simia Chickering
Nentwig & Heimer (1983)
Canal Zone
Nops ursumus Chickering
Chickering (1967)
Barro Colorado Island, Canal Zone, Santa Rosa (Colon)

Tetrablemmidae

Caraimatta cambridgei (Bry-ant)
(= Matta c.)
Bryant (1940)
Panama

(Matta cambridgei (Bryant) = Caraimatta c.)

Monoblemma unicum Gertsch
Gertsch (1941)
Barro Colorado Island

Pholcidae

Anopsicus chiriqui Gertsch
Gertsch (1982)
El Volcan

Anopsicus chickeringi Gertsch
Gertsch (1982)
Canal Zone

Anopsicus facetus Gertsch
Nentwig & Heimer (1983)
Cerro Punta

Anopsicus panama Gertsch
Gertsch (1982)
Canal Zone

Anopsicus placens (O. Pickard-Cambridge)
(= Pholcophorina p.)
(= Spermophora p.)
Petrunkevitch (1925), Chickering (1936)
Canal Zone, Barro Colorado Island

Anopsicus zeteki (Gertsch)
(= Pholcophora z.)
(= Pholcophorina z.)
Gertsch (1936), Gertsch (1982), Nentwig & Heimer (1983)
Barro Colorado Island, Canal Zone, Gamboa

Artema atlanta Walckenaer
Petrunkevitch (1925)
Colon

Blechroscelis modesta Banks
Banks (1929)
Canal Zone, Barro Colorado Island

(Metagonia caudata O. Pickard-Cambridge, misidentification of Banks (1929) of *M. panamana*)

Metagonia panama Gertsch
Banks (1929), Nentwig & Heimer (1983), Gertsch (1986)
El Valle, Carti Road, Barro Colorado Island, Gamboa

Micromerys delicatus O. Pickard-Cambridge
Deeleman-Reinhold (pers. comm. 1983)
Barro Colorado Island
(This species does not belong into the genus *Micromerys* which is an Australian genus. It represents probably a new, monotypic genus (Deeleman-Reinhold, pers. comm.).)

Modisimus caeruleolineatus Petrunkevitch
Nentwig & Heimer (1983)
Canal Zone

Modisimus chickeringi Gertsch
Gertsch (1973)
Canal Zone

Modisimus dilutus Gertsch
Gertsch (1941), Nentwig (1990)
Barro Colorado Island, Gamboa

Modisimus inornatus O. Pickard-Cambridge
Petrunkevitch (1925)
Wilcox Camp

Modisimus maculatipes O. Pickard-Cambridge
Banks (1929)
Barro Colorado Island
Modisimus montanus Petrunkevitch
Nentwig & Heimer (1983)
Canal Zone

Modisimus propinquus O. Pickard-Cambridge
Nentwig & Heimer (1983)
Gamboa

Modisimus putus O. Pickard-Cambridge
Chickering (1936)
Barro Colorado Island

Modisimus pulchellus Banks
Banks (1929), Nentwig (1990)
Canal Zone, Barro Colorado Island, Gamboa

(Pholcophorina = Anopsicus)
(Pholcophora placens (O. Pickard-Cambridge) = Anopsicus p.)
(Pholcophora zeteki Gertsch = Anopsicus z.)

Physocyclus globosus (Taczanowski)
Petrunkevitch (1925), Chickering (1936), Nentwig & Heimer (1983)
El Valle, San Juan (Chiriqui), Santiago, Remedios, Panama City, Gamboa

Smeringopus elongatus (Vinson)
Petrunkevitch (1925)
Panama
(Reported by Petrunkevitch (1939 b) as "absent")

(Spermophora placens O. Pickard-Cambridge = Anopsicus p.)

Ochyroceratidae

Speocera sp.
Nentwig & Heimer (1983)
Canal Zone, Gamboa

Theotima centralis Gertsch
(= Oonopinus c.)
Gertsch (1941), Williams (1941), Chickering (1951 c), Quintero (pers. comm. 1988)
Barro Colorado Island

Segestriidae

(Ariadna comata Banks = A. isthmica Beatty)

Ariadna isthmica Beatty
(= A. comata Banks)
Banks (1929), Beatty (1970), Nentwig & Heimer (1983)
Bella Vista, Portobelo, Canal Zone, Gamboa, Barro Colorado Island

Oonopidae

Dysderina abdita Chickering
Chickering (1968 a)
El Volcan

Dysderina belinda
Chickering (1968 a)
Boquete

Dysderina concinna Chickering
Chickering (1968 a)
El Volcan

Dysderina dura Chickering
Chickering (1951 c, 1968 a), Nentwig & Heimer (1983)
Canal Zone, Gamboa, El Valle, Portobelo, Barro Colorado Island

Dysderina humphreyi Chickering
Chickering (1968 a)
Boquete

Dysderina improvisia Chickering
Chickering (1968 a)
Barro Colorado Island, Canal Zone

Dysderina intempina Chickering
Chickering (1968 a)
Boquete

Dysderina obtina Chickering
Chickering (1968 a)
El Volcan

Dysderina plena O. Pickard-Cambridge
Banks (1929), Chickering (1951c, 1968a)
Barro Colorado Island

Dysderina potena Chickering
Chickering (1968 a)
El Volcan, Boquete

Dysderina recondita Chickering
Chickering (1951 c, 1968 a), Nentwig & Heimer (1983)
Boquete, Canal Zone

Dysderina rigida Chickering
Chickering (1968 a)
El Volcan

Dysderina seclusa Chickering
Chickering (1951 c, 1968 a)
Canal Zone, Barro Colorado Island

Dysderina silvatica Chickering
Chickering (1951 c, 1968 a)
Barro Colorado Island, Canal Zone, Gamboa, El Valle

Heteroonops spinimanus (Simon)
Chickering (1973)
Canal Zone

Ischnothyreus browni Chickering
Nentwig & Heimer (1983)
Canal Zone

Ischnothyreus peltifer (Simon)
Chickering (1951 c, 1968 d), Nentwig & Heimer (1983)
Canal Zone, Gamboa

(Oonopinus centralis Gertschn = Theotima c., Ochyroceratidae)

Oonopinus modestus Chickering
Chickering (1951 c)
Barro Colorado Island

(Oonopinus pallidulus Chickering = Oonops p.)

Oonopoides bryantae Chickering
Chickering (1951 c), Nentwig & Heimer (1983)
Barro Colorado Island, Canal Zone

Oonops amacus Chickering
Nentwig & Heimer (1983)
Canal Zone

Oonops anoxus Chickering
Chickering (1970 c)
Barro Colorado Island, Canal Zone, Boquete

Oonops balanus Chickering
Nentwig & Heimer (1983)
Canal Zone, Gamboa

Oonops castellus Chickering
Nentwig & Heimer (1983)
Gamboa

Oonops donaldi Chickering
Chickering (1951 c, 1970 c)
Canal Zone, El Valle

Oonops ornatus Chickering
Chickering (1970 c), Nentwig & Heimer (1983)
Boquete, Gamboa

Oonops pallidulus (Chickering)
(= Oonopinus p.)
Chickering (1951 c, 1970 c), Nentwig & Heimer (1983)
Barro Colorado Island, Canal Zone

Oonops persitus Chickering
Chickering (1970 c), Nentwig & Heimer (1983)
Gamboa, Barro Colorado Island, Canal Zone

Oonops reticulatus Petrunkevitch
Chickering (1951 c, 1970 c)
Panama City, Canal Zone, El Volcan

Oonops tenebus Chickering
Chickering (1970 c), Nentwig & Heimer (1983)
Canal Zone, Gamboa

Oonops zeteki Chickering
Chickering (1951 c)
Barro Colorado Island

Opopaea recondita Chickering
Chickering (1951 c)
Barro Colorado Island

Opopaea timida Chickering
Chickering (1951 c)
Canal Zone

Scaphiella barroana Gertsch
Gertsch (1941), Chickering (1951 c, 1968 c)
Barro Colorado Island

Scaphiella gertschi Chickering
Chickering (1951 c, 1968 c)
Barro Colorado Island, Canal Zone

Scaphiella kalunda Chickering
Nentwig & Heimer (1983)
Canal Zone, Gamboa

Scaphiella scutata Chickering
Nentwig & Heimer (1983)
Canal Zone

Scaphiella scutiventris Simon
Chickering (1936)
Barro Colorado Island

Scaphiella simla Chickering
Nentwig & Heimer (1983)
Canal Zone

Scaphiella williamsi Gertsch
Gertsch (1941), Chickering (1951 c, 1968 c), Nentwig & Heimer (1983)
Barro Colorado Island, Canal Zone, Gamboa

Stenoonops petrunkevitch Chickering
Chickering (1951 c, 1969 b)
Barro Colorado Island

Triaeris patellaris Bryant
Chickering (1951 c)
Barro Colorado Island

Triaeris stenaspis Simon
Chickering (1968 e), Nentwig & Heimer (1983)
Canal Zone, Gamboa

Palpimanidae

Otiothops macleayi Banks
Banks (1929), Chickering (1936, 1941), Platnick (1975 b)
Barro Colorado Island, Canal Zone, Portobelo

Mimetidae

Gelanor consequus O. Pickard-Cambridge
O. Pickard-Cambridge (1902), Petrunkevitch (1925), Chickering (1947)
Bugaba

Gelanor depressus Chickering
Chickering (1956 a)
Barro Colorado Island

Gelanor distinctus O. Pickard-Cambridge
O. Pickard-Cambridge (1899), Petrunkevitch (1925), Chickering (1947), Nentwig & Heimer (1983)
Bugaba, Barro Colorado Island, Canal Zone, Gamboa

(Gelanor errans O. Pickard-Cambridge = G. distinctus O. Pickard-Cambridge)

Gelanor gertschi Chickering
Chickering (1947)
Barro Colorado Island

Gelanor heraldicus Petrunkvitch
Petrunkevitch (1925), Chickering (1947)
Wilcox Camp

Gelanor mabelae Chickering
Chickering (1947)
Barro Colorado Island

Gelanor mixtus O. Pickard-Cambridge
O. Pickard-Cambridge (1899), Petrunkevitch (1925), Chickering (1947)
Wilcox Camp, Bugaba

Gelanor zonatus C.L.Koch
Banks (1929)
Barro Colorado Island

Mimetus banksi Chickering
Chickering (1947)
Barro Colorado Island

Mimetus bigibbosus O. Pickard-Cambridge
Banks (1929), Nentwig & Heimer (1983)
Barro Colorado Island, Carti Road

Mimetus rapax O. Pickard-Cambridge
O. Pickard-Cambridge (1899), Petrunkevitch (1925), Chickering (1947)
Bugaba

Mimetus rusticus Chickering
Chickering (1947)
Barro Colorado Island

Mimetus saetosus Chickering
Chickering (1956 a)
Barro Colorado Island

Mimetus trituberculatus O. Pickard-Cambridge
O. Pickard-Cambridge (1899), Petrunkevitch (1925), Chickering (1947)
Los Remedios

Mimetus variegatus Chickering
Chickering (1956 a)
Canal Zone, Arraijan

Mimetus verecundus Chickering
Chickering (1947)
Boquete

Oecobiidae

Oecobius concinnus Simon
Shear (1970), Nentwig & Heimer (1983)
Gamboa, Canal Zone, Taboga Island, Playa Coronado

Hersiliidae

Tama mexicana (O. Pickard-Cambridge)
Petrunkevitch (1925), Banks (1929), Nentwig & Heimer (1983)
Barro Colorado Island, Gamboa

Tama sp.
Nentwig (1990)
Gamboa

Deinopidae

Deinopis longipes F.O. Pickard-Cambridge
(= Dinopis l.)
F.O. Pickard-Cambridge (1902), Chickering (1936, 1963 c), Robinson & Robinson (1971), Lubin (1978), Nentwig & Heimer (1983)
Bugaba, Barro Colorado Island, Boquete, Canal Zone, Gamboa

(There are at least two species of *Deinopis* in Panama, probably more (Coddington, pers. comm. 1988))

Uloboridae

Ariston albicans O. Pickard-Cambridge
(= Uloborus a.)
Banks (1929)
Barro Colorado Island

Ariston aristus Opell
Opell (1979, 1981)
Barro Colorado Island

Miagrammopes alboguttatus F.O. Pickard-Cambridge
Banks (1929)
Barro Colorado Island

Miagrammopes aspinatus Chickering
Chickering (1968 b)
Barro Colorado Island

Miagrammopes intempus Chickering
Chickering (1968 b)
Portobelo

Miagrammopes larundus Chickering
Chickering (1968 b)
Barro Colorado Island

Miagrammopes licinus Chickering
Chickering (1968 b)
El Volcan

Miagrammopes simus Chamberlin & Ivie
Chamberlin & Ivie (1936), Chickering (1968 b), Lubin (1978), Lubin et al. (1978)
Barro Colorado Island

Miagrammopes unipus Chickering
Chickering (1968 b)
Barro Colorado Island

Philoponella divisa Opell
Nentwig & Heimer (1983)
Gamboa, Canal Zone

Philoponella republicana (Simon)
(= Uloborus mundior Chamberlin & Ivie)
(= Uloborus r.)
Chickering (1936), Chamberlin & Ivie (1936), Lubin (1978, 1980), Opell (1979), Hoffmaster (1982), Nentwig & Heimer (1983)
El Real, Barro Colorado Island

Philoponella semiplumosa (Simon)
(= Uloborus variegatus O. Pickard-Cambridge)
Banks (1939), Opell (1979), Nentwig & Heimer (1983)
Barro Colorado Island, Las Lajas

Philoponella signatella (Roewer)
Banks (1929), Nentwig & Heimer (1983)
El Valle, Canal Zone, Barro Colorado Island

Philoponella tingens Chamberlin & Ivie
Chamberlin & Ivie (1936), Opell (1979), Nentwig & Heimer (1983)
Barro Colorado Island, Canal Zone, Gamboa

Philoponella vicina (O. Pickard-Cambridge)
Nentwig & Heimer (1983)
Gamboa

Philoponella vittata (Keyserling)
Opell (1979)
Panama

(Uloborus aegrotus Simon = U. trilineatus Keyserling)
(Uloborus albicans (O. Pickard-Cambridge) = Ariston a.)
(Uloborus americanus Walckenaer = U. globosus Walckenaer)

Uloborus campestratus Simon
(= U. cinereus O. Pickard-Cambridge)
Chickering (1936), Opell (1979), Nentwig & Heimer (1983)
Barro Colorado Island, Gamboa

(Uloborus cinereus O. Pickard-Cambridge = U. campestratus Simon)
(Uloborus geniculatus (Olivier) = Zosis g.)

Uloborus globosus Walckenaer
(= U. americanus Walckenaer)
Petrunkevitch (1925, 1939 b), Chickering (1936)
Barro Colorado Island, Wilcox Camp

(Uloborus mundior Chamberlin & Ivie = Philoponella republicana (Simon))

Uloborus segregatus Gertsch
Opell (1979), Nentwig & Heimer (1983)
Canal Zone, Gamboa

(Uloborus penicillatus Simon = U. trilineatus Keyserling)

Uloborus trilineatus Keyserling
(= U. aegrotus Simon)
(= U. penicillatus Simon)
Banks (1929), Opell (1979), Nentwig & Heimer (1983), Eberhard (pers. comm. 1988)
Barro Colorado Island, Gamboa, Canal Zone, Fortuna

(Uloborus variegatus O. Pickard-Cambridge = Philoponella semiplumosa (Simon))

Zosis geniculatus (Olivier)
(= Uloborus g.)

Petrunkevitch (1925, 1939 b), Chickering (1936)
Barro Colorado Island, Cerro Flores

Nesticidae

Gaucelmus augustinus Keyserling
Gertsch (1984)
El Volcan

Gaucelmus pygmaeus Gertsch
Gertsch (1984)
Canal Zone

Gaucelmus tropicus Gertsch
Gertsch (1984)
Canal Zone

Theridiidae

(Achaea acutiventer Keyserling = Coleosoma a.)
(Achaea compressa Keyserling = Chrysso c.)
(Achaea ignota Keyserling = Achaearanea hirta (Taczanowski))

Achaearanea alacre (Keyserling)
(= Theridion a.)
Banks (1929)
Barro Colorado Island, Canal Zone

Achaearanea apex Levi
Levi (1959 b)
Canal Zone, Barro Colorado Island

Achaearanea caqueza Levi
Nentwig & Heimer (1983)
Gamboa

Achaearanea florendida Levi
Levi (1959 b, 1963 d), Nentwig & Heimer (1983)
Canal Zone, Barro Colorado Island

Achaearanea florens (O. Pickard-Cambridge)
(= Theridion f.)
Chickering (1936), Levi (1955 b, 1959 b), Eberhard (pers. comm. 1988)
Barro Colorado Island, Cerro Punta, Fortuna

Achaearanea hermosillo Levi
Nentwig & Heimer (1983)
Gamboa, Canal Zone

Achaearanea hirta (Taczanowski)
(= Achaea ignota Keyserling)
(= Theridion i.)
Chickering (1936), Levi (1959 b), Nentwig & Heimer (1983)
Gamboa, Barro Colorado Island, Canal Zone, El Valle, Arraijan

Achaearanea machaera Levi
Levi (1959 b)
El Volcan

Achaearanea maricaoensis (Bryant)
Levi (1959 b)
Boquete, Canal Zone, Barro Colorado Island, Portobelo

(Achaearanea mesax Levi = A. nigrovittata (Keyserling))

Achaearanea nigrovittata (Keyserling)
(= A. mesax Levi)
(= Theridion n.)
Chickering (1936), Levi (1959 b, 1963 d), Nentwig & Heimer (1983)
Barro Colorado Island, Gamboa, Canal Zone, Boqueron River

Achaearanea oblivia (O. Pickard-Cambridge)
Levi (1959 b), Nentwig & Heimer (1983)
Gamboa, El Volcan, Cerro Punta

(Achaearanea picadoi (Banks) = Achaearanea tesselata (Keyserling))

Achaearanea porteri (Banks)
Levi (1955 b, 1959 b, 1963 d)
El Volcan

Achaearanea rostrata (O. Pickard-Cambridge)
(= Theridion r.)
Banks (1939)
Barro Colorado Island

Achaearaneae schneirla Levi
Levi (1959 b)
Barro Colorado Island

Achaearanea taeniata (Keyserling)
(= Steatoda t.)
(= Theridion t.)
F.O. Pickard-Cambridge (1902), Petrunkevitch (1925), Banks (1929), Levi (1959 b), Nentwig & Heimer (1983)
Bugaba, Gamboa, Barro Colorado Island, Canal Zone, Boquete

Achaearanea tepidariorum (C.L. Koch)
(= Theridion t.)
Petrunkevitch (1925)
Panama
(Levi (1959 b) constates "no record from Central America")

Achaearanea terex Levi
Levi (1959 b)
Barro Colorado Island

Achaearanea tesselata (Keyserling)
(= A. picadoi (Banks))
(= Theridion t.)
Levi (1959 b), Nentwig & Heimer (1983)
Gamboa, Canal Zone, Barro Colorado Island

Achaearanea trapezoidales (Taczanowski)
Levi (1959 b, 1963 d)
Arraijan, Canal Zone, Barro Colorado Island

Achaearanea zonensis Levi
Levi (1959 b, 1963 d), Nentwig & Heimer (1983)
Canal Zone, Gamboa, Barro Colorado Island, Portobelo, Arraijan

Anelosimus chickeringi Levi
Levi (1956)
Barro Colorado Island

Anelosimus eximius (Keyserling)
(= Theridion e.)
Banks (1929), Chickering (1936), Levi (1956, 1963 a), Robinson (1978), Vollrath (1982, 1986), Nentwig & Heimer (1983),Christenson (1984), Nentwig (1985 c)
Barro Colorado Island, Canal Zone, Taboga Island, El Valle, Carti Road, Arraijan, Old Panama

Anelosimus jucundus (O. Pickard-Cambridge)
(= Theridion j.)
Petrunkevitch (1925), Levi (1956), Nentwig & Heimer (1983), Nentwig (1985 c), Nentwig & Christenson (1986)
El Valle, Boquete, Taboga Island, Canal Zone, Barro Colorado Island, Gamboa, Cerro Flores

Anelosimus pacificus Levi
Levi (1956)
Panama

Anelosimus studiosus (Hentz)
(= Theridion st.)
Petrunkevitch (1925), Chickering (1936), Levi (1956)
Boquete, El Armeno, Barro Colorado Island, Canal Zone

Argyrodes americanus (Taczanowski)
(= A. indignus Chamberlin & Ivie)
(= A. parvior Chamberlin & Ivie (male))
Banks (1929), Chamberlin & Ivie (1936), Exline & Levi (1962)
Barro Colorado Island, Boquete, El Valle, Canal Zone

Argyrodes amplifrons O. Pickard-Cambridge
(= A. obtusus O. Pickard-Cambridge)
Chickering (1936), Exline & Levi (1962)
Barro Colorado Island

(Argyrodes argentatus O. Pickard-Cambridge = A. nephilae Taczanowski)

Argyrodes arthuri Exline & Levi
Exline & Levi (1962), Nentwig & Heimer (1983)
Barro Colorado Island, Canal Zone

Argyrodes atopus Chamberlin & Ivie
(= A. pachysomus Chamberlin & Ivie)
Chamberlin & Ivie (1936), Exline & Levi (1962), Vollrath (1976), Nentwig & Heimer (1983)
Barro Colorado Island, Canal Zone, Gamboa, El Valle

Argyrodes attenuatus (O. Pickard-Cambridge)
(= Ariamnes approximata O. Pickard-Cambridge)
(= Ariamnes gracillimus O. Pickard-Cambridge)
O. Pickard-Cambridge (1894), F.O. Pickard-Cambridge (1902), Petrunkevitch (1925), Exline & Levi (1962), Nentwig & Heimer (1983)
Bugaba, Los Remedios, Canal Zone, Barro Colorado Island

Argyrodes bryantae Exline & Levi
Exline & Levi (1962)
Boquete, El Volcan

Argyrodes caudatus (Taczanowski)
(= A. parvior Chamberlin & Ivie (female))
(= A. vexus Chamberlin & Ivie)
Banks (1929), Chamberlin & Ivie (1936), Exline & Levi (1962), Vollrath (1976), Nentwig & Heimer (1983)
Barro Colorado Island, Canal Zone, Gamboa, Bella Vista, La Campana, Arraijan, El Valle, Taboga Island

Argyrodes chickeringi Exline & Levi
Exline & Levi (1962)
Barro Colorado Island

(Argyrodes cingulatus Petrunkevitch = A. elevatus Taczanowski)

Argyrodes dracus Chamberlin & Ivie
Chamberlin & Ivie (1936), Exline & Levi (1962), Vollrath (1976), Nentwig & Heimer (1983)
Barro Colorado Island, Canal Zone, Boquete, El Valle, Carti Road, Gamboa

Argyrodes elevatus Taczanowski
(= A. cingulatus Petrunkevitch)
(= A. jucunda O. Pickard-Cambridge)
Petrunkevitch (1925), Banks (1929), Chickering (1936), Exline & Levi (1962), Vollrath (1976, 1979), Nentwig & Heimer (1983)
Barro Colorado Island, Canal Zone, Bella Vista, Portobelo, Arraijan, Gamboa, Old Panama, El Valle, Taboga Island, Panama City

Argyrodes fictilium (Hentz)
(= Romphaea f.)
Exline & Levi (1962), Nentwig & Heimer (1983)
Barro Colorado Island, Canal Zone, Gamboa

Argyrodes furcatus (O. Pickard-Cambridge)
(= Rhomphaea bifissa (F.O. Pickard-Cambridge))
Chickering (1936)
Barro Colorado Island

Argyrodes gertschi Exline & Levi
Exline & Levi (1962)
El Volcan

(Argyrodes indiguus Chamberlin & Ivie = A. americanus (Taczanowski))
(Argyrodes jucunda O. Pickard-Cambridge = A. elevatus Taczanowski)

Argyrodes maculosus O. Pickard-Cambridge
Chickering (1936), Exline & Levi (1962)
Barro Colorado Island

Argyrodes metaltissimus (Soares & Camargo)
(= Rhomphaea m.)
Exline & Levi (1962)

Barro Colorado Island

Argyrodes nephilae Taczanowski
(= A. argentata O. Pickard-Cambridge)
Petrunkevitch (1925), Banks (1929), Chickering (1936)
Barro Colorado Island, Bella Vista, Panama City, Canal Zone, Remedios

Argyrodes obscurus Keyserling
Chickering (1936)
Barro Colorado Island

(Argyrodes obtusus O. Pickard-Cambridge = A. amplifrons O. Pickard-Cambridge)
(Argyrodes pachysomus Chamberlin & Ivie = A. atopus Chamberlin & Ivie)

Argyrodes paradoxus Taczanowski
(= Ariamnes longicaudatus Keyserling (lapsus) pro Ariamnes spinicaudatus Keyserling)
Exline & Levi (1962)
El Volcan, Cerro Punta, Barro Colorado Island, Canal Zone

Argyrodes parvus (Exline & Levi)
Exline & Levi (1962)
Boquete, Barro Colorado Island

(Argyrodes parvior Chamberlin & Ivie: male = A. americanus (Taczanowski), female = A. caudatus Taczanowski)

Argyrodes procerus (O. Pickard-Cambridge)
(= Ariamnes p.)
O. Pickard-Cambridge (1898), Petrunkevitch (1925), Exline & Levi (1962), Nentwig & Heimer (1983)
Bugaba, Arraijan, Canal Zone, Barro Colorado Island, Gamboa

Argyrodes projiciens (O. Pickard-Cambridge)
Exline & Levi (1962), Nentwig & Heimer (1983)
Barro Colorado Island, Gamboa, Canal Zone, Arraijan, Boquete, El Valle

Argyrodes quasiobtusus Exline & Levi
Nentwig & Heimer (1983)
Gamboa

(Argyrodes socius Chamberlin & Ivie = A. ululans O. Pickard-Cambridge)

Argyrodes spinosus Keyserling
Nentwig & Heimer (1983)
Gamboa

Argyrodes subdolus O. Pickard-Cambridge
Nentwig & Heimer (1983)
Cerro Punta

Argyrodes triangularis Taczanowski
Exline & Levi (1962)
Barro Colorado Island

Argyrodes trigonum (Hentz)
Nentwig & Heimer (1983)
Carti Road

Argyrodes ululans O. Pickard-Cambridge
(= A. socius Chamberlin & Ivie)
Chamberlin & Ivie (1936), Exline & Levi (1962), Nentwig & Heimer (1983)
Gamboa, Barro Colorado Island, El Valle, El Real

(Argyrodes vexus Chamberlin & Ivie = A. caudatus (Taczanowski))
(Ariamnes = Argyrodes)
(Ariamnes approximata O. Pickard-Cambridge = Argyrodes attenuatus (O. Pickard-Cambridge))
(Ariamnes gracillimus O. Pickard-Cambridge = Argyrodes attenuatus (O. Pickard-Cambridge))
(Ariamnes longicaudatus Keyserling (lapsus) pro Ariamnes spinicaudatus Keyserling = Argyrodes paradoxus Taczanowski)
(Asagena quadrimaculata O. Pickard-Cambridge = Steatoda q.)
(Chrosiothes conservaticus Chamberlin & Ivie = C. silvaticus Simon)

Chrosiothes probabilis (O. Pickard-Cambridge)
(= Euryopis p.)
(= Theridiotis p.)
Chickering (1936), Levi (1954 b)
Barro Colorado Island, Cerro Punta

Chrosiothes proximus (O. Pickard-Cambridge)
Levi (1964 d)
El Volcan, Chiriqui

Chrosiothes silvaticus Simon
(= C. conservaticus Chamberlin & Ivie)
Chamberlin & Ivie (1936), Levi (1964 d)
Barro Colorado Island, Canal Zone, Boquete, El Valle

Chrysso albomaculata O. Pickard-Cambridge
(= C. elegans (Taczanowski))
Banks (1929), Chickering (1936), Levi (1955 a, 1962), Harrison (1968), Nentwig & Heimer (1983)
Barro Colorado Island, Gamboa, Canal Zone, Bocas del Toro, Arraijan, Changuinola

Chrysso alecula Levi
Levi (1962)
El Valle

Chrysso antonio Levi
Nentwig & Heimer (1983)
Gamboa

Chrysso cambridgei (Petrunkevitch)
(= Theridion c.)
(= Theridion vittatum (O. Pickard-Cambridge))
F.O. Pickard-Cambridge (1902), Levi (1962), Nentwig & Heimer (1983)
Bugaba, Barro Colorado Island, Gamboa

Chrysso clementinae (Petrunkevitch)
(= Meotipa c.)
Chickering (1936), Levi (1962), Harrison (1968), Nentwig & Heimer (1983)
Changuinola, Barro Colorado Island, Gamboa, Canal Zone

Chrysso compressa (Keyserling)
(= Achaea c.)
(= Theridion c.)
Banks (1929)
Barro Colorado Island

Chrysso diplosticha Chamberlin & Ivie
Chamberlin & Ivie (1936), Levi (1955 a, 1962)
Barro Colorado Island, Canal Zone, Portobelo, Carti Road

(Chrysso elegans (Taczanowski) = C. albomaculata O. Pickard-Cambridge)

Chrysso indicifer Chamberlin & Ivie
Chamberlin & Ivie (1936), Levi (1955 a, 1962)
Barro Colorado Island, Canal Zone

(Chrysso lyparus Chamberlin & Ivie = C. vexabilis Keyserling)

Chrysso melba Levi
Levi (1962)
Canal Zone

Chrysso nigripalpis Banks
Banks (1929)
Barro Colorado Island

Chrysso questona Levi
Levi (1962)
El Valle, Chiriqui

Chrysso silva Levi
Levi (1962), Nentwig & Heimer (1983)
Canal Zone, Portobelo, Barro Colorado Island, Gamboa

Chrysso vallensis Levi
Levi (1955 a)
El Valle

Chrysso vexabilis Keyserling
(= C. lyparus Chamberlin & Ivie)

(= Steatoda rubicunda F.O. Pickard-Cambridge)
(= Theridion r.)
F.O. Pickard-Cambridge (1902), Petrunkevitch (1925), Banks (1929), Chamberlin & Ivie (1936), Levi (1955 a, 1960), Nentwig & Heimer (1983)
Barro Colorado Island, Gamboa, Canal Zone, Taboga Island, Portobelo, Old Panama, El Valle, Boquete, Arraijan, Carti Road, Bugaba

Chrysso volcanensis Levi
Levi (1962)
El Volcan

Coleosoma acutiventer (Keyserling)
(= Achaea a.)
(= Theridion a.)
Chickering (1936), Levi (1959 a), Harrison (1968)
Boquete, Canal Zone, Changuinola

Coleosoma floridanum Banks
(= Theridion f.)
Chickering (1936), Levi (1959 a)
Barro Colorado Island

Dipoena alta Keyserling
(= D. furtiva Chickering)
(= D. pallida Chickering)
Chickering (1943), Levi (1963 b), Roberts (1979)
Portobelo, Canal Zone, Barro Colorado Island, El Volcan

Dipoena anas Levi
Levi (1963 b)
Boquete

Dipoena atlantica Chickering
Chickering (1943, 1948), Levi (1963 b)
Barro Colorado Island, Canal Zone

Dipoena balboae Chickering
Chickering (1943), Levi (1963 b)
Canal Zone, Arraijan

Dipoena banksi Chickering
Chickering (1943, 1948)
Canal Zone, Boquete, Arraijan, El Valle

Dipoena barro Levi
Levi (1963 b)
Barro Colorado Island

Dipoena bequaerti Chickering
Chickering (1948), Levi (1963 b)
Canal Zone

Dipoena boquete Levi
Levi (1963 b)
Boquete

Dipoena bryantae Chickering
Chickering (1943, 1948), Levi (1963 b), Nentwig & Heimer (1983)
Barro Colorado Island, Gamboa, Canal Zone

Dipoena chickeringi Levi
Levi (1953)
El Volcan

Dipoena conica Chickering
Chickering (1943, 1948), Levi (1963 b)
Barro Colorado Island, Gamboa, Canal Zone

Dipoena copiosa Levi
Levi (1963 b), Nentwig & Heimer (1983)
Canal Zone

Dipoena cordiformis Keyserling
Chickering (1936), Levi (1963 b)
Barro Colorado Island, Canal Zone

Dipoena cornuta Chickering
Chickering (1943), Levi (1963 b)
Barro Colorado Island, Canal Zone

Dipoena crassiventris Keyserling
Chickering (1936)
Barro Colorado Island

Dipoena donaldi Chickering
Chickering (1943), Levi (1963 b)
Canal Zone, Barro Colorado Island, Portobelo

Dipoena dorsata Muma
Levi (1963 b)
Barro Colorado Island, Canal Zone

Dipoena duodecimpunctata Chickering
Levi (1963 b)
Canal Zone

Dipoena eatoni Chickering
Chickering (1943), Levi (1963 b)
El Valle, Canal Zone

(Dipoena furtiva Chickering = D. alta Keyserling)

Dipoena hortoni Chickering
Chickering (1943), Nentwig & Heimer (1983)
Canal Zone, Gamboa

Dipoena insulana Chickering
Chickering (1943), Levi (1963 b), Nentwig & Heimer (1983)
Barro Colorado Island, Canal Zone, Gamboa

Dipoena isthmia Chickering
Chickering (1943), Levi (1963 b)
Canal Zone

Dipoena jamesi Roberts
Roberts (1979)
El Volcan

Dipoena josephus Levi
Levi (1963 b)
Boquete, Barro Colorado Island

Dipoena lana Levi
Levi (1963 b)
Boquete

Dipoena lineatipes Bryant
Roberts (1979)
Canal Zone

Dipoena mertoni Levi
Levi (1963 b)
Boquete

Dipoena militaris Chickering
(= D. striatipes Simon)
Banks (1929), Chickering (1943), Levi (1963 b)
Canal Zone, Barro Colorado Island

Dipoena orvillei Chickering
Chickering (1943, 1948), Levi (1963 b), Nentwig & Heimer (1983)
El Valle, Barro Colorado Island, Boquete, Gamboa

Dipoena pacifica Chickering
Chickering (1943), Levi (1963 b)
El Valle

(Dipoena pallida Chickering = D. alta Keyserling)

Dipoena parki Chickering
Chickering (1943), Levi (1963 b)
Barro Colorado Island, Canal Zone

Dipoena perimenta Levi
Levi (1963 b)
Canal Zone

(Dipoena petrunkevitchi Chickering = D. rubellum (Keyserling))

Dipoena proterva Chickering
Chickering (1943), Levi (1963 b)
Barro Colorado Island, Canal Zone

Dipoena pulchella Petrunkevitch
(unidentified species, Levi 1963 b)
Petrunkevitch (1925)
Wilcox Camp

Dipoena rubellum (Keyserling)
(= D. petrunkevitchi Chickering)
Chickering (1943)
Barro Colorado Island, Canal Zone

Dipoena seclusa Chickering
Chickering (1948), Levi (1963 b)
Barro Colorado Island

Dipoena standleyi Levi
Levi (1963 b)

Barro Colorado Island, Canal Zone

(Dipoena striatipes Simon = D. militaris Chickering)

Dipoena superba Chickering
Chickering (1948)
Canal Zone

Dipoena tropica Chickering
Chickering (1943), Levi (1963 b)
Arraijan, Canal Zone

Dipoena venusta Chickering
Chickering (1948), Levi (1963 b)
Canal Zone

Dipoena zeteki Chickering
Chickering (1943), Levi (1963 b)
Canal Zone

Episinus bigibbosus O. Pickard-Cambridge
O. Pickard-Cambridge (1896)
Bugaba

Episinus bruneoviridis (Mello-Leitao)
Levi (1964 a), Nentwig & Heimer (1983)
Gamboa, Barro Colorado Island, Canal Zone,

Episinus cognatus O. Pickard-Cambridge
Banks (1929), Levi (1964 a),
Barro Colorado Island, Canal Zone, El Valle

Episinus emanus Levi
Levi (1964 a)
Barro Colorado Island

Episinus erythrophthalmus (Simon)
Levi (1964 a), Nentwig & Heimer (1983)
Canal Zone, Gamboa, Arraijan, El Valle

Episinus panamensis Levi
Levi (1964 a)
Barro Colorado Island

Episinus putus O. Pickard-Cambridge
F.O. Pickard-Cambridge (1902), Petrunkevitch (1925)
Bugaba

Episinus pyrus Levi
Levi (1964 a)
Canal Zone

Episinus vaticus Levi
Levi (1964 a), Nentwig & Heimer (1983)
Canal Zone, El Valle

Euryopis lineatipes O. Pickard-Cambridge
F.O. Pickard-Cambridge (1902), Petrunkevitch (1925), Banks (1929), Levi (1954 a)
Gamboa, Barro Colorado Island, Bugaba

(Euryopis nigripes Banks = E. taczanowski Keyserling)

Euryopis pickardi Levi
Levi (1963 b)
Boquete

(Euryopis probabilis O. Pickard-Cambridge = Theridiotis p.)

Euryopis spinigera O. Pickard-Cambridge
Levi (1954 a), Nentwig & Heimer (1983)
Canal Zone, Gamboa, Carti Road, El Valle

Euryopis taczanowski Keyserling
(= E. nigripes Banks)
Banks (1929), Chickering (1936), Levi (1954 a, 1963 b), Nentwig & Heimer (1983)
Barro Colorado Island, Gamboa, El Valle, Boquete, Canal Zone

(Helvibis keyserlingi Banks = H. longistylum (F.O. Pickard-Cambridge))

Helvibis longistylum (F.O. Pickard-Cambridge)

(= H. keyserlingi Banks)
Banks (1929), Levi (1964 b)
Canal Zone, Barro Colorado Island

(Hubba insignis O. Pickard-Cambridge = Thymoites caracasanus (Simon))

Latrodectus geometricus C.L.Koch
Petrunkevitch (1925)
Panama

Latrodectus mactans (Fabricius)
F.O. Pickard-Cambridge (1902), Petrunkevitch (1925, 1939 b)
Bugaba, Las Palmas

(Lithyphantes lugubris F.O. Pickard-Cambridge = in part Steatoda saltensis Levi; in part Steatoda autumnalis (Banks) (probably erroneous record of Panama (Levi 1957))
(Lithyphantes nigrofemoratus Keyserling = Steatoda n.)
(Meotipa clementinae Petrunkevitch = Chrysso c.)

Paratheridula perniciosa (Keyserling)
(= Theridion p.)
Chickering (1936)
Barro Colorado Island

Paratheridula quadrimaculata (Banks)
Levi (1957)
Boquete, Old Panama, Barro Colorado Island, Canal Zone

Phoroncidia sp.
Nentwig & Heimer (1983)
Gamboa

(Romphaea = Argyrodes)
(Rhomphaea anomala Chamberlin & Ivie = Synotaxus turbinatus Simon)
(Rhomphaea bifissa (F.O. Pickard-Cambridge) = Argyrodes furcatus (O. Pickard-Cambridge))
(Sphyrotinus = Thymoites)

Spintharus flavidus Hentz
F.O. Pickard-Cambridge (1902), Petrunkevitch (1925), Banks (1929), Levi (1963 e)
Bugaba, Barro Colorado Island, Canal Zone, El Volcan, Boquete

(Steatoda forda Keyserling = Tidarren f.)
(Steatoda longistylum F.O. Pickard-Cambridge = Theridion l.)

Steatoda moesta (O. Pickard-Cambridge)
Levi (1957 b), Nentwig & Heimer (1983)
Gamboa, El Valle, Barro Colorado Island, Canal Zone

Steatoda nigrofemorata (Keyserling)
(= Lithyphantes n.)
Banks (1929), Levi (1957 b), Nentwig & Heimer (1983)
Barro Colorado Island, Gamboa, Canal Zone, Boquete, La Campana, Santa Rosa (Colon)

Steatoda quadrimaculata (O. Pickard-Cambridge)
(= Asagena q.)
Chickering (1936), Levi (1957 b)
Barro Colorado Island

(Steatoda rubicunda F.O. Pickard-Cambridge = Chrysso vexabilis Keyserling)

Steatoda saltensis Levi
(= Lithyphantes lugubris F.O. Pickard-Cambridge (in part))
F.O. Pickard-Cambridge (1902), Petrunkevitch (1925), Levi (1957 b)
Bugaba

Steatoda septemmaculata (Keyserling)
Levi (1957 b), Nentwig & Heimer (1983)
Taboga Island, Ensenada, Santa Rosa (Colon)

(Steatoda taeniata Keyserling = Achaearanea t.)

Stemmops bicolor O. Pickard-Cambridge
Levi (1964 d)
Boquete, Canal Zone, Barro Colorado Island

Stemmops mellus Levi
Levi (1964 d), Nentwig & Heimer (1983), Heimer (1987)
Barro Colorado Island, Canal Zone, Gamboa

Stemmops orsus Levi
Levi (1964 d), Nentwig & Heimer (1983), Heimer (1987)
Canal Zone, Gamboa, Boquete

Stemmops questa Levi
Levi (1964 d), Nentwig & Heimer (1983), Heimer (1987)
Bocas del Toro (Changuinola District), Canal Zone, Gamboa

Stemmops servus Levi
Levi (1964 d), Nentwig & Heimer (1983), Heimer (1987)
Canal Zone

Styposis chickeringi Levi
Levi (1960)
El Valle

Styposis clausis Levi
Levi (1960)
Canal Zone

Styposis flavescens Simon
Nentwig & Heimer (1983)
Gamboa

Styposis scleropsis Levi
Levi (1960)
Canal Zone

Synotaxus ecuadorensis Exline
Exline & Levi (1965), Lubin (1978)
Barro Colorado Island
(*Synotaxus* is considered to be a separate family Synotaxidae by Forster et al. 1990)

Synotaxus turbinatus Simon
(= Rhomphaea anomala Chamberlin & Ivie)
Chamberlin & Ivie (1936), Exline & Levi (1965)
Barro Colorado Island
(see note above)

(Theridiella zygops Chamberlin & Ivie = Theridiosomatidae)
(Theridion acutiventer (Keyserling) = Coleosoma a.)

Theridion adamsoni Berland
Levi (1959 b)
Canal Zone

Theridion adjacens (O. Pickard-Cambridge)
Levi (1959 b), Nentwig & Heimer (1983)
Conception, Boquete, El Volcan, Cerro Punta

Theridion akme Levi
Levi (1959 b)
Canal Zone

Theridion akron Levi
Levi (1959 b)
Barro Colorado Island, Canal Zone

(Theridion alacre Keyserling = Achaearanea a.)

Theridion albulum O. Pickard-Cambridge
O. Pickard-Cambridge (1898), Petrunkevitch (1925), Levi (1959 b)
Bugaba, Canal Zone, Boquete, Arraijan

Theridion analyticum Chamberlin
Chickering (1936)
Barro Colorado Island

Theridion armouri Levi
Levi (1959 b, 1963 c), Nentwig & Heimer (1983)
Canal Zone, Gamboa
Theridion artum Levi

Levi (1959 b, 1963 c)
Canal Zone

Theridion atropunctatum Petrunkevitch
(= Wamba congener O. Pickard-Cambridge)
Banks (1929), Levi (1959 b)
Barro Colorado Island, Canal Zone, Arraijan

Theridion barroanum Levi
Levi (1959 b, 1963 c)
Canal Zone, Barro Colorado Island

(Theridion cambridgei Petrunkevitch = Chrysso c.)

Theridion centrum Levi
Levi (1959 b)
Chiriqui

Theridion chiriqui Levi
Levi (1959 b)
El Volcan

Theridion cobanum Levi
Nentwig & Heimer (1983)
Cerro Punta

Theridion cocosense Strand
Nentwig & Heimer (1983)
Gamboa

(Theridion compressum (Keyserling) = Chrysso c.)

Theridion costaricaense Levi
Levi (1963 c)
Canal Zone

Theridion crispulum Simon
Levi (1959 b), Nentwig & Heimer (1983)
Cerro Punta, El Volcan, Canal Zone, Barro Colorado Island

(Theridion defunctum Petrunkevitch = T. dilucidum Simon)

Theridion dilucidum Simon
(= T. defunctum Petrunkevitch)
Petrunkevitch (1925), Levi (1959 b), Nentwig & Heimer (1983)
Gamboa, Barro Colorado Island, Canal Zone, Wilcox Camp

Theridion dotanum (Banks)
Levi (1959 b)
El Volcan

Theridion elisabethae Roewer
Levi (1959 b)
Playa Grande

Theridion evexum Keyserling
Levi (1959 b)
El Volcan

(Theridion eximium Keyserling = Anelosimus e.)
(Theridion florens O. Pickard-Cambridge = Achaearanea f.)
(Theridion floridanum (Banks) = Coleosoma f.)
(Theridion fordum Keyserling = Tidarren f.)

Theridion fuessleyi Simon
Chickering (1936)
Barro Colorado Island

Theridion galerum Levi
Levi (1959 b), Nentwig & Heimer (1983)
Canal Zone, Barro Colorado Island

Theridion grecia Levi
Levi (1959 b)
El Valle

Theridion hispidum O. Pickard-Cambridge
Levi (1959 b), Nentwig & Heimer (1983)
El Valle, Boquete, Portobelo, Canal Zone, Barro Colorado Island

(Theridion ignotum (Keyserling) = Achaearanea hirta (Taczanowski))

Theridion indicatum Banks
Banks (1929)
Barro Colorado Island

Theridion istokpoga Levi
Levi (1959 b)
Canal Zone

Theridion jamaicense Levi
Nentwig & Heimer (1983)
El Valle

(Theridion jucundum O. Pickard-Cambridge = Anelosimus j.)

Theridion lathropi Levi
Levi (1959 b), Nentwig & Heimer (1983)
Canal Zone, Barro Colorado Island, Portobelo

Theridion longistylum (F.O. Pickard-Cambridge)
(= Steatoda l.)
F.O. Pickard-Cambridge (1902), Petrunkevitch (1925)
Bugaba

Theridion marvum Levi
Levi (1959 b, 1963 c)
Canal Zone, Barro Colorado Island, Boquete

Theridion metabolum Chamberlin & Ivie
Chamberlin & Ivie (1936), Levi (1959 b), Nentwig & Heimer(1983)
Barro Colorado Island, Canal Zone, Gamboa, El Valle

Theridion micheneri Levi
Levi (1963 c)
Canal Zone

Theridion minutissimum Keyserling
Levi (1959 b)
Canal Zone

(Theridion mixtum O. Pickard-Cambridge = Tidarren m.)

Theridion morulum O. Pickard-Cambridge
Nentwig & Heimer (1983)
Chiriqui

(Theridion nigrovittatum Keyserling = Achaearanea n.)

Theridion niveum O. Pickard-Cambridge
O. Pickard-Cambridge (1898), Petrunkevitch (1925), Banks (1929)
Bugaba, Barro Colorado Island, Canal Zone

Theridion nudum Levi
Levi (1959 b)
El Volcan, El Valle, Boquete

Theridion oswaldocruzi Levi
Nentwig & Heimer (1983)
Gamboa

Theridion paidiscum Levi
Levi (1959 b)
Boquete

Theridion panamense Levi
Levi (1959 b, 1963 c)
El Volcan, El Valle, Boquete

Theridion parvum Levi
Levi (1959 b)
El Volcan

(Theridion perniciosum Keyserling = Paratheridula p.)

Theridion petrum Levi
Levi (1959 b, 1963 c)
Canal Zone, Barro Colorado Island

(Theridion picadoi Banks = Achaearanea p.)

Theridion plantatum Levi
Levi (1959 b)
Barro Colorado Island

Theridion progum Levi
Levi (1959 b)
El Valle

Theridion quantum Levi
Levi (1959 b), Nentwig & Heimer (1983)
Canal Zone

Theridion reservum Levi
Levi (1959 b)
Canal Zone, Barro Colorado Island

Theridion resum Levi
Levi (1959 b)
Canal Zone

(Theridion rostratum O. Pickard-Cambridge = Achaearanea r.)
(Theridion rubicundum (F.O. Pickard-Cambridge) = Chrysso vexabilis Keyserling)

Theridion rufipes Lucas
(= Steatoda r.)
F.O. Pickard-Cambridge (1901), Petrunkevitch (1925), Levi (1959 b), Nentwig & Heimer (1983)
Bugaba, Wilcox Camp, El Valle, Boquete, Gamboa

Theridion rufipunctum Levi
Levi (1959 b, 1963 c)
El Volcan, Boquete

Theridion sexmaculatum Keyserling
Levi (1959 b)
Canal Zone, Barro Colorado Island, El Volcan

Theridion signaculum Levi
Levi (1959 b, 1963 c)
Canal Zone, Barro Colorado Island, Portobelo

Theridion signum Levi
Levi (1959 b)
Boquete

(Theridion studiosus Hentz = Anelosimus st.)
(Theridion taeniatum Keyserling = Achaearanea t.)

Theridion tempum Levi
Levi (1959 b, 1963 c)
Canal Zone

(Theridion tepidariorum C.L.Koch = Achaearanea t.)
(Theridion tesselatum Keyserling = Achaearanea t.)

Theridion trepidum O. Pickard-Cambridge
Levi (1959 b)
El Volcan

Theridion valleculum Levi
Levi (1959 b)
El Valle

(Theridion vittatum O. Pickard-Cambridge = Chrysso cambridgei (Petrunkevitch))

Theridion vituperabile Petrunkevitch
Petrunkevitch (1925)
Panama

Theridion vulvum Levi
Levi (1959 b)
El Valle, El Volcan

(Theridiotis probabilis (O. Pickard-Cambridge) = Chrosiothes p.)
(Theridium = Theridion)

Theridula gonygaster (Simon)
(= T. opulenta (Simon))
(= T. triangularis Keyserling)
Petrunkevitch (1925), Banks (1929), Chickering (1936), Levi (1954 c, 1966), Harrison (1968), Eberhard (pers. comm. 1988)
Barro Colorado Island, Canal Zone, Fortuna, Changuinola

Theridula puebla Levi
Levi (1966)
Barro Colorado Island, Canal Zone, Boquete

(Theridula opulenta (Simon) = T. gonygaster (Simon))
(Theridula triangularis Keyserling = T. gonygaster (Simon))

Thwaitesia affinis O. Pickard-Cambridge
O. Pickard-Cambridge (1896), Petrunkevitch (1925), Banks (1929), Levi (1963 e), Nentwig & Heimer (1983)
Bugaba, Barro Colorado Island, Gamboa, Canal Zone, Arraijan, Las Sabanas, Wilcox Camp

Thymoites amprus Levi
Levi (1964 c)
Canal Zone

Thymoites bogus (Levi)
(= Sphyrotinus b.)
Levi (1959 b)
Boquete

Thymoites boquete (Levi)
(= Sphyrotinus b.)
Levi (1959 b), Nentwig & Heimer (1983)
Gamboa, Canal Zone, Boquete, El Volcan

Thymoites caracasanus (Simon)
(= Hubba insignis O. Pickard-Cambridge)
(= Sphyrotinus i.)
Banks (1929), Levi (1959 b)
Barro Colorado Island, Canal Zone

Thymoites chickeringi (Levi)
(= Sphyrotinus c.)
Levi (1959 b), Nentwig & Heimer (1983)
Canal Zone, Barro Colorado Island, Gamboa

Thymoites confraternus (Banks)
(= Sphyrotinus deprus Levi)
Levi (1959 b, 1964 c)
Canal Zone, Barro Colorado Island

Thymoites delicatulus (Levi)
(= Sphyrotinus d.)
Levi (1959 b, 1964 c)
Canal Zone, Barro Colorado Island, Old Panama

(Thymoites deprus (Levi) = T. confraternus (Banks))

Thymoites illudens (Gertsch & Archer)
(= Sphyrotinus i.)
Levi (1959 b)
Canal Zone

Thymoites indicatus (Banks)
(= Sphyrotinus i.)
Levi (1959 b)
Barro Colorado Island, Canal Zone

(Thymoites insignis O. Pickard-Cambridge = T. caracasanus (Simon))

Thymoites luculentus (Simon)
(= Sphyrotinus l.)
Levi (1959 b)
Barro Colorado Island

Thymoites maderae (Gertsch & Archer)
(= Sphyrotinus m.)
Levi (1959 b, 1964 c), Nentwig & Heimer (1983)
El Valle, Canal Zone, Gamboa

Thymoites notabilis (Levi)
Levi (1959 b, 1964 c), Nentwig & Heimer (1983)
Canal Zone

Thymoites prolatus (Levi)
(= Sphyrotinus p.)
Levi (1959 b)
Barro Colorado Island, Canal Zone

Thymoites reservatus (Levi)
Levi (1959 b)
Canal Zone

Thymoites stylifrons (Simon)
(= Sphyrotinus s.)
Levi (1959 b)
Arraijan, Canal Zone, Barro Colorado Island

Tidarren fordum (Keyserling)
(= Steatoda f.)
(= Theridion f.)

(= T. haemorrhoidale (Bertkau))
F.O. Pickard-Cambridge (1902), Petrunkevitch (1925), Banks (1929), Levi (1955 a), Harrison (1968), Lubin (1978), Nentwig & Heimer (1983)
Bugaba, Canal Zone, Gamboa, Barro Colorado Island, Taboga Island, Changuinola

(Tidarren haemorrhoidale (Bertkau) = T. fordum (Keyserling))

Tidarren mixtum (O. Pickard-Cambridge)
(= Theridion m.)
Chickering (1936)
Barro Colorado Island

Tidarren sisyphoides (Walckenaer)
Levi (1955 a), Nentwig & Heimer (1983)
Barro Colorado Island, Canal Zone, Gamboa

(Wamba congener O. Pickard-Cambridge = Theridion atropunctatum Petrunkevitch)

Theridiosomatidae

(Andrasta globosa O. Pickard-Cambridge = Epilineutes g.))

Baalzebub baubo Coddington
Coddington (1986)
Canal Zone, Barro Colorado Island

Epeirotypus chavarria Coddington
Eberhard (pers. comm. 1988)
Fortuna

Epilineutes globosus (O. Pickard-Cambridge)
(= Andrasta g.)
(= Theridiosoma g.)
Petrunkevitch (1925), Coddington (1986)
Canal Zone, Barro Colorado Island

Naatlo sutila Coddington
Coddington (1986), Eberhard (pers. comm. 1988)
Fortuna

Naatlo fauna (Simon)
Coddington (1986)
Canal Zone, Barro Colorado Island

Ogulnius gertschi Archer
Archer (1953)
El Volcan

Theridiosoma argentiolunulatum Simon
Chickering (1936)
Barro Colorado Island

(Theridiosoma globosum (O. Pickard-Cambridge) = Epilineutes g.)

Theridiosoma zygops (Chamberlin & Ivie)
(= Theridiella z.)
Chamberlin & Ivie (1936), Levi (1962)
Barro Colorado Island

Wendilgarda clara Keyserling
(= W. panamica Archer)
Chickering (1936), Archer (1953), Coddington (1986)
Barro Colorado Island, Canal Zone, Portobelo

(Wendilgarda panamica Archer = W. clara Keyserling)

Symphytognathidae

Anapistula secreta Gertsch
Gertsch (1941), Forster & Platnick (1977), Nentwig & Heimer (1983)
Barro Colorado Island, Canal Zone, Gamboa

Curimagua bayano Forster
Forster & Platnick (1977), Vollrath (1978)
Bayano

Anapidae

Anapis keyserlingi Gertsch
Gertsch (1941), Platnick & Shadab (1978 b)
Barro Colorado Island, Canal Zone

Anapis monteverde Platnick & Shadab
Nentwig & Heimer (1983)
Cerro Punta

Anapisona furtiva Gertsch
Gertsch (1941), Platnick & Shadab (1979)
Barro Colorado Island

Anapisona hamigera (Simon)
Platnick & Shadab (1979), Nentwig & Heimer (1983)
Barro Colorado Island, Cerro Punta

Anapisona kethleyi Platnick & Shadab
Nentwig & Heimer (1983)
Chiriqui

Anapisona simoni Gertsch
Gertsch (1941), Platnick & Shadab (1979)
Barro Colorado Island, Canal Zone

Pseudanapis gertschi (Forster)
Platnick & Shadab (1979)
Barro Colorado Island, Boquete

Mysmenidae

(Lucarachne beebei Gertsch = Mysmenopsis b.)

Mysmena incredula (Gertsch & Davis)
Levi (1956 b), Nentwig & Heimer (1983)
Canal Zone, Gamboa, Barro Colorado Island

Mysmenopsis beebei (Gertsch)
(= Lucarachne b.)
Chickering (1960), Platnick & Shadab (1978)
El Valle

Mysmenopsis dipluramigo Platnick & Shadab
Platnick & Shadab (1978 a), Vollrath (1978), Nentwig & Heimer (1983)
Gamboa, El Valle, Bayano

Mysmenopsis gamboa Platnick & Shadab
Platnick & Shadab (1978 a), Vollrath (1978)
Gamboa, Bayano

Mysmenopsis ischnamigo Platnick & Shadab
Platnick & Shadab (1978 a), Vollrath (1978), Nentwig & Heimer (1983)
Canal Zone

Linyphiidae

Bathyphantes plagiatus Banks
Banks (1929)
Canal Zone, Barro Colorado Island

Ceraticelus similis (Banks)
Nentwig & Heimer (1983)
Gamboa

Eperigone caelebs
Millidge (1987)
Panama

Erigone antegona Chickering
Chickering (1970 b)
Canal Zone

Erigone aptuna Chickering
Chickering (1970 b)
Barro Colorado Island

Erigone bereta Chickering
Chickering (1970 b)
Canal Zone

Erigone bimaculata Keyserling
F.O. Pickard-Cambridge (1902), Petrunkevitch (1925)
Isthmus of Darien

Erigone digena Chickering
Chickering (1970 b)
Barro Colorado Island, Canal Zone

Erigone dipona Chickering
Chickering (1970 b)
Boquete

(Frontinella calcarifera (Keyserling) = Linyphia c.)

(Frontinella uncata F.O. Pickard-Cambridge = Novofrontina u.)

Gongylidiellum panamanum Petrunkevitch
Petrunkevitch (1925)
Wilcox Camp

Grammonota dalunda Chickering
Chickering (1970 a)
Canal Zone

Grammonota innota Chickering
Chickering (1970 a)
Barro Colorado Island, Canal Zone

Grammonota lutacola Chickering
Chickering (1970 a)
Boquete

Grammonota secata Chickering
Chickering (1970 a)
Barro Colorado Island, Canal Zone

Grammonota tabuna Chickering
Chickering (1970 a), Nentwig & Heimer (1983)
Barro Colorado Island, Canal Zone, Gamboa

Grammonota terresta Chickering
Chickering (1970 a)
Canal Zone, Barro Colorado Island, Arraijan, Boquete

Laminacauda dentichelis Millidge
Millidge (1985)
Panama

Lepthyphantes sp.
Chickering (1936)
Barro Colorado Island

(Linyphia bipunctata (Keyserling) = Novofrontina b.)

Linyphia bifasciata (F.O. Pickard-Cambridge)
(= Neriene b.)
Nentwig & Heimer (1983)
Cerro Punta

Linyphia calcarifera Keyserling
(= Frontinella c.)
Petrunkevitch (1925), Banks (1929)
Barro Colorado Island, Canal Zone, Las Sabanas, Wilcox Camp

Linyphia lambda (F.O. Pickard-Cambridge)
(= Neriene l.)
Nentwig & Heimer (1983)
El Valle

Linyphia nigrita (F.O. Pickard-Cambridge)
(= Neriene n.)
Nentwig & Heimer (1983)
Cerro Punta

Linyphia octopunctata (Chamberlin & Ivie)
(= Neriene o.)
Chamberlin & Ivie (1936)
Barro Colorado Island

Linyphia trifalcata (F.O. Pickard-Cambridge)
(= Neriene t.)
Nentwig & Heimer (1983)
Cerro Punta

(Linyphia uncata (F.O. Pickard-Cambridge) = Novofrontina u.)
(= Frontinella u.)

Microneta semiatra (Keyserling)
Chickering (1936)
Barro Colorado Island

(Neriene bipunctata Keyserling = Novofrontina b.)
(Neriene bifasciata F.O. Pickard-Cambridge = Novofrontina b.)
(Neriene lambda F.O. Pickard-Cambridge = Linyphia l.)
(Neriene nigrita F.O. Pickard-Cambridge = Linyphia n.)
(Neriene octopunctata Chamberlin & Ivie = Linyphia o.)
(Neriene trifalcata F.O. Pickard-Cambridge = Linyphia t.)

Novofrontina bipunctata (Keyserling)
(=Lynyphia b.)
(=Neriene b.)
Banks (1929)
Taboga Island
("The record for Panama may refer to *N. uncata*", Millidge (1991).)

Novofrontina uncata (F.O. Pickard-Cambridge)
(=Frontinella u.)
(=Linyphia u.)
F.O. Pickard-Cambridge (1902), Banks (1929), Nentwig & Heimer (1983), Millidge (1991)
Bugaba, Barro Colorado Island, Gamboa, Canal Zone, El Valle, Carti Road

Pocobletus coroniger Simon
Chickering (1969 a), Nentwig & Heimer (1983)
Barro Colorado Island, Canal Zone, Gamboa

Sthelota albonotata (Keyserling)
F.O. Pickard-Cambridge (1902), Petrunkevitch (1925), Chickering (1936)
Bugaba, Barro Colorado Island

Tetragnathidae

(Agriognatha insolita Chickering = Cyrtognatha i.)
(Agriognatha lepida (O. Pickard-Cambridge) = Cyrtognatha l.)
(Argyroepeira = Leucauge)
(Argyroepeira lepida O. Pickard-Cambridge = Cyrtognatha l.)

Azilia guatemalensis O. Pickard-Cambridge
O. Pickard-Cambridge (1889), Petrunkevitch (1925)
Bugaba

Chrysometa alajuela Levi
Levi (1986 a), Eberhard (pers. comm. 1988)
El Volcan, Fortuna

Chrysometa alboguttata (O. Pickard-Cambridge)
(= Meta a.)
(= Pseudometa a.)
(= Pseudometa superans (= O. Pickard-Cambridge))
O. Pickard-Cambridge (1889), Petrunkevitch (1925), Banks (1929), Harrison (1968), Levi (1986 a)
Barro Colorado Island, Canal Zone, Gamboa, Bugaba, Boquete, Changuinola

Chrysometa boquete Levi
Levi (1986 a)
Boquete

Chrysometa flava (O. Pickard-Cambridge)
(= Pseudometa f.)
Nentwig & Heimer (1983)
Cerro Punta

Chrysometa poas Levi
Levi (1986 a)
Boquete, Cerro Punta

Chrysometa universitaria Levi
Levi (1986 a)
Boquete, El Volcan, Rio Sereno

Chrysometa zelotypa (Keyserling)
Levi (1986 a)
Cerro Punta

Cyrtognatha globosa Petrunkevitch
Petrunkevitch (1925)
Wilcox Camp

Cyrtognatha lepida (O. Pickard-Cambridge)
(= Agriognatha l.)
(= Argyroepeira l.)
O. Pickard-Cambridge (1889), Petrunkevitch (1925)
Bugaba

Cyrtognatha pachygnathoides (O. Pickard-Cambridge)
(= Agriognatha p.)

(= Tetragnatha p.)
O. Pickard-Cambridge (1894), Petrunkevitch (1925)
El Volcan

Dolichognatha pentagona (Hentz)
Nentwig & Heimer (1983)
Canal Zone

Dolichognatha spinosa (Petrunkevitch)
(= Landana sp.)
Petrunkevitch (1939 a), Chickering (1942)
Barro Colorado Island

Dolichognatha tuberculata (Keyserling)
Chickering (1936)
Barro Colorado Island

(Eucta lacerta (Walckenaer) = Tetragnatha caudata Emerton)

Glenognatha centralis Chamberlin
Chamberlin (1925), Banks (1929), Hormiga & Döbel (1990)
Panama

Glenognatha emertoni Simon
Nentwig & Heimer (1983)
Gamboa

Glenognatha foxi (McCook)
(= Mimognatha f.)
Banks (1929), Levi (1980), Nentwig & Heimer (1983)
Gamboa, Carti Road, Bella Vista

Homalometa chiriqui Levi
Levi (1986 a)
El Volcan

Homalometa nigritarsis Simon
Levi (1986 a)
Barro Colorado Island

(Landana spinosa Petrunkevitch = Dolichognatha s.)

Leucauge acuminata (O. Pickard-Cambridge)
(= Argyroepeira a.)
O. Pickard-Cambridge (1889), Petrunkevitch (1925)
Bugaba, Tolé

Leucauge argyra (Walckenaer)
(= Plesiometa a.)
F.O. Pickard-Cambridge (1903), Petrunkevitch (1925), Banks (1929), Chickering (1936), Harrison (1968), Nentwig & Heimer (1983)
Barro Colorado Island, Canal Zone, Colon, Gamboa, Taboga Island, Las Lajas, Conception, Bugaba, Changuinola

Leucauge aurostriata (O. Pickard-Cambridge)
(= Argyroepeira a.)
Banks (1929)
Gamboa

Leucauge curta (O. Pickard-Cambridge)
(= Argyroepeira c.)
O. Pickard-Cambridge (1889), Petrunkevitch (1925)
Bugaba

Leucauge geminipunctata Chamberlin & Ivie
Chamberlin & Ivie (1936)
Barro Colorado Island

Leucauge globosa (O. Pickard-Cambridge)
(= Mecynometa g.)
Petrunkevitch (1925), Banks (1929), Chickering (1963 a), Nentwig & Heimer (1983)
Barro Colorado Island, Canal Zone, Gamboa, El Valle, Wilcox Camp

Leucauge idonea (O. Pickard-Cambridge)
(= Argyroepeira i.)
O. Pickard-Cambridge (1889), Petrunkevitch (1925), Banks (1929), Nentwig & Heimer (1983)
Bugaba, Canal Zone

Leucauge lugens O. Pickard-Cambridge
Banks (1929), Nentwig & Heimer (1983)
Barro Colorado Island, Gamboa

(Leucauge maerens (O. Pickard-Cambridge) = L. moerens)

Leucauge mandibulata F.O. Pickard-Cambridge
Banks (1929), Nentwig & Heimer (1983)
Barro Colorado Island, El Valle

Leucauge moerens (O. Pickard-Cambridge)
(= L. maerens)
F.O. Pickard-Cambridge (1903), Petrunkevitch (1925), Banks (1929), Nentwig & Heimer (1983)
Bugaba, Barro Colorado Island, Canal Zone, Gamboa, El Valle

Leucauge saphes Chamberlin & Ivie
Chamberlin & Ivie (1936)
Barro Colorado Island

Leucauge simplex F.O. Pickard-Cambridge
Nentwig & Heimer (1983)
Canal Zone, Gamboa, El Real

Leucauge synthetica Chamberlin & Ivie
Chamberlin & Ivie (1936)
Barro Colorado Island

Leucauge venusta (Walckenaer)
F.O. Pickard-Cambridge (1903), Petrunkevitch (1925), Hoffmaster (1982), Nentwig & Heimer (1983)
Bugaba, Canal Zone, Gamboa, Carti Road, El Valle, El Real

(Mecynometa globosa (O. Pickard-Cambridge) = Leucauge g.)

Mecynometa quadrituberculata (Keyserling)
Chickering (1936)
Barro Colorado Island

(Meta gravida (O. Pickard-Cambridge) = Metabus g.)

Metabus gravidus (O. Pickard-Cambridge)
(= Meta g.)
Petrunkevitch (1925), Eberhard (pers. comm. 1988)
Cerro Flores, Fortuna

(Mimognatha foxi McCook = Glenognatha f.)

Nephila clavipes (Linnaeus)
Petrunkevitch (1925, 1939 b), Banks (1929), Harrison (1968), Robinson & Mirick (1971), Robinson & Robinson (1973, 1974, 1980), Vollrath (1976, 1979, 1980, 1985), Lubin (1978), Hoffmaster (1982), Nentwig & Heimer (1983), Nentwig (1985 e), Levi & von Eickstedt (1989)
Barro Colorado Island, Gamboa, Canal Zone, Taboga Island, El Real, Changuinola

(Plesiometa argyra (Walckenaer) = Leucauge a.)
(Pseudometa alboguttata (O. Pickard-Cambridge) = Chrysometa a.)
(Pseudometa flava (O. Pickard-Cambridge) = Chrysometa f.)
(Pseudometa superans (O. Pickard-Cambridge) = Chrysometa alboguttata (O. Pickard-Cambridge))

Tetragnatha alba F.O. Pickard-Cambridge
F.O. Pickard-Cambridge (1903), Petrunkevitch (1925)
Bugaba

Tetragnatha amplidens Chamberlin & Ivie
Chamberlin & Ivie (1936)
Barro Colorado Island

(Tetragnatha antillana Simon = T. nitens (Audouin in Savigny)

Tetragnatha caudata Emerton
(= Eucta lacerta (Walckenaer))
Chickering (1957 c)
Boquete

Tetragnatha cognata O. Pickard-Cambridge
Chickering (1957 c)
Boquete, El Volcan, Canal Zone

Tetragnatha confraterna Banks
Chickering (1936, 1957 c)
Barro Colorado Island

Tetragnatha ethodon Chamberlin & Ivie
Chamberlin & Ivie (1936), Chickering (1957 c)
Barro Colorado Island

Tetragnatha fragilis Chickering
Chickering (1957 c)
Barro Colorado Island, Canal Zone, Gamboa, Arraijan, Portobelo

Tetragnatha gertschi Chickering
Chickering (1957 c)
El Volcan, Boquete

Tetragnatha guatemalensis O. Pickard-Cambridge
Chickering (1957 c), Levi (1981)
Barro Colorado Island, Gamboa

Tetragnatha laboriosa Hentz
Chickering (1957 c), Levi (1981), Nentwig & Heimer (1983)
El Volcan, Boquete, Canal Zone, Barro Colorado Island, Gamboa

Tetragnatha mabelae Chickering
Chickering (1957 c)
Barro Colorado Island, Canal Zone

Tetragnatha mexicana Keyserling
Chickering (1936, 1957 c)
Barro Colorado Island

Tetragnatha nitens (Audouin in Savigny) (= T. antillana Simon)
Petrunkevitch (1925), Chickering (1957 c), Levi 1981
Boquete, El Valle, Wilcox Camp, El Volcan, Canal Zone

(Tetragnatha pachygnathoides O. Pickard-Cambridge = Cyrtognatha p.)

Tetragnatha pallescens F.O. Pickard-Cambridge
Chickering (1936, 1957 c), Levi (1981), Nentwig & Heimer (1983)
Barro Colorado Island, El Volcan, Chiriqui, Gamboa, Canal Zone

Tetragnatha pallida O. Pickard-Cambridge
O. Pickard-Cambridge (1889), Banks (1929), Chickering (1957 c), Nentwig & Heimer (1983)
Barro Colorado Island, Canal Zone, Bugaba

Tetragnatha siduo Chamberlin & Ivie
Chamberlin & Ivie (1936)
Barro Colorado Island

Tetragnatha sinuosa Chickering
Chickering (1957 c)
Canal Zone

Tetragnatha tenuis O. Pickard-Cambridge
O. Pickard-Cambridge (1889), Petrunkevitch (1925), Chickering (1957 c), Nentwig & Heimer (1983)
Bugaba, Barro Colorado Island, Gamboa

Tetragnatha tenuissima O. Pickard-Cambridge
O. Pickard-Cambridge (1889), Petrunkevitch (1925), Banks (1929), Chickering (1957 c), Nentwig & Heimer (1983)
Bugaba, Canal Zone, Barro Colorado Island, El Valle, Boquete, Wilcox Camp, Gamboa

Tetragnatha tropica O. Pickard-Cambridge
Chickering (1936, 1957 c)
Barro Colorado Island, Canal Zone

Tetragnatha vermiformis Emerton
Levi (1981)
Barro Colorado Island

Araneidae

Acacesia alticeps Chamberlin & Ivie
Chamberlin & Ivie (1936)
Barro Colorado Island

Acacesia cornigera Petrunkevitch
Petrunkevitch (1925)
Wilcox Camp

(Acacesia foliata Hentz = A. hamata)

Acacesia hamata (Hentz)
(= A. foliata Hentz)
F.O. Pickard-Cambridge (1904), Petrunkevitch (1925), Banks (1929), Chickering (1936), Levi (1976), Nentwig & Heimer (1983), Eberhard (pers. comm. 1988)
Bugaba, Las Sabanas, El Valle, Carti Road, Gamboa, Barro Colorado Island, Canal Zone

(Acrosoma = Micrathena)
(Agriognatha insolita Chickering = Cyrtognatha i., Tetragnathidae)

Aculepeira azul Levi
Levi (1991 b)
Cerro Azul (Panama City)

Aculepeira gravabilis (O. Pickard-Cambridge)
(= Araneus g.)
(= Epeira g.)
O. Pickard-Cambridge (1889), Petrunkevitch (1925), Nentwig & Heimer (1983), Levi (1991 b)
Cerro Punta, El Volcan, Cerro Jefe

Aculepeira travassosi (Soares & Camargo)
Levi (1991 b)
Barro Colorado Island

Alpaida acuta (Keyserling)
(= Epeira a.)
Levi (1988)
Canal Zone

Alpaida bicornuta (Taczanowski)
(= Araneus designatus Chamberlin & Ivie)
(= Araneus spinigerus O. Pickard-Cambridge)
(= Edricus spinigerus (O. Pickard-Cambridge)
(= Epeira spinigera O. Pickard-Cambridge)
O. Pickard-Cambridge (1889), Petrunkevitch (1925), Chickering (1936), Chamberlin & Ivie (1936), Levi (1988)
Barro Colorado Island, Colon, Canal Zone, Bugaba

Alpaida championi (O. Pickard-Cambridge)
(= Epeira ch.)
Petrunkevitch (1925), Banks (1929), Levi (1988)
Bugaba, Barro Colorado Island

Alpaida chickeringi Levi
(Levi (1988)
Barro Colorado Island

Alpaida conica O. Pickard-Cambridge
O. Pickard-Cambridge (1889), Levi (1988)
Bugaba, Canal Zone

Alpaida graphica (O. Pickard-Cambridge)
(= Araneus g.)
(= Epeira g.)
O. Pickard-Cambridge (1889), Petrunkevitch (1925), Levi (1988)
Bugaba

Alpaida leucogramma (White)
(= Araneus albostriata (Keyserling))
(= Epeira a.)
(= Parepeira a.)
Banks (1929), Chickering (1936), Petrunkevitch (1939 b), (Levi 1988)

Coclé, Santa Rosa (Colon), Canal Zone, El Real (Darien), Barro Colorado Island, Las Sabanas

Alpaida moata (Chamberlin & Ivie)
(= Araneus m.)
Chamberlin & Ivie (1936), Eberhard (pers. comm. 1988), Levi (1988)
Canal Zone, Barro Colorado Island, Gamboa, Bocas del Toro, Rio Changuinola, Corriente Grande, El Valle

Alpaida septemmammata (O. Pickard-Cambridge)
(= Araneus s.)
Nentwig & Heimer (1983), Levi (1988)
Alto Limon Lo Yegueda (Veraguas), Barro Colorado Island, Canal Zone, Gamboa, Guayabalito / Gatuncillo / Chagres River (Colon),

Alpaida trispinosa (Keyserling)
(= Araneus t.)
Banks (1929), Petrunkevitch (1939 b), Levi (1988)
Playa Las Lajas (Chiriqui), Colon, Barro Colorado Island, Canal Zone

Alpaida truncata (Keyserling)
(= Araneus mutatus Chamberlin & Ivie)
(= Araneus t.)
(= Edricus t.)
Banks (1929), Chamberlin & Ivie (1936), Petrunkevitch (1939 b), Levi (1988)
Arraijan, Barro Colorado Island, Canal Zone

Alpaida tuonabo (Chamberlin & Ivie)
(= Araneus t.)
Chamberlin & Ivie (1936), Lubin (1978), Shelly (1983), Nentwig & Heimer (1983), Eberhard (pers. comm. 1988), Levi (1988)
Gamboa, Barro Colorado Island, Canal Zone, Portobelo, El Valle

Alpaida variabilis (Keyserling)
(= Araneus v.)
Chickering (1936), Levi (1988)
Barro Colorado Island

Alpaida veniliae (Keyserling)
(= Araneus v.)
F.O. Pickard-Cambridge (1904), Petrunkevitch (1925), Chickering (1936), Nentwig & Heimer (1983), Levi (1988)
Bugaba, Barro Colorado Island, Gamboa, Desierto de Sarigua (Herrera), Santa Rosa (Colon), Canal Zone, El Real (Darien)

(Araneus albostriatus (Keyserling) = Alpaida leucogramma (White))

Araneus bispinosus (Keyserling)
F.O. Pickard-Cambridge (1904), Petrunkevitch (1925)
Panama

(Araneus championi (O. Pickard-Cambridge) = Alpaida ch.)
(Araneus clivosus (O. Pickard-Cambridge) = Wixia c.)

Araneus concoloratus F.O. Pickard-Cambridge
F.O. Pickard-Cambridge (1904), Petrunkevitch (1925), Levi (1991 b)
Veraguas

(Araneus consequus (O. Pickard-Cambridge) = Wixia c.)
(Araneus designatus Chamberlin & Ivie = Alpaida bicornuta (Taczanoweski))

Araneus detrimentosus (O. Pickard-Cambridge)
Banks (1929), Levi (1973, 1991 b)
Barro Colorado Island, Boquete, El Valle, Taboga Islands, Canal Zone

Araneus expletus (O. Pickard-Cambridge)
(= Neosconella styligera F.O. Pickard-Cambridge)
Nentwig & Heimer (1983), (Levi 1991 b)
Canal Zone, Bambito (Chiriqui), Volcan

(Araneus fasciolatus (O. Pickard-Cambridge) = Mastophora f.)
(Araneus glabratus F.O. Pickard-Cambridge = A. guttatus (Keyserling))

Araneus galero Levi
Levi (1991 b)
Arraijan, Playa Coronado, Barro Colorado Island, Canal Zone

(Araneus globosus Petrunkevitch = Wixia g. F.O. Pickard-Cambridge)
(Araneus graphicus (O. Pickard-Cambridge) = Alpaida g.)
(Araneus gravabilis (O. Pickard-Cambridge) = Aculepeira g.)

Araneus guttatus (Keyserling)
(= A. glabratus F.O. Pickard-Cambridge)
F.O. Pickard-Cambridge (1904), Petrunkevitch (1925), Banks (1929), Nentwig & Heimer (1983), Levi (1991 b)
Bugaba, Barro Colorado Island, Canal Zone, Gamboa, Las Sabanas, El Valle, Santiago, Villa Darien (Darien), El Salto

(Araneus hoxaeus (O. Pickard-Cambridge) = Parawixia h.)
(Araneus hypocritus (O. Pickard-Cambridge) = Parawixia h.)
(Araneus incertus O. Pickard-Cambridge = Metazygia i.)

"*Araneus*" *laticeps* (O. Pickard-Cambridge)
(= Epeira l.)
O. Pickard-Cambridge (1889), Petrunkevitch (1925), Nentwig & Heimer (1983)
Bugaba, Gamboa
("belongs into an unnamed genus close to Metazygia", Levi (1991 b))

(Araneus moatus Chamberlin & Ivie = Alpaida moata)
(Araneus mutatus Chamberlin & Ivie = Alpaida m.)
(Araneus nauticus (L.Koch) = Neoscona n.)

"*Araneus*" *nigrocinctus* F.O. Pickard-Cambridge
F.O. Pickard-Cambridge (1904), Petrunkevitch (1925)
Bugaba
("belongs in an unnamed genus", Levi (1991 b))

(Araneus oaxacensis (Keyserling) = Neoscona o.)
(Araneus pallidulus (Keyserling) = Metazygia p.)

Araneus pegnia (Walckenaer)
(= A. solersioides (O. Pickard-Cambridge)
(= Epeira s.)
O. Pickard-Cambridge (1889), Petrunkevitch (1925), Banks (1929), Levi (1973, 1991 b)
Bugaba, Canal Zone, Panama City, Taboga Island, Rio Changuinola, El Valle, Barro Colorado Island, Aguadulce

(Araneus septemmammatus (O. Pickard-Cambridge) = Alpaida s.)
(Araneus setospinosus Chamberlin & Ivie = Cyrtophora nympha Simon)

Araneus sextus (Chamberlin)
Banks (1929), Levi (1991 b)
El Valle, Arraijan, Canal Zone

(Araneus sinuoscapus Chamberlin & Ivie = A. venatrix (C.L.Koch))
(Araneus solersioides (O. Pickard-Cambridge) = Araneus pegnia (Walckenaer))
(Araneus spinigerus (O. Pickard-Cambridge) (female) = Alpaida bicornuta (Taczanowski)

Araneus stellatus (Walckenaer)
(= Marxia st.)
Chickering (1936)
Barro Colorado Island

(Araneus trispinosus (Keyserling) = Alpaida t.)

(Araneus truncatus (Keyserling) = Alpaida t.)
(Araneus tuonabo Chamberlin & Ivie = Alpaida t.)
(Araneus variabilis (Keyserling) = Alpaida v.)

Araneus venatrix (C.L.Koch)
(= Araneus sinuoscapus Chamberlin & Ivie)
(= A. venator)
Chamberlin & Ivie (1936), Chickering (1936), Levi (1991 b)
Barro Colorado Island, Canal Zone, Arraijan

(Araneus veniliae (Keyserling) = Alpaida v.)

Argiope argentata Fabricius
F.O. Pickard-Cambridge (1903), Petrunkevitch (1925, 1939 b), Banks (1929), Levi (1968), Harrison (1969), Robinson (1969), Robinson & Olazzari (1971), Robinson & Robinson (1970 a, b, 1980), Lubin (1978), Vollrath (1979), Hoffmaster (1982), Nentwig & Heimer (1983), Nentwig (1985 e), Eberhard (pers. comm. 1988)
El Volcan, Barro Colorado Island, Canal Zone, Gamboa, Las Sabanas, Panama City, Carti Road, Conception, El Valle, Punta Chame, Changuinola

Argiope savignyi Levi
Levi (1968), Robinson & Robinson (1970 a, 1980), Nentwig & Heimer (1983), Nentwig (1985 e), Eberhard (pers. comm. 1988)
Barro Colorado Island, Gamboa, Canal Zone, Bocas del Toro, Boquete

Argiope trifasciata (Forskal)
Petrunkevitch (1925, 1939 b)
Panama
(Doubtful record; "We have never seen *A. trifasciata* in Panama during 13 years of field studies."; Robinson & Robinson (1980))

(Argyroepeira = Leucauge)
(Bertrana hieroglyphica Petrunkevitch = B. striolata Keyserling)

Bertrana striolata Keyserling
(= B. hieroglyphica Petrunkevitch)
Petrunkevitch (1925), Chickering (1963 a), Nentwig & Heimer (1983), Eberhard (pers. comm. 1988), Levi (1989)
Wilcox Camp, Barro Colorado Island, Gamboa, Canal Zone

(Bion brevis O. Pickard-Cambridge = Witica crassicauda (Keyserling))
(Caira = Kaira)

Carepalxis solobrensis O. Pickard-Cambridge
(=C. gibbosa O. Pickard-Cambridge)
(=Scoloderus gibber)
O. Pickard-Cambridge (1899), Petrunkevitch (1925), Chickering (1936), Levi (1991 c)
Bugaba, Barron Colorado Island, Canal Zone

Cyclosa caroli Hentz
(= C. lacerta O. Pickard-Cambridge)
O. Pickard-Cambridge (1889), Petrunkevitch (1925), Banks (1929), Levi (1977 b), Lubin (1978), Robinson & Robinson (1980), Hoffmaster (1982), Nentwig & Heimer (1983)
Bugaba, Barro Colorado Island, Canal Zone, Gamboa, Carti Road, El Valle

(Cyclosa lacerta O. Pickard-Cambridge = C. caroli Hentz)

Cyclosa v-notata Petrunkevitch
Petrunkevitch (1925)
Wilcox Camp

(Cyclosa tuberculata (Keyserling) = C. turbinata (Walckenaer))
(Cyclosa tuberculifera O. Pickard-Cambridge = C. turbinata (Walckenaer))

Cyclosa turbinata (Walckenaer)
(= C. tuberculata (Keyserling))
(= C. tuberculifera O. Pickard-Cambridge)
Chickering (1936), Levi (1977 b)
Barro Colorado Island

Cyclosa walckenaeri (O. Pickard-Cambridge)
Banks (1929), Levi (1977 b), Nentwig & Heimer (1983)
Bella Vista, Playa Coronado

Cyrtophora nympha Simon
(= Araneus setospinosus Chamberlin & Ivie)
Chamberlin & Ivie (1936), Robinson & Robinson (1980), Levi (1991 b)
Barro Colorado Island

Cyrtophora sellata Simon
Robinson & Robinson (1980)
Panama
(Drexelia directa (Hentz) = Larinia d.)
(Edricus crassicaudus (Keyserling) = Witica c.)

Edricus spinigerus O. Pickard-Cambridge
O. Pickard-Cambridge (1890), Levi (1991 a)
Bugaba

(Edricus truncatus (Keyserling) = Alpaida t.)
(Epeira = Araneus)
(Epeira acuta Keyserling) = Alpaida a.)
(Epeira albostriata Keyserling = Alpaida leucogramma (White))
(Epeira armata O. Pickard-Cambridge = Parawixia rigida)
(Epeira bimaculata O. Pickard-Cambridge = Mangora b.)
(Epeira championi O. Pickard-Cambridge = Alpaida ch.)
(Epeira consequa O. Pickard-Cambridge = Wixia c.)
(Epeira destricta O. Pickard-Cambridge = Parawixia d.)
(Epeira fragilis O. Pickard-Cambridge = Eustala f.)
(Epeira graphica O. Pickard-Cambridge) = Alpaida g.)
(Epeira gravabilis O. Pickard-Cambridge = Aculepeira g.)
(Epeira gregalis O. Pickard-Cambridge = Metazygia g.)
(Epeira hoxaea O. Pickard-Cambridge = Parawixia h.)
(Epeira hypocrita O. Pickard-Cambridge = Parawixia h.)
(Epeira incerta O. Pickard-Cambridge = Metazygia i.)
(Epeira intercisa O. Pickard-Cambridge = Larinia directa (Hentz))
(Epeira latebricola O. Pickard-Cambridge = Eustala l.)
(Epeira laticeps O. Pickard-Cambridge = "Araneus" l.)
(Epeira pallidula Keyserling = Metazygia p.)
(Epeira purpurascens O. Pickard-Cambridge = Eriophora fuliginea (C.L.Koch))
(Epeira rigida O. Pickard-Cambridge = Parawixia r.)
(Epeira rufa O. Pickard-Cambridge = Wixia r.)
(Epeira solersioides O. Pickard-Cambridge = Araneus pegnia (Walckenaer))
(Epeira spinigera O. Pickard-Cambridge = Alpaida bicornuta (Taczanowski))
(Epeira tauricornis O. Pickard-Cambridge = Wagneriana t.)
(Epeira tetragnathoides O. Pickard-Cambridge = Larinia directa (Hentz))
(Epeira volucripes Keyserling = Neoscona nautica (L. Koch))

Epeiroides bahiensis Keyserling
Nentwig & Heimer (1983)
Carti Road

(Epeiroides fasciolata O. Pickard-Cambridge = Mastophora f.)

Eriophora edax (Blackwall)
F.O. Pickard-Cambridge (1903), Petrunkevitch (1925), Banks (1929), Chickering (1936), Harrison (1968), Levi (1970), Nentwig & Heimer (1983)
Bugaba, Barro Colorado Island, Canal Zone, Gamboa, El Valle, Changuinola

Eriophora fuliginea (C.L.Koch)
(= Epeira purpurascens O. Pickard-Cambridge)
(= E. purpurascens (O. Pickard-Cambridge))
O. Pickard-Cambridge (1889), Petrunkevitch (1925), Banks (1929), Levi (1970), Robinson et al (1971), Robinson & Robinson (1980), Nentwig & Heimer (1983), Nentwig (1985 e), Graf & Nentwig (1987)
Bugaba, Barro Colorado Island, Gamboa, Canal Zone, El Valle, Conception, El Real

Eriophora nephiloides (O. Pickard-Cambridge)
Banks (1929), Levi (1970), Lubin (1978), Nentwig & Heimer (1983)
Barro Colorado Island, Gamboa, Canal Zone

(Eriophora purpurascens (O. Pickard-Cambridge) = E. fuliginea (C.L. Koch))

Eriophora ravilla (C.L.Koch)
Levi (1970), Nentwig & Heimer (1983)
Gamboa

Eustala anastera (Walckenaer)
Nentwig & Heimer (1983)
Cerro Punta

Eustala bifida F.O. Pickard-Cambridge
Banks (1929), Nentwig & Heimer (1983)
Barro Colorado Island, Gamboa

Eustala bucolica Chickering
Chickering (1955 a)
Boquete, El Volcan

Eustala conformans Chamberlin
Chamberlin (1925), Banks (1929), Chickering (1955 a)
Barro Colorado Island

Eustala delecta Chickering
Chickering (1955 a)
Barro Colorado Island, Canal Zone, Boquete

Eustala devia (Gertsch & Mulaik)
(= E. minima Chickering)
Chickering (1955 a), Levi (1977 b), Nentwig & Heimer (1983)
Barro Colorado Island, Gamboa, Canal Zone, Boquete, El Valle, Arraijan, Portobelo

Eustala eleuthera Levi
Nentwig & Heimer (1983)
Gamboa

Eustala exigua Chickering
Chickering (1955 a)
Barro Colorado Island, Canal Zone, El Valle, Arraijan

Eustala fragilis (O. Pickard-Cambridge)
(= Epeira f.)
O. Pickard-Cambridge (1889), Petrunkevitch (1925)
Bugaba

Eustala fuscovittata (Keyserling)
Petrunkevitch (1925), Banks (1929), Chickering (1955 a), Harrison (1968), Nentwig & Heimer (1983), Eberhard (pers. comm. 1988)
Barro Colorado Island, Gamboa, Canal Zone, Changuinola

Eustala gertschi Chickering
Chickering (1955 a)
Barro Colorado Island

Eustala guttata F.O.P.C
F.O. Pickard-Cambridge (1904), Petrunkevitch (1925), Banks (1929), Chickering (1955 a), Nentwig & Heimer (1983)

Bugaba, Barro Colorado Island, Canal Zone, Gamboa, El Valle

Eustala illicita (O. Pickard-Cambridge)
Chickering (1955 a), Nentwig & Heimer (1983)
Canal Zone, Gamboa, El Valle

Eustala inconstans Chickering
Chickering (1955 a)
Barro Colorado Island

Eustala ingenua Chickering
Chickering (1955 a)
Canal Zone

Eustala innoxia Chickering
Chickering (1955 a)
Canal Zone

Eustala lata Chickering
Chickering (1955 a)
Barro Colorado Island

Eustala latebricola (O. Pickard-Cambridge)
(= Epeira l.)
O. Pickard-Cambridge (1889), Petrunkevitch (1925), Chickering (1936), Chickering (1955 a)
Bugaba, Barro Colorado Island, Boquete

Eustala longembola Chickering
Chickering (1955 a)
Barro Colorado Island, Canal Zone, Arraijan, El Valle

Eustala maxima Chickering
Chickering (1955 a)
Barro Colorado Island, Canal Zone

Eustala mimica Chickering
Chickering (1955 a)
Barro Colorado Island (?), Canal Zone

(Eustala minima Chickering = E. devia (Gertsch & Mulaik))

Eustala montana Chickering
Chickering (1955 a)
El Volcan

Eustala montivaga Chickering
Chickering (1955 a)
El Volcan

Eustala oblonga Chickering
Chickering (1955 a), Nentwig & Heimer (1983)
Barro Colorado Island, Gamboa, Canal Zone

Eustala panamana Chickering
Chickering (1955 a)
Cerro Punta

Eustala redundans Chickering
Chickering (1955 a)
Canal Zone

Eustala rustica Chickering
Chickering (1955 a)
El Valle

Eustala scitula Chickering
Chickering (1955 a)
Barro Colorado Island, Canal Zone

Eustala scutigera (O. Pickard-Cambridge)
F.O. Pickard-Cambridge (1904), Petrunkevitch (1925), Chickering (1936, 1955 a)
Bugaba, Barro Colorado Island, Canal Zone, El Valle, Boquete, Portobelo

Eustala sedula Chickering
Chickering (1955 a)
Barro Colorado Island

Eustala semifoliata (O. Pickard-Cambridge)
Chickering (1955 a)
Barro Colorado Island

Eustala tantula Chickering
Chickering (1955 a)
Barro Colorado Island, Canal Zone, El Valle, Boquete

(Eustala tuceps Chamberlin = Metazygia gregalis (O. Pickard-Cambridge))

Eustala tumida Chickering
Chickering (1955 a)
Canal Zone
Eustala vegeta (Keyserling)
Chickering (1955 a)
Portobelo

Eustala venusta Chickering
Chickering (1955 a), Nentwig & Heimer (1983)
Gamboa, Barro Colorado Island, Canal Zone

Gasteracantha cancriformis (Linn)
(= G. kochii Butler)
Petrunkevitch (1925), Harrison (1968), Levi (1978), Robinson & Robinson (1980), Nentwig & Heimer (1983)
Canal Zone, Gamboa, Barro Colorado Island, El Real, El Valle, Almirante, Changuinola

Gea heptagon (Hentz)
Banks (1929), Levi (1968)
Barro Colorado Island

(Gea panamensis Chamberlin = Argiope argentata Fabricius)

Glyptogona decemtuberculata (O. Pickard-Cambridge)
(= Cyrtarachne d.)
O. Pickard-Cambridge (1890), Petrunkevitch (1925)
Bugaba

Hypognatha elaborata Chickering
Chickering (1953 a)
Barro Colorado Island

Hypognatha testudinaria (Taczanowski)
Chickering (1936)
Barro Colorado Island

(Ildibaha = Micrathena)
(Ildibaha mutilloides Simon = Micrathena vigorsi (Perty))
(Ildibaha myrmeciaeformis (Taczanowski) = Micrathena schreibersi (Perty))

Kaira altiventer O. Pickard-Cambridge
O. Pickard-Cambridge (1889), Petrunkevitch (1925), Levi (1977 a)
Veraguas

Kaira gibberosa O. Pickard-Cambridge
O. Pickard-Cambridge (1889), Petrunkevitch (1925)
Veraguas

(Keyserlingia cornigera O. Pickard-Cambridge = Micrathena sexspinosa (Hahn))
(Landana spinosa Petrunkevitch = Dolichognatha s.)

Larinia directa (Hentz)
(= Drexelia d.)
(= Epeira intercisa O. Pickard-Cambridge)
(= Epeira tetragnathoides O. Pickard-Cambridge)
O. Pickard-Cambridge (1889), Petrunkevitch (1925), Chickering (1936), Harrison (1968), Levi (1975, 1990), Harrod et al. (1990)
Barro Colorado Island, Sona, Bugaba, Changuinola, Fortuna, Canal Zone

(Mahadeva, Mahadiva = Verrucosa)

Mangora belligerens Chamberlin & Ivie
Chamberlin & Ivie (1936)
Barro Colorado Island

Mangora bimaculata (O. Pickard-Cambridge)
(= Epeira b.)
O. Pickard-Cambridge (1889), Petrunkevitch (1925), Banks (1929), Chickering (1954), Robinson & Robinson (1980), Nentwig & Heimer (1983)
Barro Colorado Island, Gamboa, Canal Zone, Veraguas

Mangora candida Chickering
Chickering (1954)
Barro Colorado Island, Canal Zone

Mangora dentembolus Chamberlin & Ivie
Chamberlin & Ivie (1936)
Barro Colorado Island

Mangora mobilis (O. Pickard-Cambridge)
Chickering (1954)
Barro Colorado Island, Canal Zone, El Cermeno

Mangora montana Chickering
Chickering (1954)
Barro Colorado Island, Canal Zone, Taboga Island

Mangora pia Chamberlin & Ivie
Chamberlin & Ivie (1936), Chickering (1954), Lubin (1978), Nentwig & Heimer (1983)
Barro Colorado Island, Gamboa, Canal Zone, Taboga Island

Mangora picta O. Pickard-Cambridge
Banks (1929), Nentwig & Heimer (1983)
Canal Zone, Gamboa, Barro Colorado Island

Mangora schneirlai Chickering
Chickering (1954)
Barro Colorado Island

Mangora spinula F.O. Pickard-Cambridge
Banks (1929), Chamberlin & Ivie (1936), Chickering (1954), Harrison (1968), Nentwig & Heimer (1983)
Barro Colorado Island, Gamboa, Canal Zone, Arraijan, El Valle, Boquete, El Cermeno, Portobelo, El Volcan, Carti Road, Conception, Changuinola

Mangora sufflava Chickering
Chickering (1963 d)
Boquete

Mangora trilineata O. Pickard-Cambridge
O. Pickard-Cambridge (1889), Petrunkevitch (1925), Banks (1929)
Bugaba, Barro Colorado Island, Wilcox Camp, Canal Zone, Gamboa

(Marxia stellata (Walckenaer) = Araneus st.)

Mastophora fasciolatus (O. Pickard-Cambridge)
(= Araneus f.)
(= Epeiroides f.)
O. Pickard-Cambridge (1889), Petrunkevitch (1925), Levi 1991 b)
Bugaba

Mecynogea lemniscata (Walckenaer)
Levi (1980), Robinson & Robinson (1980), Nentwig & Heimer (1983)
Canal Zone, Panama

Metargyra debilis (O. Pickard-Cambridge)
(= Argyroepeira d.)
O. Pickard-Cambridge (1889), Petrunkevitch (1925)
Bugaba, Wilcox Camp

Metazygia gregalis (O. Pickard-Cambridge)
(= Epeira g.)
(= Eustala tuceps Chamberlin)
O. Pickard-Cambridge (1889), Chamberlin (1925), Petrunkevitch (1925), Banks (1929), Chickering (1936)
Veraguas, Barro Colorado Island

Metazygia incerta O. Pickard-Cambridge
(= Araneus i.)
Banks (1929), Levi (1991 b)
Barro Colorado Island

(Metazygia keyserlingi Banks = M. zilloides (Banks))

Metazygia pallidula (Keyserling)
(= Araneus p.)
F.O. Pickard-Cambridge (1904), Petrunkevitch (1925), Banks (1929), Levi (1991 b)

Bugaba, Barro Colorado Island, Gamboa

Metazygia wittfeldae (McCook)
Harrison (1968), Nentwig & Heimer (1983)
Changuinola, Gamboa, El Valle

Metazygia zilloides (Banks)
(= M. keyserlingi Banks)
Banks (1929), Levi (1977 b), Nentwig & Heimer (1983)
Barro Colorado Island, Gamboa, Carti Road, El Valle

Metepeira labyrinthea Hentz
F.O. Pickard-Cambridge (1903), Petrunkevitch (1925), Banks (1929)
Bugaba, Barro Colorado Island

Micrathena bimucronata (O. Pickard-Cambridge)
Levi (1985)
Bocas del Toro

Micrathena brevipes (O. Pickard-Cambridge)
(= Acrosoma b.)
O. Pickard-Cambridge (1890), Petrunkevitch (1925), Chickering (1961), Levi (1985)
Cerro Campana / Capira, Bugaba

Micrathena brevispina (Keyserling)
Levi (1985)
Volcan

(Micrathena calcaratum O. Pickard-Cambridge = M. sexspinosa (Hahn))

Micrathena clypeata (Walckenaer)
(= M. planata Chamberlin & Ivie)
Chamberlin & Ivie (1936), Chickering (1936, 1961), Lubin (1978), Robinson & Robinson (1980), Levi (1985)
Barro Colorado Island, Gamboa

(Micrathena coleophora Chamberlin & Ivie = M. schreibersi (Perty))

(Micrathena cornigera (F.O. Pickard-Cambridge) = M. sexspinosa (Hahn))

Micrathena crassa (Keyserling)
Petrunkevitch (1925), Chickering (1961)
Cerro Flores

(Micrathena disjuncta Chickering = M. duodecimspinosa (O. Pickard-Cambridge))

Micrathena donaldi Chickering
Chickering (1961), Levi (1985)
Barro Colorado Island, Canal Zone, Capira

Micrathena duodecimspinosa (O. Pickard-Cambridge)
(= Acrosoma d.)
(= M. disjuncta Chickering)
(= M. macilenta Chickering)
(= M. subspinosa F.O. Pickard-Cambridge)
O. Pickard-Cambridge (1890), Petrunkevitch (1925), Chickering (1936, 1961), Lubin (1978), Robinson & Robinson (1980), Nentwig & Heimer (1983), Levi (1985)
Barro Colorado Island, Canal Zone, Boquete, Bugaba, Bocas del Toro, El Real

Micrathena evansi Chickering
(= M. insolita Chickering)
Chickering (1961, 1969), Levi (1985)
Portobelo

Micrathena excavata (C.L.Koch)
(= M. mirifica Chickering)
(= M. subflava Chickering)
Chickering (1961), Nentwig & Heimer (1983), Levi (1985)
Canal Zone, Barro Colorado Island, Gamboa

(Micrathena fericula (O. Pickard-Cambridge) = M. flaveola (C.L. Koch))

Micrathena fidelis (Banks)
Levi (1985)
Cerro Punta

Micrathena flaveola (C.L. Koch)
(= Acrosoma vitiosa O. Pickard-Cambridge)
(= M. fericula (O. Pickard-Cambridge))
(= M. gladiola Walckenaer)
(= M. vitiosa (O. Pickard-Cambridge))
O. Pickard-Cambridge (1890), F.O. Pickard-Cambridge (1904), Petrunkevitch (1925), Chickering (1961), Nentwig & Heimer (1983), Levi (1985)
Bugaba, Canal Zone, El Real, Gamboa

Micrathena furcula (O. Pickard-Cambridge)
(= Acrosoma f.)
Banks (1929), Chickering (1961), Nentwig & Heimer (1983), Levi (1985)
Barro Colorado Island, Canal Zone, Gamboa

(Micrathena gladiola Walckenaer = M. flaveola (C.L.Koch))

Micrathena gracilis (Walckenaer)
(= M. nigrior Chamberlin & Ivie)
Chamberlin & Ivie (1936), Chickering (1961), Levi (1978)
Barro Colorado Island

Micrathena horrida (Taczanowski)
(= Acrosoma longicauda O. Pickard-Cambridge)
(= Acrosoma mammillata Butler)
(= Micrathena longicauda (O. Pickard-Cambridge))
(= Micrathena mammilata (Butler))
O. Pickard-Cambridge (1890), F.O. Pickard-Cambridge (1897), Petrunkevitch (1925), Banks (1929), Chickering (1961), Lubin (1978), Nentwig & Heimer (1983), Levi (1985)
Barro Colorado Island, Gamboa, Canal Zone, El Valle, Bugaba

(Micrathena inaequalis (F.O. Pickard-Cambridge) = M. lucasi (Keyserling))
(Micrathena insolita Chickering = M. evansi Chickering)

Micrathena lucasi (Keyserling)
(= M. inaequalis (F.O. Pickard-Cambridge)
Chickering (1936, 1961), Nentwig & Heimer (1983), Levi (1985)
El Volcan, Boquete, La Fortuna, El Valle, Barro Colorado Island

Micrathena macfarlanei Chickering
Chickering (1961), Levi (1985)
Barro Colorado Island, Canal Zone

(Micrathena macilenta Chickering = M. duodecimspinosa (O. Pickard-Cambridge))
(Micrathena mammilata (Butler) = M. horrida (Taczanowski))
(Micrathena mirifica Chickering = M. excavata (C.L.Koch))

Micrathena mitrata (Hentz)
(= M. patruelis F.O. Pickard-Cambridge)
Petrunkevitch (1925), Banks (1929), Chamberlin & Ivie (1936), Chickering (1961), Levi (1977, 1985)
Canal Zone

(Micrathena modica Chickering = M. quadriserrata F.O. Pickard-Cambridge)

Micrathena molesta Chickering
Levi (1985)
Corriente Grande (Bocas del Toro), La Fortuna

(Micrathena nigrior Chamberlin & Ivie = M. gracilis (Walckenaer))
(Micrathena obtusospina (Keyserling) = M. sexspinosa (Hahn))

Micrathena parallela (O. Pickard-Cambridge)
(= Acrosoma p.)
(= M. serrata F.O. Pickard-Cambridge)
O. Pickard-Cambridge (1890), F.O. Pickard-Cambridge (1904), Petrunkevitch (1925), Levi (1985)
El Volcan, Cerro Punta, Conception, Barro Colorado Island, Bugaba

(Micrathena patruelis C.L.Koch (Petrunkevitch 1925) probably misidentification of Micrathena mitrata (Hentz), (Levi 1985))
(Micrathena patruelis F.O. Pickard-Cambridge = M. mitrata (Hentz))
(Micrathena retracta Chamberlin & Ivie = M. saccata (C.L.Koch))

Micrathena plana (C.L.Koch)
Levi (1985)
David

(Micrathena planata Chamberlin & Ivie = M. clypeata (Walckenaer))

Micrathena quadriserrata F.O. Pickard-Cambridge
(= M. modica Chickering)
(= M. uncata F.O. Pickard-Cambridge)
Petrunkevitch (1925), Chickering (1961), Nentwig & Heimer (1983), Levi (1985), Eberhard (pers. comm. 1988)
Barro Colorado Island, Canal Zone, Gamboa, El Volcan

Micrathena saccata (C.L.Koch)
(= M. retracta Chamberlin & Ivie)
Petrunkevitch (1925), Banks (1929), Chamberlin & Ivie (1936), Chickering (1961), Levi (1985)
El Valle, Barro Colorado Island, Canal Zone, Gamboa, Taboga Island

Micrathena sagittata (Walckenaer)
Robinson & Robinson (1980), Levi (1985)
Barro Colorado Island, Canal Zone

Micrathena schreibersi (Perty)
(= Acrosoma s.)
(= Ildibaha myrmeciaeformis (Taczanowski))
(= M. coleophora Chamberlin & Ivie)
Banks (1929), Chamberlin & Ivie (1936), Petrunkevitch (1939 b), Chickering (1936, 1961), Lubin (1978), Robinson & Robinson (1980), Nentwig & Heimer (1983), Levi (1985), Eberhard (pers. comm. 1988)
Almirante, Arraijan, Barro Colorado Island, Canal Zone, Gamboa, El Real

(Micrathena serrata F.O. Pickard-Cambridge = M. parallela (O. Pickard-Cambridge))

Micrathena sexspinosa (Hahn)
(= Acrosoma calcaratum O. Pickard-Cambridge)
(= Acrosoma petersii Taczanowski)
(= Keyserlingia cornigera O. Pickard-Cambridge)
(= M. cornigera F.O. Pickard-Cambridge)
(= M. obtusospina (Keyserling))
O. Pickard-Cambridge (1890), Petrunkevitch (1925), Banks (1929), Chickering (1961), Robinson & Robinson (1980), Nentwig & Heimer (1983), Levi (1985)
Barro Colorado Island, Canal Zone, Gamboa, Almirante, Corozal (Los Santos), El Valle, El Volcan, David, El Real, Bugaba

Micrathena striata F.O. Pickard-Cambridge
Petrunkevitch (1925)
Canal Zone

(Micrathena subflava Chickering = M. excavata (C.L.Koch))
(Micrathena subspinosa F.O. Pickard-Cambridge = M. duodecimspinosa (O. Pickard-Cambridge))
(Micrathena uncata F.O. Pickard-Cambridge = M. quadriserrata F.O. Pickard-Cambridge)

Micrathena vigorsi (Perty)
(= Ildibaha mutilloides Simon)
Chickering (1936)
Barro Colorado Island

(Micrathena vitiosa (O. Pickard-Cambridge) = M. flaveola (C.L. Koch))

Mimicosa spinosa Petrunkevitch
Petrunkevitch (1925)
Canal Zone

Neoscona nautica (L.Koch)
(= Araneus n.)
(= Epeira volucripes Keyserling)
(= N. volucripes (Keyserling))
Keyserling (1884), F.O. Pickard-Cambridge (1904), Petrunkevitch (1925), Berman & Levi (1971)
Panama

Neoscona oaxacensis (Keyserling)
(= Araneus o.)
F.O. Pickard-Cambridge (1904), Petrunkevitch (1925), Berman & Levi (1971)
Bugaba

(Neoscona volucripes (Keyserling) = N. nautica (L.Koch)
(Neosconella solersioides (O. Pickard-Cambridge) = Araneus pegnia (Walckenaer))
(Neosconella styligera F.O. Pickard-Cambridge = Araneus expletus (O. Pickard-Cambridge)
(Nephila see Tetragnathidae)

Nuctenea cornuta (Clerck)
(Levi (1974)
Panama

(Parawixia armata (O. Pickard-Cambridge) = P. rigida)

Parawixia destricta (O. Pickard-Cambridge)
(= Epeira d.)
O. Pickard-Cambridge (1889), Petrunkevitch (1925), Banks (1929)
Bugaba, Barro Colorado Island, Taboga Island

Parawixia guatemalensis (O. Pickard-Cambridge)
Petrunkevitch (1925), Chickering (1936)
Wilcox Camp, Barro Colorado Island

Parawixia hoxaea (O. Pickard-Cambridge)
(= Araneus h.)
(= Epeira h.)
O. Pickard-Cambridge (1889), Petrunkevitch (1925), Levi (1992)
Tolé, El Volcan

Parawixia hypocrita (O. Pickard-Cambridge)
(= Araneus h.)
(= Epeira h.)
O. Pickard-Cambridge (1889), Petrunkevitch (1925), Levi (1992)
Bugaba, Barro Colorado Island, Canal Zone, Gamboa

Parawixia nesophila Chamberlin & Ivie
Chamberlin & Ivie (1936), Levi (1992)
Barro Colorado Island, Canal Zone

Parawixia rigida (O. Pickard-Cambridge)
(= E. armata)
(= Epeira r.)
(= P. armata)
O. Pickard-Cambridge (1889), Petrunkevitch (1925), Chickering (1936), Levi (1992)
Bugaba, Barro Colorado Island

Parawixia rimosa (Keyserling)
Chickering (1936)
Barro Colorado Island

(Parepeira albostriata (Keyserling) = Alpaida leucogramma (White))
(Pronous beatus Simon = P. tuberculifer Keyserling)

Pronous tuberculifer Keyserling
(= P. beatus Simon)
Banks (1929), Chickering (1936), Lubin (1978), Nentwig & Heimer (1983)
Barro Colorado Island, Canal Zone, Gamboa, Chiriqui, El Valle

(Salassina crassicauda Keyserling = Witica c.)

Scoloderus americanus (O. Pickard-Cambridge)
(= Carepalxis a.)
O. Pickard-Cambridge (1889), Petrunkevitch (1925), Banks (1929)
Bugaba, Barro Colorado Island

Scoloderus cordatus (Taczanowski)
Nentwig & Heimer (1983)
Gamboa

(*Scoloderus gibber* (O. Pickard-Cambridge) = Carepalxis solobrensis O. Pickard-Cambridge)

Scoloderus tuberculifer (O. Pickard-Cambridge)
(= Carepalxis t.)
O. Pickard-Cambridge (1889), Petrunkevitch (1925)
Bugaba

Spilasma sp.
Canal Zone
Eberhard (pers. comm.)

Verrucosa arenata (Walckenaer)
(= Mahadeva r.)
(= V. reticulata (O. Pickard-Cambridge))
O. Pickard-Cambridge (1889), Petrunkevitch (1925), Banks (1929), Chickering (1936), Levi (1976), Nentwig & Heimer (1983)
Barro Colorado Island, El Volcan, Gamboa, Almirante

Verrucosa mexicana (Lucas)
(= Mahadeva m.)
(= V. undecimvariolata O. Pickard-Cambridge)
O. Pickard-Cambridge (1889), Petrunkevitch (1925), Banks (1929), Harrison (1968), Nentwig & Heimer (1983)
Bugaba, Barro Colorado Island, Gamboa, Canal Zone, Changuinola

(Verrucosa reticulata (O. Pickard-Cambridge) = V. arenata (Walckenaer))

Wagneriana spicata (O. Pickard-Cambridge)
Banks (1929), Levi (1991 a)
Canal Zone

Wagneriana taboga Levi
Levi (1991 a)
Gamboa, Canal Zone, Taboga Island, Porto Belo, Panama City, Barro Colorado Island, David

Wagneriana tauricornis (O. Pickard-Cambridge)
(= Epeira t.)
O. Pickard-Cambridge (1889), Petrunkevitch (1925), Banks (1929), Nentwig & Heimer (1983), Levi (1991 a)
Bugaba, El Volcan, Barro Colorado Island, Gamboa, Canal Zone, El Valle

Wagneriana undecimtuberculata (Keyserling)
F.O. Pickard-Cambridge (1904), Petrunkevitch (1925), Chickering (1936), Harrison (1968), Nentwig & Heimer (1983), Levi (1991 a)
Bugaba, Barro Colorado Island, Gamboa, Canal Zone, Changuinola, David

Wittica crassicauda (Keyserling)
(= Bion brevis O. Pickard-Cambridge)
(= Edricus c.)
(= Salassina c.)
(= W. talis O. Pickard-Cambridge)
F.O. Pickard-Cambridge (1904), Petrunkevitch (1925), Banks (1929), Harrison (1968), Lubin (1978), Nentwig & Heimer (1983), Levi (1986 b), Eberhard (pers. comm. 1988)
Bugaba, Barro Colorado Island, Canal Zone, Gamboa, El Valle, Bocas del Toro, Chiriqui, Changuinola

Wixia clivosa (O. Pickard-Cambridge)
(= Araneus clivosus (O. Pickard-Cambridge))
Petrunkevitch (1925)
Canal Zone

Wixia consequa (O. Pickard-Cambridge)
(= Araneus c.)
(= Epeira c.)
O. Pickard-Cambridge (1889), Petrunkevitch (1925), Levi (1991 b)
Bugaba

Wixia gibbifera (O. Pickard-Cambridge)
Nentwig & Heimer (1983)
Gamboa

Wixia globosa F.O. Pickard-Cambridge
(= Araneus g.)
Petrunkevitch (1925)
Wilcox Camp

Wixia rufa (O. Pickard-Cambridge)
(= Epeira r.)
O. Pickard-Cambridge (1889), Petrunkevitch (1925)
Bugaba

Zigana wixoides Chamberlin & Ivie
Chamberlin & Ivie (1936)
Barro Colorado Island

Lycosidae

Allocosa panamena Chamberlin
(= A. panamana)
(= Arctosa p.)
Chamberlin (1925), Petrunkevitch (1925), Banks (1929), Chickering (1936), Dondale & Redner (1983 a)
Barro Colorado Island, Remedios, Canal Zone, Boquete

(Allocosa minuta F.O. Pickard-Cambridge = Arctosa m.)
(Arctosa cheluncata Petrunkevitch = A. minuta F.O. Pickard-Cambridge)

Arctosa littoralis (Hentz)
Petrunkevitch (1925), Dondale & Redner (1983 b)
Remedios
Arctosa minuta (F.O. Pickard-Cambridge)
(= Allocosa m.)
(= A. cheluncata Petrunkevitch)
Petrunkevitch (1925), Dondale & Redner (1983 a)
Santiago, Remedios, La Mesa, Wilcox Camp

(Arctosa panamena Chamberlin = Allocosa p.)
(Arctosa trifida F.O. Pickard-Cambridge = A. littoralis (Hentz))

Lycosa brevitarsis F.O. Pickard-Cambridge
F.O. Pickard-Cambridge (1902)
Chiriqui

Lycosa cerrofloresiana Petrunkevitch
Petrunkevitch (1925)
Cerro Flores

Lycosa chickeringi Chamberlin & Ivie
Chamberlin & Ivie (1936)
Barro Colorado Island

Lycosa dilatata F.O. Pickard-Cambridge
Petrunkevitch (1925)
Canal Zone, Boca del Monte, Cerro Flores

(Lycosa eutypa Chamberlin = L. longitarsis F.O. Pickard-Cambridge)

Lycosa fusca (Keyserling)
Petrunkevitch (1925)
Panama

Lycosa longitarsis F.O. Pickard-Cambridge
(= L. eutypa Chamberlin)
F.O. Pickard-Cambridge (1902), Petrunkevitch (1925), Banks (1929)
Chiriqui, Barro Colorado Island, Canal Zone

Lycosa rabida Walckenaer
Petrunkevitch (1925)
Canal Zone, Boca del Monte
Lycosa sanlorenziana Petrunkevitch
Petrunkevitch (1925)
Wilcox Camp

Lycosa tivior Chamberlin & Ivie
Chamberlin & Ivie (1936)
Barro Colorado Island

Lycosa tristani Banks
(= Pardosa albopilosa Petrunkevitch)
(= Schizogyna t.)
Petrunkevitch (1925), Chickering (1936), Gertsch & Wallace (1937)
Barro Colorado Island, Panama City, Remedios

(Pardosa albopilosa Petrunkevitch = Lycosa tristani Banks)

Pardosa fastosa (Keyserling)
(= P. uncatula F.O. Pickard-Cambridge)
Petrunkevitch (1925)
Wilcox Camp

Pardosa sagei Gertsch & Wallace
Gertsch & Wallace (1937), Dondale & Redner (1984)
El Volcan

(Pardosa uncatula F.O. Pickard-Cambridge = P. fastosa (Keyserling))

Pirata pagicola Chamberlin
Chamberlin (1925), Banks (1929), Wallace & Exline (1978)
Barro Colorado Island

Pirata turrialbicus Wallace
Wallace & Exline (1978)
Boquete, Chanquinola, Canal Zone, El Valle

(Schizocosa tristani (Banks) = Lycosa t.)
(Schizogyna tristani (Banks) = Lycosa t.)

Pisauridae

(including Trechaleidae, see remark p. 21)
Ancylometes bogotensis (Keyserling)
(= Lycoctenus b.)
F.O. Pickard-Cambridge (1901), Petrunkevitch (1925), Banks (1929)
Bugaba, Canal Zone

Ancylometes valentinei Petrunkevitch
Petrunkevitch (1925)
Wilcox Camp

Architis nitidopilosa Simon
Carico (1981), Nentwig & Heimer (1983), Nentwig (1985 d)
Gamboa

Architis robusta Carico
Carico (1981)
Canal Zone

Architis tenuis Simon
Banks (1929)
Barro Colorado Island

Dyrines lineatipes Petrunkevitch
Petrunkevitch (1925)
Wilcox Camp

Enna approximata (O. Pickard-Cambridge)
(= Perissoblemma a.)
O. Pickard-Cambridge (1893), Petrunkevitch (1925)
Bugaba

Enna julliení (Simon)
(= Hesydrus j.)
F.O. Pickard-Cambridge (1901), Simon (1898), Petrunkevitch (1925)
Panama

Enna minor Petrunkevitch
Petrunkevitch (1925)
Santiago

Enna nesiotes Chamberlin
Chamberlin (1925)
Barro Colorado Island

Hesydrus habilis (O. Pickard-Cambridge)
Petrunkevitch (1925)
Remedios

(Hesydrus jullieni Simon = Enna j.)

(Lycoctenus bogotensis (Keyserling) = Ancylometes b.)

Paradossenus nigricans F.O. Pickard-Cambridge
Chickering (1936)
Barro Colorado Island

(Perissoblemma approximatum O. Pickard-Cambridge = Enna a.)
(Staberius aculeatus Simon = S. spinipes (Taczanowski))

Staberius spinipes (Taczanowski))
(= S. aculeatus Simon)
Banks (1929), Chickering (1936), Carico (1981), Nentwig & Heimer (1983)
Gamboa, Canal Zone, Barro Colorado Island, Las Sabanas

Syntrechalea tenuis F.O. Pickard-Cambridge
F.O. Pickard-Cambridge (1902), Petrunkevitch (1925), Banks (1929)
Bugaba, Barro Colorado Island

Thaumasia argenteonotata (Simon)
(= T. uncata F.O. Pickard-Cambridge)
F.O. Pickard-Cambridge (1901), Petrunkevitch (1925), Banks (1929), Harrison (1968), Nentwig & Heimer (1983), Nitzsche(1988)
Bugaba, Barro Colorado Island, Canal Zone, Gamboa, El Valle, Carti Road, Changuinola

Thaumasia argyrotypa Chamberlin & Ivie
Chamberlin & Ivie (1936)
Barro Colorado Island

Thaumasia heterogyna Chamberlin & Ivie (1936)
Chamberlin & Ivie (1936)
Barro Colorado Island
(Thaumasia uncata F.O. Pickard-Cambridge = T. argenteonotatus (Simon))

Thaumasia velox Simon
Banks (1929), Chickering (1936)
Gamboa, Barro Colorado Island, Canal Zone

Tinus sp.
Nentwig & Heimer (1983)
Gamboa

Trechalea connexa (O. Pickard-Cambridge)
Nentwig & Heimer (1983)
Gamboa

Trechalea magnifica Petrunkevitch
Petrunkevitch (1925), Banks (1929)
Barro Colorado Island, Wilcox Camp, Boca del Monte, Santiago, La Mesa

Agelenidae

(Agelena panamana Petrunkevitch = Melpomene p.)

Melpomene chiricana Chamberlin & Ivie
Chamberlin & Ivie (1942)
Cerro Punta
(Probably a synonym of *M. penetralis* (F.O. Pickard-Cambridge), V. Roth pers. comm. 1987)

Melpomene panamana (Petrunkevitch)
Petrunkevitch (1925), Chamberlin & Ivie (1942)
Boca del Monte, Wilcox Camp

Melpomene plesia Chamberlin & Ivie
Chamberlin & Ivie (1942)
El Volcan

Tegenaria domestica (Clerck)
Petrunkevitch (1925)
Panama

Dictynidae

Dictyna alyaceae Chickering
Chickering (1950 a)
Panama
Dictyna donaldi Chickering
Chickering (1950 a)
Boquete

Dictyna lecta Chickering

Chickering (1951 b)
El Volcan

Dictyna meditata Gertsch
Gertsch (1946), Chickering (1950 a)
Barro Colorado Island, Canal Zone, El Valle, Portobelo

Dictyna nivea (O. Pickard-Cambridge)
(= Mallos n.)
Petrunkevitch (1925)
Sona

Dictyna parietalis O. Pickard-Cambridge
Chickering (1936)
Barro Colorado Island

Mallos margaretae Gertsch
Gertsch (1946)
El Volcan

(Mallos niveus O. Pickard-Cambridge = Dictyna n.)

Thallumetus dulcineus Gertsch
Gertsch (1946)
El Volcan

Thallumetus pullus Chickering
Chickering (1951 b)
Canal Zone

Thallumetus pusillus Chickering
Chickering (1950 a, 1951 b)
Canal Zone

Amaurobiidae

Barrisca nannella Chamberlin & Ivie
Platnick (1978)
Barro Colorado Island
(genus of unknown family relationship within the amaurobioid complex, maintained here for reasons of convenience, see Platnick (1978))

Titanoecidae

Goeldia mexicana (O. Pickard-Cambridge)
(= Temecula m.)
Leech (1972), Nentwig & Heimer (1983)
Gamboa

(Temecula mexicana O. Pickard-Cambridge = Goeldia m.)

Tengellidae

(Metafecenia albolineata F.O. Pickard-Cambridge = Tengella a.)

Tengella albolineata (F.O. Pickard-Cambridge)
(= Metafecenia a.)
Nentwig & Heimer (1983)
El Valle

Zorocrates fuscus Simon
F.O. Pickard-Cambridge (1902), Petrunkevitch (1925)
Bugaba

Oxyopidae

Hamataliwa barroana (Chamberlin & Ivie)
(= Oxyopes b.)
Chamberlin & Ivie (1936), Brady (1970)
Barro Colorado Island, Canal Zone, Gamboa

Hamataliwa bufo Brady
Brady (1970)
Barro Colorado Island, Canal Zone

Hamataliwa cavata (Kraus)
Nentwig & Heimer (1983)
Gamboa

(Hamataliwa clypeata (F.O. Pickard-Cambridge) = H. triangularis (Kraus))

Hamataliwa crocata Brady
Brady (1970)
Barro Colorado Island, Canal Zone

Hamataliwa flebilis (O. Pickard-Cambridge)
(= Oxyopeidon f.)
O. Pickard-Cambridge (1894), Brady (1970)
Bugaba, Barro Colorado Island, Gamboa, Canal Zone

Hamataliwa globosa (F.O. Pickard-Cambridge)
(= Oxyopes g. (in part))
F.O. Pickard-Cambridge (1902), Petrunkevitch (1925), Banks (1929), Brady (1970)
Bugaba, Barro Colorado Island

Hamataliwa hista Brady
Brady (1970)
Boquete

Hamataliwa puta (O. Pickard-Cambridge)
(= Oxyopeidon p.)
O. Pickard-Cambridge (1894), Petrunkevitch (1925), Brady (1970), Nentwig & Heimer (1983)
Bugaba, Barro Colorado Island, Gamboa, Wilcox Camp

Hamataliwa triangularis (Kraus)
(= Hamataliwa clypeata (F.O. Pickard-Cambridge))
(= Oxyopes c. (in part))
(= Oxyopes globosus F.O. Pickard-Cambridge (in part))
F.O. Pickard-Cambridge (1902), Petrunkevitch (1925), Banks (1929), Brady (1970), Nentwig & Heimer (1983)
Bugaba, Barro Colorado Island, Canal Zone, Gamboa

Hamataliwa tricuspidata (F.O. Pickard-Cambridge)
(= Oxyopes clypeatus F.O. Pickard-Cambridge)
(= Oxyopes t.)
F.O. Pickard-Cambridge (1902), Petrunkevitch (1925), Banks (1929), Brady (1970), Nentwig & Heimer (1983)
Arraijan, Bugaba, Barro Colorado Island, Canal Zone

Hamataliwa ursa Brady
Brady (1970)
Barro Colorado Island, Canal Zone

(Oxyopeidon flebile O. Pickard-Cambridge = Hamataliwa f.)
(Oxyopeidon putum O. Pickard-Cambridge = Hamataliwa p.)

Oxyopes acleistus Chamberlin
(= O. aureus Brady)
Nentwig & Heimer (1983)
Canal Zone, Gamboa

Oxyopes aglossus Chamberlin
Nentwig & Heimer (1983)
Gamboa

Oxyopes apollo Brady
Nentwig & Heimer (1983)
Gamboa

(Oxyopes aureus Brady = O. acleistus Chamberlin)
(Oxyopes barroanus Chamberlin & Ivie = Hamataliwa b.)

Oxyopes bifidus F.O. Pickard-Cambridge
Brady (1975)
Arraijan, Canal Zone

Oxyopes bifissus F.O. Pickard-Cambridge
Chickering (1936)
Barro Colorado Island

(Oxyopes clypeatus F.O. Pickard-Cambridge = Hamataliwa triangularis (Kraus))
(Oxyopes globosus F.O. Pickard-Cambridge = (in part) Hamataliwa g., (in part) Hamataliwa triangularis (Kraus)
(Oxyopes gracilis Keyserling = O. salticus Hentz)

Oxyopes salticus Hentz
(= O. gracilis Keyserling)
Petrunkevitch (1925, 1939 b), Banks (1929), Chickering (1936), Harrison (1968), Nentwig & Heimer (1983)

Barro Colorado Island, Canal Zone, Gamboa, Los Sabanas, Carti Road, El Valle, Changuinola

Oxyopes summus Brady
Brady (1975)
Canal Zone, Gamboa

(Oxyopes tricuspidatus F.O. Pickard-Cambridge = Hamataliwa t.)

Peucetia bibranchiata F.O. Pickard-Cambridge
Banks (1929)
Canal Zone, Taboga Island

Peucetia longipalpis F.O. Pickard-Cambridge
Nentwig & Heimer (1983)
Carti Road, El Valle

Peucetia rubricapilla Petrunkevitch
Petrunkevitch (1925)
Sona

Peucetia thalassina (C.L.Koch)
Petrunkevitch (1925)
Sona

Peucetia viridans (Hentz)
Brady (1964), Nentwig & Heimer (1983)
El Valle, Boquete

Schaenioscelis leucochlorus Mello-Leitao
Nentwig (1990)
Gamboa

Tapinillus longipes (Taczanowski)
Banks (1929), Chickering (1936), Nentwig & Heimer (1983)
Barro Colorado Island, Canal Zone, Gamboa

Senoculidae

(Labdacus = Senoculus)

Senoculus barroanus Chickering
Chickering (1940 b)
Canal Zone, Barro Colorado Island, Gamboa

Senoculus bucolicus Chickering
Chickering (1940 b)
Barro Colorado Island

Senoculus canaliculatus F.O. Pickard-Cambridge
(= S. valentinei Petrunkevitch)
F.O. Pickard-Cambridge (1902), Petrunkevitch (1925), Chickering (1940 b, 1953)
Bugaba, Wilcox Camp, Barro Colorado Island, Canal Zone, Portobelo

(Senoculus parallelus (Simon) = S. purpureus (Simon))

Senoculus rubicundus Chickering
Chickering (1953 b)
Barro Colorado Island, Canal Zone

Senoculus purpureus (Simon)
(= Labdacus parallelus Simon)
(= Senoculus p.)
Simon (1880), F.O. Pickard-Cambridge (1902), Petrunkevitch (1925)
Panama

Senoculus robustus Chickering
Chickering (1940 b)
Barro Colorado Island

Senoculus silvaticus Chickering
Chickering (1940 b)
Barro Colorado Island

Senoculus tigrinus Chickering
Chickering (1940 b)
Barro Colorado Island

(Senoculus valentinei Petrunkevitch = S. canaliculatus F.O. Pickard-Cambridge)

Senoculus zeteki Chickering
Chickering (1953 b)
Barro Colorado Island

Anyphaenidae

Anyphaena excepta O. Pickard-Cambridge
Nentwig & Heimer (1983)
Gamboa

Anyphaena judicata O. Pickard-Cambridge
Nentwig & Heimer (1983)
Cerro Punta

Anyphaena lanta Chickering
Chickering (1940 a)
Panama

Anyphaena nexuosa Chickering
Chickering (1940 a)
Panama

Anyphaena plana F.O. Pickard-Cambridge
F.O. Pickard-Cambridge (1900), Petrunkevitch (1925), Platnick & Lau (1975)
Bugaba, Chiriqui

Anyphaena stylifer F.O. Pickard-Cambridge
F.O. Pickard-Cambridge (1900), Petrunkevitch (1925), Chickering (1940 a)
Bugaba

Arachosia puta O. Pickard-Cambridge
O. Pickard-Cambridge (1892), Petrunkevitch (1925)
El Volcan

Aysha insulana Chickering
Chickering (1937 a, 1940 a)
Barro Colorado Island

Aysha similaris Banks
Banks (1929), Chickering (1937 a, 1940 a)
Barro Colorado Island, Canal Zone, El Valle

(Aysha simplex O. Pickard-Cambridge = A. tenuis (L.Koch))

Aysha tenuis (L. Koch)
(= A. simplex O. Pickard-Cambridge)
O. Pickard-Cambridge (1897), Petrunkevitch (1925), Banks (1929)
Taboga Island, Veraguas

Aysha valvula F.O. Pickard-Cambridge
F.O. Pickard-Cambridge (1900), Petrunkevitch (1925)
Bugaba

(Cragus = Wulfila)

Macrophyes attenuata (O. Pickard-Cambridge)
Nentwig & Heimer (1983)
Canal Zone

Macrophyes elongata Chickering
Chickering (1937 a, 1940 a)
Barro Colorado Island

Pelayo montanus Chickering
Chickering (1940 a)
Panama

Sillus armatus Chickering
Chickering (1940 a)
Panama

Sillus coloratus Chickering
Chickering (1937 a, 1940 a)
Barro Colorado Island, El Valle, Arraijan

Sillus curvispinus F.O. Pickard-Cambridge
F.O. Pickard-Cambridge (1900), Petrunkevitch (1925)
Bugaba

Sillus longispinus F.O. Pickard-Cambridge
(= Wulfila dubia Chickering)
Petrunkevitch (1925), Banks (1929), Chickering (1940), Nentwig & Heimer (1983)
Barro Colorado Island, El Valle, Canal Zone, Wilcox Camp, Remedios

Sillus placens (O. Pickard-Cambridge) (= Clubiona p.)
O. Pickard-Cambridge (1896), Petrunkevitch (1925)
Panama

Sillus ravus Chickering
Chickering (1940 a)
Panama

Teudis bucolicus Chickering
Chickering (1940 a)
Panama

Teudis cambridgei Chickering
Chickering (1940 a)
Panama

Teudis donaldi Chickering
Chickering (1940 a)
Panama

Teudis juradoi Chickering
Chickering (1940 a), Nentwig & Heimer (1983)
Panama, El Valle, Gamboa

Teudis lycosoides Chickering
Chickering (1937 a)
Barro Colorado Island

Teudis lectus Chickering
Chickering (1940 a)
Panama

Teudis mordax (O. Pickard-Cambridge)
Platnick (1974), Nentwig & Heimer (1983)
Barro Colorado Island, Gamboa

Teudis roseus F.O. Pickard-Cambridge
F.O. Pickard-Cambridge (1900), Petrunkevitch (1925), Banks (1929), Nentwig & Heimer (1983)
Bugaba, Canal Zone, Gamboa

Teudis ruber F.O. Pickard-Cambridge
Nentwig & Heimer (1983)
El Valle
Wulfila arraijanica Chickering
Chickering (1940 a)
Arraijan

(Wulfila dubia Chickering = Sillus longispinus F.O. Pickard-Cambridge)

Wulfila fragilis Chickering
Chickering (1937 a)
Barro Colorado Island

Wulfila innoxia Chickering
Chickering (1940)
El Valle

Wulfila inornata O. Pickard-Cambridge
Nentwig & Heimer (1983)
Gamboa

Wulfila maculata Chickering
Chickering (1937 a, 1940 a)
Barro Colorado Island

Wulfila mandibulata Petrunkevitch
Petrunkevitch (1925), (Banks 1929), Chickering (1940 a)
Wilcox Camp, Canal Zone, Arraijan,

Wulfila modesta Chickering
Chickering (1937 a, 1940 a)
Barro Colorado Island

Wulfila pallida O. Pickard-Cambridge
Petrunkevitch (1925)
Wilcox Camp

Wulfila pellucida Chickering
Chickering (1937 a)
Barro Colorado Island

Wulfila pulverulenta Chickering
Chickering (1937 a, 1940 a)
Barro Colorado Island

Wulfila spinosa Chickering
Chickering (1937)
Barro Colorado Island

Wulfila sublesta Chickering
Chickering (1940 a), Nentwig & Heimer (1983)

Gamboa, El Valle

Wulfila tantilla Chickering
Chickering (1940 a), Platnick (1974)
Canal Zone

Wulfila tenella Chickering
Chickering (1940 a)
Panama

Miturgidae

Eutichurus longipes Chickering
Chickering (1937 b)
Panama

Eutichurus putus O. Pickard-Cambridge
O. Pickard-Cambridge (1898), Petrunkevitch (1925)
Veraguas

Eutichurus sigillatus Chickering
Chickering (1937 b)
Panama

(Syrisca hirsuta Petrunkevitch = Teminius h.)
(Syrisca pulchra Petrunkevitch = Teminius hirsutus Petrunkevitch)

Teminius hirsutus (Petrunkevitch)
(= Syrisca h.)
(= Syrisca pulchra Petrunkevitch)
Petrunkevitch (1925), Platnick & Shadab (1989)
Canal Zone, BCI, Gamboa, Cerro Iglesia, Remedios, Santa Rosa, Old Panama City

Liocranidae

Parachemmis fuscus Chickering
Chickering (1937 b)
Panama

Parachemmis trilineatus Chickering
Chickering (1937 b)
Panama

Clubionidae

Cheiracanthium debile O. Pickard-Cambridge
O. Pickard-Cambridge (1897)
Veraguas

Cheiracanthium ferum O. Pickard-Cambridge
F.O. Pickard-Cambridge (1899), Petrunkevitch (1925), Banks (1929), Chickering (1937 b), Nentwig & Heimer (1983)
Gamboa, Colon, Veraguas

Cheiracanthium inclusum (Hentz)
Harrison (1968)
Changuinola

Clubiona aliceae Chickering
Chickering (1937 b)
Panama

Clubiona balboae Chickering
Chickering (1937 b)
Panama

Clubiona barroana Chickering
Chickering (1937 b)
Panama

Clubiona hortani Chickering
Chickering (1937 b)
Panama

Clubiona multinotata Chickering
Chickering (1937 b)
Panama

Clubiona orvillei Chickering
Chickering (1937 b)
Panama
(Clubiona placens O. Pickard-Cambridge = Sillus p., Anyphaenidae)

Clubiona sericea (O. Pickard-Cambridge)
Banks (1929), Chickering (1937 b), Nentwig & Heimer (1983)
Barro Colorado Island, Carti Road

Clubiona sigillata Petrunkevitch
Petrunkevitch (1925)
Wilcox Camp

Clubiona silvatica Chickering
Chickering (1937 b)
Panama

Clubiona tigrina (O. Pickard-Cambridge)
Chickering (1936, 1937 b)
Barro Colorado Island

Corinnidae

Castianeira abuelita Reiskind
Reiskind (1969)
Barro Colorado Island, Canal Zone

Castianeira alba Reiskind
Reiskind (1969), Nentwig & Heimer (1983)
Boquete, Canal Zone, Gamboa

Castianeira cincta (Banks)
(= Myrmecotypus c.)
Banks (1929), Reiskind (1969)
Barro Colorado Island, Canal Zone

Castianeira cubana (Banks)
Reiskind (1969)
Arraijan, Boquete, Barro Colorado Island, Canal Zone

Castianeira dentata Chickering
(= C. pusilla Chickering)
Chickering (1937 b), Reiskind (1969, 1977)
Barro Colorado Island, Portobelo, El Valle, Canal Zone, Gamboa

Castianeira dubia (O. Pickard-Cambridge)
(= C. nitida Chickering)
Chickering (1937 b), Reiskind (1969, 1977)
Barro Colorado Island, Portobelo, Bocas del Toro

Castianeira guapa Reiskind
Reiskind (1969)
Canal Zone, Barro Colorado Island

Castianeira lachrymosa (O. Pickard-Cambridge)
Banks (1929), Chickering (1937 b)
Barro Colorado Island

Castianeira luctuosa O. Pickard-Cambridge
Banks (1929)
Almirante

Castianeira memnonia (C.L. Koch)
Banks (1929), Chickering (1937 b), Reiskind (1969, 1977), Nentwig & Heimer (1983)
Barro Colorado Island, Gamboa, Canal Zone

(Castianeira nitida Chickering = C. dubia (O. Pickard-Cambridge))
(Castianeira pusilla Chickering = C. dentata Chickering)

Castianeira similis (Banks)
(= Myrmecotypus s.)
Banks (1929), Reiskind (1969)
Barro Colorado Island

Castianeira trimac Reiskind
Reiskind (1969)
Canal Zone, Barro Colorado Island

Corinna bulbosa F.O.P.C
(= C. eutypa Chamberlin)
Chamberlin (1925), Petrunkevitch (1925), Banks (1929), Chickering (1937 b, 1977)
Barro Colorado Island, San Lorenzo

Corinna bulbula F.O. Pickard-Cambridge
F.O. Pickard-Cambridge (1899), Petrunkevitch (1925)
Bugaba

(Corinna eutypa Chamberlin = C. bulbosa F.O. Pickard-Cambridge)

Corinna casueta Chickering
Chickering (1972)

Barro Colorado Island, Canal Zone

Corinna crassipalpis Chickering
Chickering (1937 b, 1972)
Barro Colorado Island, Canal Zone

Corinna grassima Chickering
Chickering (1972)
Barro Colorado Island, Canal Zone

Corinna luctuosa F.O. Pickard-Cambridge
Nentwig & Heimer (1983)
Gamboa

Corinna luteomaculata Petrunkevitch
Petrunkevitch (1925)
Cerro Flores

Corinna mucronata F.O. Pickard-Cambridge
F.O. Pickard-Cambridge (1899), Petrunkevitch (1925), Banks (1929), Chickering (1937 b, 1972),
Barro Colorado Island, Canal Zone, Bugaba, Boquete, El Valle, Portobelo

Corinna nervosa F.O. Pickard-Cambridge
F.O. Pickard-Cambridge (1899), Petrunkevitch (1925), Chickering (1972)
Bugaba

Corinna robusta Chickering
Chickering (1937 b, 1972), Nentwig & Heimer (1983)
Barro Colorado Island, Canal Zone, El Valle, Portobelo

Corinna saga F.O. Pickard-Cambridge
Nentwig & Heimer (1983)
Gamboa

Corinna sylvatica Chickering
Chickering (1937 b, 1972), Nentwig & Heimer (1983)
Gamboa, Barro Colorado Island, Canal Zone

Corinna perida Chickering
Chickering (1972)
Barro Colorado Island

Corinna tonubra Chickering
Chickering (1972)
Barro Colorado Island

Corinna variegata F.O. Pickard-Cambridge
Nentwig & Heimer (1983)
Canal Zone

Mazax ajax Reiskind
Nentwig & Heimer (1983)
Canal Zone, Gamboa

Mazax pax Reiskind
(= M. spinosa O. Pickard-Cambridge)
Reiskind (1969, 1977)
Barro Colorado Island, Canal Zone

(Mazax plana F.O. Pickard-Cambridge = M. spinosa (Simon))
(Mazax segregata Chamberlin = M. spinosa (Simon))

Mazax spinosa (Simon)
(= M. plana F.O. Pickard-Cambridge)
(= M. segregata Chamberlin)
Chamberlin (1925), Banks (1929), Chickering (1937 b), Reiskind (1969, 1977)
Barro Colorado Island, Canal Zone, Gamboa

Megalostrata mordicans (F.O. Pickard-Cambridge)
Petrunkevitch (1925)
Cerro Flores
(Myrmecotypus cinctus Banks = Castianeira c.)

Myrmecotypus dubius (O. Pickard-Cambridge)
Petrunkevitch (1925)
Boca del Monte, Wilcox Camp

Myrmecotypus lineatipes Chickering
Chickering (1937 b), Reiskind (1969)
Barro Colorado Island

Myrmecotypus niger Chickering
Chickering (1937 b), Reiskind (1969)
Barro Colorado Island, Canal Zone

(Myrmecotypus obscurus Chickering = M. pilosus (O. Pickard-Cambridge))

Myrmecotypus olympus Reiskind
Reiskind (1969)
Barro Colorado Island

Myrmecotypus orpheus Reiskind
Reiskind (1969)
Barro Colorado Island, Canal Zone

Myrmecotypus rettenmeyeri Unzicker
(= M. discreta Unzicker)
Reiskind (1969, 1977)
Barro Colorado Island, Canal Zone

Myrmecotypus pilosus (O. Pickard-Cambridge)
(= M. obscurus Chickering)
Chickering (1937 b), Reiskind (1969, 1977)
Barro Colorado Island, El Valle, Canal Zone

(Myrmecotypus similis Banks = Castianeira s.)

Sphecotypus niger (Perty)
(= S. formicarius O. Pickard-Cambridge)

O. Pickard-Cambridge (1895), Petrunkevitch (1925), Reiskind (1969)
Bugaba

Trachelas barroanus Chamberlin
Chamberlin (1925), Banks (1929), Chickering (1937 b), Platnick & Shadab (1974 b), Nentwig & Heimer (1983)
Barro Colorado Island, Canal Zone

Trachelas bispinosus F.O. Pickard-Cambridge
F.O. Pickard-Cambridge (1899), Petrunkevitch (1925), Platnick & Shadab (1974 b), Nentwig & Heimer (1983)
Cerro Flores, Canal Zone, Boquete, Bugaba, El Valle, Barro Colorado Island

Trachelas cambridgei Kraus
Platnick & Shadab (1974 a)
Canal Zone

Trachelas ecudobus Chickering
Platnick & Shadab (1974 b)
Canal Zone, Barro Colorado Island, Gamboa

Trachelas panamanus Chickering
Platnick & Shadab (1974 b)
Portobelo, Canal Zone, Barro Colorado Island

Trachelas prominens Platnick & Shadab
Platnick & Shadab (1974 b)
Arraijan, Boquete, El Valle, Canal Zone, Barro Colorado Island

Trachelas transversus F.O. Pickard-Cambridge
Nentwig & Heimer (1983)
Gamboa

Trachelas triangulus Platnick & Shadab
Platnick & Shadab (1974 b)
Canal Zone, Barro Colorado Island

Trachelas trifidus Platnick & Shadab
Platnick & Shadab (1974 b)
Canal Zone

Zodariidae

(Leprolochus decoratus Chickering = L. spinifrons Simon (Jocqué 1988)

Ishania certa Jocqué & Ubick
Jocqué & Ubick (in 1991)
Chiriqui

Leprolochus spinifrons Simon
(= L. decoratus Chickering)
Chickering (1957 b), Jocqué (1991)
Canal Zone

(Naibena barroana Chamberlin = Tenedos b.)
(Storena barroana (Chamberlin) = Tenedos b.)

Tenedos barroanus (Chamberlin)
(= Naibena b.)
(= Storena b.)
Banks (1929), Chickering (1945 b), Nentwig & Heimer (1983), Jocqué (1991)
Barro Colorado Island, Gamboa, Canal Zone, Portobelo, El Valle

Gnaphosidae

Camillina balboa Platnick & Shadab
Platnick & Shadab (1982)
Barro Colorado Island, Canal Zone

Camillina colon Platnick & Shadab
Platnick & Shadab (1982)
Colon

Camillina cordifera (Tullgren)
Platnick & Shadab (1982)
Panama

Cesonia boca Platnick & Shadab
Platnick & Shadab (1980)
Boquete

(Cesonia cincta Banks = Gertschosa c.)

Cesonia lacertosa Chickering
Chickering (1949), Platnick & Shadab (1980)
Canal Zone
Cesonia nigra Chickering
Chickering (1949)
El Valle

Cesonia notata Chickering
Chickering (1949), Platnick & Shadab (1980)
Canal Zone, Portobelo, Barro Colorado Island

Cesonia pudica Chickering
Chickering (1949), Platnick & Shadab (1980)
Barro Colorado Island, Canal Zone, El Valle, Coclé

(Echemus iotus Banks = Zimirimus i.)
(Echemus montanus Chickering = Zimiromis tropicalis (Banks))

Eilica chickeringi Platnick
Platnick (1975 a)
El Valle, Coclé

Gertschosa cincta (Banks)
(= Cesonia c.)
Banks (1929), Chickering (1949), Platnick & Shadab (1981 a)
Las Sabanas, Canal Zone, Barro Colorado Island, El Valle, Coclé

Lygromma chamberlini Gertsch
Gertsch (1941), Chickering (1949), Platnick & Shadab (1976 a)
Canal Zone, Barro Colorado Island, Portobelo

Lygromma dybasi Platnick & Shadab
Platnick & Shadab (1976 a, 1981 c)
Cerro Pando, Hato del Volcan, Finca Lerida (Chiriqui)

Lygromma volcan Platnick & Shadab
Platnick & Shadab (1981 c)
Las Lagunas, Hato del Volcan (Chiriqui)

Neozimiris chickeringi Platnick & Shadab
Platnick & Shadab (1976 a)
Canal Zone
Zimiromus aduncus Platnick & Shadab
Platnick & Shadab (1976 c)
Canal Zone

Zimiromus chickeringi Platnick & Shadab
Platnick & Shadab (1976 c)
Barro Colorado Island, Canal Zone

Zimiromus iotus (Banks) (= Echemus i.)
Banks (1929), Platnick & Shadab (1976 c)

Canal Zone, Barro Colorado Island, Las Sabanas, El Cermeno

Zimiromus tropicalis (Banks)
(= Echemus montanus Chickering)
Platnick & Shadab (1976 c)
El Valle

Ctenidae

Acanthoctenus maculatus Petrunkevitch
Petrunkevitch (1925)
Santiago, Wilcox Camp, Canal Zone

Acanthoctenus spinigerus Keyserling
Chickering (1936)
Barro Colorado Island

Acanthoctenus spinipes Keyserling
Banks (1929)
Almirante

Celaetycheus paradoxus F.O. Pickard-Cambridge
F.O. Pickard-Cambridge (1900), Petrunkevitch (1925)
Bugaba

Ctenus calcaratus F.O. Pickard-Cambridge
Nentwig & Heimer (1983)
Gamboa

Ctenus curvipes Keyserling
(= Microctenus c.)
(= Oligoctenus c.)
Keyserling (1881), F.O. Pickard-Cambridge (1900), Petrunkevitch (1925)
Panama
Ctenus dolomedes Chamberlin
Chamberlin (1925)
Barro Colorado Island

Ctenus medius Keyserling
Banks (1929), Nentwig & Heimer (1983)
Barro Colorado Island, Gamboa

Ctenus rubripes Keyserling
Keyserling (1881), F.O. Pickard-Cambridge (1900), Petrunkevitch (1925)
Panama

Ctenus sinuatipes F.O. Pickard-Cambridge
Petrunkevitch (1925), Banks (1929), Chickering (1936)
Barro Colorado Island, San Pablo, Cerro Flores, Canal Zone

Ctenus w-notatus Petrunkevitch
Petrunkevitch (1925), Chickering (1936)
Barro Colorado Island, Wilcox Camp

Cupiennius coccineus F.O. Pickard-Cambridge
F.O. Pickard-Cambridge (1901), Petrunkevitch (1925), Banks (1929), Lachmuth et al. (1985)
Barro Colorado Island, Canal Zone, El Volcan

Cupiennius foliatus F.O. Pickard-Cambridge
F.O. Pickard-Cambridge (1901), Petrunkevitch (1925), Banks (1929), Williams (1941), Nentwig & Heimer (1983), Lachmuth et al. (1985)
Bugaba, Barro Colorado Island, Canal Zone, Gamboa, Las Palmas, Veraguas, Carate, Los Santos

Cupiennius getazi Simon
Banks (1929), Lachmuth et al.(1985)
Almirante, Barro Colorado Island, Portobelo, Canal Zone, San Pablo

Cupiennius granadensis (Keyserling)
(= C. medius Chamberlin & Ivie)
Chamberlin & Ivie (1936), Lachmuth et al. (1985)
Barro Colorado Island, Canal Zone, Arraijan, Gamboa, El Valle, Boquete

(Cupiennius medius Chamberlin & Ivie = C. granadensis (Keyserling))

Cupiennius panamensis Lachmuth, Grasshoff & Barth
Lachmuth et al. (1985)
El Volcan, Boquete, El Valle

Cupiennius salei (Keyserling)
F.O. Pickard-Cambridge (1901), Petrunkevitch (1925), Chickering (1936), Nentwig & Heimer (1983)
Barro Colorado Island, Chiriqui, Gamboa

(Lycoctenus bogotensis Keyserling = Ancylometes b., Pisauridae)
(Microctenus curvipes Keyserling = Ctenus c.)
(Oligoctenus curvipes (Keyserling) = Ctenus c.)

Phoneutria sp.
Nentwig & Heimer (1983)
Carti Road

Selenopidae

Selenops aissus Walckenaer
Petrunkevitch (1925)
Remedios, Sona

Selenops banksi Muma
Muma (1953)
Barro Colorado Island

Selenops buschi Muma
Muma (1953)
Taboga Island

Selenops galapagoensis Banks
Muma (1953)
El Volcan, Panama City, Boquete, Canal Zone

Selenops marginalis F.O. Pickard-Cambridge
Nentwig & Heimer (1983)
Canal Zone

Selenops mexicanus Keyerling
F.O. Pickard-Cambridge (1900), Petrunkevitch (1925), Banks (1929), Nentwig & Heimer (1983)
David, Barro Colorado Island, Gamboa, Canal Zone, Chiriqui

Selenops minutus F.O. Pickard-Cambridge
Banks (1929)
Barro Colorado Island

Heteropodidae

(Chemmis barroanus Chamberlin = Sparianthis b.)

Defectrix defectrix Petrunkevitch
Petrunkevitch (1925)
Panama City

Heteropoda venatoria (L.)
Petrunkevitch (1925)
Panama
(Reported by Petrunkevitch (1939 b) as "absent")

(Macrinus formosus (Banks) = Olios f.)

Olios formosus Banks
(= Macrinus f.)
Banks (1929), Chickering (1936)
Barro Colorado Island, Canal Zone

Olios manifestus O. Pickard-Cambridge
(= Sparassus m.)
Nentwig & Heimer (1983)
Canal Zone

Olios obscurus (Keyserling)
(= Sparassus o.)
F.O. Pickard-Cambridge (1900), Petrunkevitch (1925)
Tolé

Olios simoni (O. Pickard-Cambridge)
(= Sadala s.)
Nentwig & Heimer (1983)
Gamboa

Olios stylifer (F.O. Pickard-Cambridge)
(= Sparassus st.)
Harrison (1968), Nentwig & Heimer (1983)
Canal Zone, Gamboa, Changuinola

Prusias nugalis O. Pickard-Cambridge
O. Pickard-Cambridge (1892), Petrunkevitch (1925), Banks (1929)

Barro Colorado Island, Bugaba

Prusias semotus (O. Pickard-Cambridge)
(= Ramnes s.)
O. Pickard-Cambridge (1892), Petrunkevitch (1925)
Bugaba

(Pseudosparianthis sp. (Chickering (1936)) = Tentabunda chickeringi Gertsch)
(Ramnes semotus O. Pickard-Cambridge = Prusias s.)
(Sadala simoni O. Pickard-Cambridge = Olios s.)
(Sparassus manifestus (O. Pickard-Cambridge) = Olios m.)
(Sparassus obscurus Keyserling = Olios o.)
(Sparassus stylifer F.O. Pickard-Cambridge = Olios st.)

Sparianthis barroana Chamberlin
(= Chemmis b.)
Chamberlin (1925), Banks (1929)
Barro Colorado Island

Sparianthina selenopoides Banks
Banks (1929)
Barro Colorado Island, Canal Zone

Tentabunda chickeringi Gertsch
Chickering (1936), Gertsch (1941), Quintero (pers. comm. 1988)
Barro Colorado Island

Philodromidae

Apollophanes crispus Dondale & Redner
Dondale & Redner (1975)
Canal Zone, Barro Colorado Island

Apollophanes punctipes (O. Pickard-Cambridge)
Banks (1929), Chickering (1936), Dondale & Redner (1975), Nentwig & Heimer (1983)
Canal Zone, Gamboa, Taboga Island, Boquete, El Volcan, Barro Colorado Island

Philodromus traviatus Banks
Banks (1929), Dondale & Redner (1969
Barro Colorado Island

Thomisidae

(Diaea puta O. Pickard-Cambridge = Isaloides p.)

Epicadinus trifidus (O. Pickard-Cambridge)
Nentwig & Heimer (1983)
Canal Zone

Epicadinus trispinosus (Taczanowski)
Nentwig & Heimer (1983)
Gamboa

Epicadus heterogaster (Guerin)
Chickering (1936), Nentwig & Heimer (1983)
Canal Zone, Gamboa, Barro Colorado Island

Isaloides puta (O. Pickard-Cambridge)
(= Diaea p.)
O. Pickard-Cambridge (1891), Petrunkevitch (1925)
Bugaba

Misumena conjuncta O. Pickard-Cambridge
O. Pickard-Cambridge (1891)
Veraguas

Misumena nigripes (Taczanowski)
Banks (1929)
Canal Zone

(Misumena ornata O. Pickard-Cambridge = Misumenops dubius (Keyserling))
(Misumena pallens Keyserling = Misumenops p.)
(Misumena pallida O. Pickard-Cambridge = Misumenops damnosus (Keyserling))
(Misumena particeps O. Pickard-Cambridge = Misumenops damnosus (Keyserling))
(Misumena pascalis O. Pickard-Cambridge = Misumenops p.)

Misumenoides annulipes O. Pickard-Cambridge
Nentwig & Heimer (1983)
El Valle

Misumenoides blanda (O. Pickard-Cambridge)
(= Runcinia b.)
O. Pickard-Cambridge (1889), Petrunkevitch (1925), Nentwig & Heimer (1983)
Bugaba, El Volcan, Canal Zone

Misumenoides magna (Keyserling)
(= Runcinia propinqua O. Pickard-Cambridge)
O. Pickard-Cambridge (1891), Petrunkevitch (1925), Nentwig & Heimer (1983)
Bugaba, Canal Zone

Misumenoides rugosa (O. Pickard-Cambridge)
(= Runcinia r.)
O. Pickard-Cambridge (1891), Petrunkevitch (1925), Harrison (1968), Nentwig & Heimer (1983)
Bugaba, Gamboa, Canal Zone, Changuinola

Misumenoides tibialis (O. Pickard-Cambridge)
(= Runcinia t.)
O. Pickard-Cambridge (1891), Petrunkevitch (1925), Nentwig & Heimer (1983)
Bugaba, Canal Zone

Misumenops americanus (Keyserling)
Nentwig & Heimer (1983)
El Valle

Misumenops celer (Hentz)
(= M. spinosus F.O. Pickard-Cambridge)
Petrunkevitch (1925), Nentwig & Heimer (1983)
Canal Zone

Misumenops damnosus (Keyserling)
(= Misumena pallida O. Pickard-Cambridge)
(= Misumena particeps O. Pickard-Cambridge)
O. Pickard-Cambridge (1891), Petrunkevitch (1925), Nentwig & Heimer (1983)
Canal Zone, Veraguas, Gamboa

Misumenops dubius (Keyserling)
(= Misumena ornata O. Pickard-Cambridge)
Harrison (1968), Nentwig & Heimer (1983)
Changuinola, Gamboa

Misumenops pallens (Keyserling)
(= Misumena p.)
Petrunkevitch (1925), Banks (1929)
Barro Colorado Island, Gamboa, Bella Vista, Las Sabanas

Misumenops pascalis (O. Pickard-Cambridge)
(= Misumena p.)
O. Pickard-Cambridge (1891), Petrunkevitch (1925), Nentwig & Heimer (1983)
Bugaba, Veraguas, Gamboa

Misumenops prosper O. Pickard-Cambridge
Nentwig & Heimer (1983)
El Valle

(Misumenops spinosus F.O. Pickard-Cambridge = Misumenops celer (Hentz))

Onocolus episcopus Mello-Leitao
Nentwig & Heimer (1983)
Canal Zone

Onocolus pentagonus (Keyserling)
F.O. Pickard-Cambridge (1900), Petrunkevitch (1925), Lise (1981), Nentwig & Heimer (1983)
Bugaba, Gamboa

Parastephanops cognatus (O. Pickard-Cambridge)
(= Stephanopoides c.)

O. Pickard-Cambridge (1892), Petrunkevitch (1925)
Bugaba, Veraguas

Parasynaema civipes (O. Pickard-Cambridge)
Nentwig & Heimer (1983)
Canal Zone

Paratobias championi F.O. Pickard-Cambridge
F.O. Pickard-Cambridge (1900)
Bugaba

Platyarachne episcopalis (Taczanowski)
(= Thomisus e.)
Nentwig & Heimer (1983)
Gamboa

(Runcinia blanda O. Pickard-Cambridge = Misumenoides b.)
(Runcinia rugosa O. Pickard-Cambridge = Misumenoides r.)
(Runcinia tibialis O. Pickard-Cambridge = Misumenoides t.)

Runcinia uragua Keyserling
Banks (1929), Chickering (1936)
Barro Colorado Island, Canal Zone

Runcinia parva Keyserling
Banks (1929)
Barro Colorado Island

(Runcinia propinqua O. Pickard-Cambridge = Misumenoides magna (Keyserling))
Runcinia sagittata O. Pickard-Cambridge
O. Pickard-Cambridge (1891)
Bugaba

(Stephanopoides cognata O. Pickard-Cambridge = Parastephanops c.)

Strophius hirsutus O. Pickard-Cambridge
O. Pickard-Cambridge (1891), Petrunkevitch (1925), Chickering (1936), Nentwig & Heimer (1983)
Bugaba, Barro Colorado Island, Canal Zone, Gamboa

Strophius signatus O. Pickard-Cambridge
Chickering (1936)
Barro Colorado Island

Synaema adjunctum O. Pickard-Cambridge
O. Pickard-Cambridge (1891), Petrunkevitch (1925)
El Volcan

Synaema affinitatum O. Pickard-Cambridge
(= S. profuga O. Pickard-Cambridge)
O. Pickard-Cambridge (1891), Petrunkevitch (1925)
Bugaba

(Synaema bimaculatum O. Pickard-Cambridge = S. haenschi Dahl)

Synaema fuscomandibulatum Petrunkevitch
Petrunkevitch (1925)
Cerro Flores

Synaema haenschi Dahl
(= Synaema bimaculatum O. Pickard-Cambridge)
Nentwig & Heimer (1983)
Canal Zone

Synaema jocosum Banks
Banks (1929), Chickering (1936)
Barro Colorado Island
Synaema maculosum O. Pickard-Cambridge
O. Pickard-Cambridge (1891), Petrunkevitch (1925)
Bugaba, Cerro Flores

Synaema palliatum O. Pickard-Cambridge
O. Pickard-Cambridge (1891), Petrunkevitch (1925)
Bugaba

(Syneama profuga O. Pickard-Cambridge = S. affinitatum O. Pickard-Cambridge)

Synaema socium O. Pickard-Cambridge
O. Pickard-Cambridge (1891), Petrunkevitch (1925)
Bugaba

(Thomisus episcopalis (Taczanowski) = Platyarachne e.)

Thomisus odiosus O. Pickard-Cambridge
Nentwig & Heimer (1983)
Canal Zone, Gamboa

Tmarus aculeatus Chickering
Chickering (1950 b)
Barro Colorado Island, Canal Zone, Arraijan

Tmarus bucculentus Chickering
Chickering (1950 b), Nentwig & Heimer (1983)
Barro Colorado Island, Canal Zone

Tmarus cognatus Chickering
Chickering (1950 b)
Canal Zone

Tmarus contortus Chickering
Chickering (1950 b), Nentwig & Heimer (1983)
Canal Zone

Tmarus corruptus O. Pickard-Cambridge
O. Pickard-Cambridge (1892), Petrunkevitch (1925), Chickering (1936, 1950 b)
Barro Colorado Island, Bugaba

Tmarus curvus Chickering
Chickering (1950 b)
Barro Colorado Island, Canal Zone, Portobelo, Arraijan

Tmarus decens O. Pickard-Cambridge
O. Pickard-Cambridge (1892), Petrunkevitch (1925), Chickering (1950 b)
Bugaba

Tmarus ineptus O. Pickard-Cambridge
O. Pickard-Cambridge (1892), Petrunkevitch (1925)
Bugaba

Tmarus intentus O. Pickard-Cambridge
O. Pickard-Cambridge (1892), Petrunkevitch (1925), Banks (1929), Chickering (1936, 1950 b)
Bugaba, Barro Colorado Island, Canal Zone, El Valle, Portobelo, Arraijan

Tmarus jocosus O. Pickard-Cambridge
Nentwig & Heimer (1983)
Canal Zone, Gamboa, Carti Road

Tmarus morosus Chickering
Chickering (1950 b)
Barro Colorado Island, Canal Zone

Tmarus mundulus O. Pickard-Cambridge
O. Pickard-Cambridge (1892), Petrunkevitch (1925), Chickering (1950 b), Nentwig & Heimer (1983)
Bugaba, Barro Colorado Island, Canal Zone, Gamboa

Tmarus parki Chickering
Chickering (1950 b)
Barro Colorado Island

Tmarus pauper O. Pickard-Cambridge
O. Pickard-Cambridge (1892), Petrunkevitch (1925), Chickering (1950 b)
Bugaba

Tmarus peregrinus Chickering
Chickering (1950 b)
Barro Colorado Island

Tmarus probus Chickering
Chickering (1950 b), Nentwig & Heimer (1983)

Canal Zone, El Valle, Arraijan, Taboga Island

Tmarus productus Chickering
Chickering (1950 b)
Portobelo

Tmarus sigillatus Chickering
Chickering (1950 b)
Canal Zone

Tmarus studiosus O. Pickard-Cambridge
O. Pickard-Cambridge (1892), Petrunkevitch (1925), Banks (1929), Chickering (1950 b), Nentwig & Heimer (1983)
Barro Colorado Island, Canal Zone, Bugaba, Gamboa

Tobias rugosus (Taczanowski)
F.O. Pickard-Cambridge (1900), Petrunkevitch (1925), Chickering (1936), Nentwig & Heimer (1983)
Bugaba, Barro Colorado Island, Gamboa

Xysticus pellax O. Pickard-Cambridge
Nentwig & Heimer (1983)
Canal Zone

Aphantochilidae

Aphantochilus rogersi O. Pickard-Cambridge
Banks (1929)
Barro Colorado Island

(Bucranium spinigerum O. Pickard-Cambridge = Majella s.)

Bucranium taurifrons O. Pickard-Cambridge
Chickering (1936)
Barro Colorado Island

Majella spinigera (O. Pickard-Cambridge)
(= Bucranium spinigera (O. Pickard-Cambridge)
O. Pickard-Cambridge (1891)
Bugaba

Salticidae

Acragas fallax (Peckham & Peckham)
(= Amycus f.)
Peckham & Peckham (1896), F.O. Pickard-Cambridge (1901), Petrunkevitch (1925), Chickering (1946), Nentwig & Cutler (1983)
Barro Colorado Island, Canal Zone, Gamboa

Acragas hieroglyphicus (Peckham & Peckham)
(= Amycus h.)
Peckham & Peckham (1896), F.O. Pickard-Cambridge (1901), Petrunkevitch (1925), Chickering (1936, 1946), Nentwig & Cutler (1983)
Barro Colorado Island, Gamboa

Acragas pacatus (Peckham & Peckham)
(= Amycus p.)
Petrunkevitch (1925), Chickering (1946), Galiano (1968)
Barro Colorado Island

Acragas peckhami (Chickering)
(= Amycus p.)
Chickering (1946), Nentwig & Cutler (1983)
Barro Colorado Island, Canal Zone, Gamboa, Portobelo

Acragas quadriguttatus (F.O. Pickard-Cambridge)
(= Amycus qu.)
F.O. Pickard-Cambridge (1901), Petrunkevitch (1925), Banks (1929), Chickering (1946), Galiano (1968)
Bugaba, Barro Colorado Island

Acragas zeteki (Chickering)
(= Amycus z.)
Chickering (1946), Galiano (1968)
Barro Colorado Island

Akela charlottae Peckham & Peckham
(= Freya striata Chickering)
Chickering (1946)
Barro Colorado Island

Albionella propria Chickering
Chickering (1946), Nentwig & Cutler (1983)
Barro Colorado Island, Gamboa

(Amycus benignus (Peckham & Peckham) = Hypaeus b.)
(Amycus fallax Peckham & Peckham = Acragas f.)
(Amycus hieroglyphicus Peckham & Peckham = Acragas h.)
(Amycus pacatus Peckham & Peckham = Acragas p.)

Amycus palpinalis F.O. Pickard-Cambridge
F.O. Pickard-Cambridge (1901), Petrunkevitch (1925), Banks (1929)
Barro Colorado Island, Wilcox Camp, Bugaba

(Amycus peckhami Chickering = Acragas p.)
(Amycus quadriguttatus F.O. Pickard-Cambridge = Acragas qu.
(Amycus spiralifer F.O. Pickard-Cambridge = Noegus s.)
(Amycus zeteki Chickering = Acragas z.)

Ashtabula dentata Chickering
Chickering (1946), Nentwig & Cutler (1983)
Barro Colorado Island, Gamboa, El Valle, Canal Zone
Ashtabula montana Chickering
Chickering (1946)
El Valle

(Atelurius incertus Banks = Zygoballus i.)

Attus infuscatus Peckham & Peckham
Peckham & Peckham (1896), F.O. Pickard-Cambridge (1901)
Panama

Avitus diolenii Peckham & Peckham
Peckham & Peckham (1896), F.O. Pickard-Cambridge (1901), Petrunkevitch (1925), Chickering (1946), Nentwig & Cutler (1990)
Barro Colorado Island, Gamboa

(Balmaceda peckhami Chickering = Hentzia tibialis Bryant (partim))

Balmaceda picta Peckham & Peckham
Chickering (1946), Nentwig & Cutler (1983)
Barro Colorado Island, Gamboa

Balmaceda reducta Chickering
Chickering (1946)
Canal Zone

Balmaceda turneri Chickering
Chickering (1946)
Portobelo

Banketosa dubia Chickering
Chickering (1946)
Boquete

Banketosa notata Chickering
Chickering (1946)
El Valle

Beata cephalica F.O. Pickard-Cambridge
F.O. Pickard-Cambridge (1901), Petrunkevitch (1925), Chickering (1946)
Bugaba

Beata flavolineata (F.O. Pickard-Cambridge)
(= Metaphidippus f.)
F.O. Pickard-Cambridge (1901), Petrunkevitch (1925), Chickering (1946)
Boquete, Bugaba

Beata levispina (F.O. Pickard-Cambridge)
Chickering (1946), Nentwig & Cutler (1983)
El Valle

Beata longipes (F.O. Pickard-Cambridge)
Petrunkevitch (1925), Chickering (1946)
Panama

Beata magna Peckham & Peckham
(= Homalattoides roseus F.O. Pickard-Cambridge)
(= Partona roseus (F.O. Pickard-Cambridge)
F.O. Pickard-Cambridge (1901), Petrunkevitch (1925)
Bugaba, Wilcox Camp

Beata munda Chickering
Chickering (1946)
El Valle, Boquete

Beata striata Petrunkevitch
Petrunkevitch (1925), Chickering (1946)
Cerro Iglesia

Beata variegata (F.O. Pickard-Cambridge)
Petrunkevitch (1925), Chickering (1946)
Canal Zone, Boquete

Beata venusta Chickering
Chickering (1946)
El Valle

Beata zeteki Chickering
Chickering (1946)
Canal Zone

Breda milvina (C.L.Koch)
Petrunkevitch (1925), Chickering (1946)
Cerro Iglesia, Canal Zone, Barro Colorado Island

Breda notata Chickering
Chickering (1946), Nentwig & Cutler (1983)
Barro Colorado Island, Canal Zone, Portobelo, Arraijan, El Valle, Gamboa

Breda punctata (Peckham & Peckham)
(= Fuentes p.)
Banks (1929), Chickering (1946)
Las Sabanas

Bryantella speciosa Chickering
Chickering (1946), Scioscia (1988)
Barro Colorado Island, Canal Zone

Capidava nigropicta (F.O. Pickard-Cambridge)
(= Sidusa n.)
F.O. Pickard-Cambridge (1901), Petrunkevitch (1925)
Bugaba

Carabella banksi Chickering
Chickering (1946)
Barro Colorado Island

Carabella insignis (Banks)
(= Paramarpissa i.)
Banks (1929), Chickering (1946)
Barro Colorado Island

Chapoda festiva (Peckham & Peckham)
Chickering (1946), Nentwig & Cutler (1983)
Gamboa, Barro Colorado Island

Chapoda inermis (F.O. Pickard-Cambridge)
Petrunkevitch (1925), Chickering (1946)
Barro Colorado Island, Canal Zone, El Valle, Cerro Flores

Chapoda panamana Chickering
Chickering (1946), Nentwig & Cutler (1983)
Barro Colorado Island, Canal Zone

Chapoda peckhami Banks
Banks (1929), Chickering (1946), Nentwig & Cutler (1983)
Barro Colorado Island, Canal Zone, El Valle, Gamboa, Carti Road

Cheliferoides planus Chickering
Chickering (1946)
Barro Colorado Island

Chinoscopus flavus (Peckham, Peckham & Wheeler)
Banks (1929), Chickering (1936), Nentwig & Cutler (1983)
Barro Colorado Island, Canal Zone, Gamboa

(Chira designata Peckham & Peckham = C. trivittata (Taczanowski))
(Panamanian specimens of designata are trivittata according to Galiano (1961), but the type should remain pending further investigation (Cutler, pers. comm. 1987))

Chira trivittata (Taczanowski)
(See C. designata)
(= Shira t.)
Peckham & Peckham (1896), F.O. Pickard-Cambridge (1901), Petrunkevitch (1925), Chickering (1946), Galiano (1961), Nentwig & Cutler (1983)
Barro Colorado Island, Canal Zone, Arraijan, Portobelo, Gamboa

Chirothecia modesta Chickering
Chickering (1946)
Boquete

Cobanus bifurcatus Chickering
Chickering (1946), Nentwig & Cutler (1983)
Canal Zone, El Valle

Cobanus cambridgei Chickering
Chickering (1946)
Boquete

Cobanus electus Chickering
Chickering (1946), Nentwig & Cutler (1990)
Barro Colorado Island, El Valle, Gamboa

Cobanus erythrocras Chamberlin & Ivie
Chamberlin & Ivie (1946), Chickering (1946), Nentwig & Cutler (1990)
Barro Colorado Island, Gamboa

Cobanus extensus (Peckham & Peckham)
(= Helorus e.)
Peckham & Peckham (1896), F.O. Pickard-Cambridge (1900), Petrunkevitch (1925), Banks (1929), Chickering (1946)
Bugaba, Barro Colorado Island

Cobanus flavens (Peckham & Peckham)
(= Helorus f.)
Peckham & Peckham (1896), F.O. Pickard-Cambridge (1900), Petrunkevitch (1925), Chickering (1936, 1946)
Barro Colorado Island

Cobanus incurvus Chickering
Chickering (1946)
Barro Colorado Island, El Valle

Cobanus mandibularis (Peckham & Peckham)
(= Helorus m.)
Peckham & Peckham (1896), F.O. Pickard-Cambridge (1900), Petrunkevitch (1925), Banks (1929), Chickering (1946), Nentwig & Cutler (1990)
Barro Colorado Island, Canal Zone, El Valle, Gamboa

Cobanus obscurus Chickering
Chickering (1946)
El Valle

Cobanus sedusus Chickering
Chickering (1946)
Barro Colorado Island, Canal Zone

Cobanus subfuscus F.O. Pickard-Cambridge
Petrunkevitch (1925), Banks (1929), Chickering (1946)
Barro Colorado Island, Cerro Flores

Cobanus unicolor F.O. Pickard-Cambridge
F.O. Pickard-Cambridge (1900), Petrunkevitch (1925), Chickering (1936, 1946)
Bugaba, Barro Colorado Island

(Coccorchestes scarabaeoides O. Pickard-Cambridge = Cylistella s.)
(Colonus = Thiodina)

Compsodecta montana Chickering

Chickering (1946)
Boquete

Corythalia albicincta (F.O. Pickard-Cambridge)
(= Sidusa a.)
F.O. Pickard-Cambridge (1901), Petrunkevitch (1925), Chickering (1936, 1946)
Barro Colorado Island, Bugaba

Corythalia bicincta Petrunkevitch
(= C. sitticus Chamberlin & Ivie)
(= Sidusa tulineba Chamberlin & Ivie)
Petrunkevitch (1925), Banks (1929), Chamberlin & Ivie (1936), Chickering (1946), Nentwig & Cutler (1983)
Barro Colorado Island, Canal Zone, Gamboa, El Valle, Portobelo, Wilcox Camp, Carti Road

Corythalia bryantae Chickering
Chickering (1946)
Boquete

Corythalia canalis (Chamberlin)
(= Saitis c.)
Chamberlin (1925), Banks (1929), Chickering (1946)
Barro Colorado Island, Canal Zone

Corythalia clara Chamberlin & Ivie
Chamberlin & Ivie (1936), Chickering (1946)
Barro Colorado Island, Canal Zone, Portobelo

(Corythalia conformans Chamberlin & Ivie = C. sulphurea F.O. Pickard-Cambridge)

Corythalia conspecta (Peckham & Peckham)
(= C. diffusa Chamberlin & Ivie)
Banks (1929), Chamberlin & Ivie (1936), Chickering (1946)
Barro Colorado Island

(Corythalia diffusa Chamberlin & Ivie = C. conspecta Peckham & Peckham)

Corythalia modesta Chickering
Chickering (1946)
Boquete

Corythalia murcida F.O. Pickard-Cambridge
Banks (1929)
Barro Colorado Island

Corythalia nigriventer (F.O. Pickard-Cambridge)
(= Sidusa n.)
F.O. Pickard-Cambridge (1901), Petrunkevitch (1925), Chickering (1946)
Bugaba

Corythalia obsoleta Banks
Banks (1929), Chickering (1946), Nentwig & Cutler (1983)
Barro Colorado Island, Canal Zone, Gamboa, Portobelo

Corythalia opima (Peckham & Peckham)
Chickering (1936, 1946), Nentwig & Cutler (1983)
Barro Colorado Island, El Valle, Gamboa

Corythalia panamana Petrunkevitch
Petrunkevitch (1925), Banks (1929), Chickering (1946)
Bella Vista, Remedios, Panama City

Corythalia parvula (Peckham & Peckham)
Chickering (1946)
Portobelo

Corythalia pulchra Petrunkevitch
Petrunkevitch (1925), Chickering (1946), Nentwig & Cutler (1983)
Gamboa, Canal Zone, Panama City

Corythalia quadriguttata (F.O. Pickard-Cambridge)
(= Sidusa q.)
F.O. Pickard-Cambridge (1901), Petrunkevitch (1925)
Bugaba

Corythalia serrapophysis (Chamberlin & Ivie)
(= Freya s.)
Chamberlin & Ivie (1936), Chickering (1946), Nentwig & Cutler (1983)
Barro Colorado Island, Canal Zone, Gamboa, Portobelo, Arraijan

(Corythalia sitticus Chamberlin & Ivie = C. bicincta Petrunkevitch)

Corythalia spiralis (F.O. Pickard-Cambridge)
(= Sidusa s.)
F.O. Pickard-Cambridge (1901), Petrunkevitch (1925), Banks (1929), Harrison (1968), Nentwig & Cutler (1983)
Bugaba, Barro Colorado Island, Gamboa, Canal Zone, Portobelo, El Valle, Arraijan, Wilcox Camp, Changuinola

Corythalia spirorbis (F.O. Pickard-Cambridge)
(= Sidusa s.)
F.O. Pickard-Cambridge (1901), Petrunkevitch (1925), Chickering (1946)
Bugaba

Corythalia sulphurea (F.O. Pickard-Cambridge)
(= Corythalia conformans Chamberlin & Ivie)
(= Sidusa s.)
F.O. Pickard-Cambridge (1901), Petrunkevitch (1925), Chickering (1946, 1936), Nentwig & Cutler (1983)
Bugaba, Barro Colorado Island, Canal Zone, Gamboa, El Valle, Portobelo, Cerro Flores, Wilcox Camp

Corythalia voluta (F.O. Pickard-Cambridge)
(= Sidusa v.)
F.O. Pickard-Cambridge (1901), Petrunkevitch (1925), Chickering (1946)
Bugaba, Boquete

Cotinusa bryantae Chickering
Chickering (1946)
Barro Colorado Island, Canal Zone

Cylistella castanea Petrunkevitch
Petrunkevitch (1925), Chickering (1946)
Wilcox Camp

Cylistella fulva Chickering
Chickering (1946)
Canal Zone

Cylistella scarabaeoides (O. Pickard-Cambridge)
(= Coccorchestes s.)
O. Pickard-Cambridge (1894), Petrunkevitch (1925), Chickering (1946)
Bugaba, Canal Zone, Barro Colorado Island

(Cyrene = Freya, Phiale)
(Cyrene delecta Peckham & Peckham = Nycerella d.)
(Cyrene interrupta F.O. Pickard-Cambridge = Nycerella delecta Peckham & Peckham)
(Dendryphantes centralis Peckham & Peckham = Metaphidippus c.)
(Dendryphantes cupreus (F.O. Pickard-Cambridge) = Metaphidippus c.)
(Dendryphantes dentiger F.O. Pickard-Cambridge = misidentification by Banks (1929) of *Paraphidippus perpolitus* Chickering (Chickering 1946))
(Dendryphantes laetabilis Peckham & Peckham = Metaphidippus l.)
(Dendryphantes longipalpus (F.O. Pickard-Cambridge) = Metaphidippus l.)
(Dendryphantes momus F.O. Pickard-Cambridge = Metaphidippus m.)
(Dendryphantes quadrinotatus (F.O. Pickard-Cambridge) = Metaphidippus q.)

Dendryphantes perplexans Chamberlin & Ivie
Chamberlin & Ivie (1936)
Barro Colorado Island

Descanso insolitus Chickering
Chickering (1946)
Barro Colorado Island, Canal Zone, El Valle

Descanso peregrinus Chickering
Chickering (1946)
Barro Colorado Island, Canal Zone

(Deza sumptuosa (Perty) = misidentification by Banks (1929) of *Psecas euoplus* Chamberlin & Ivie (Chickering (1946))

Donaldius lucidus Chickering
Chickering (1946)
El Valle

(Donatinus niger Chickering = Simprulla n.)

Dryphias maccuni (Peckham & Peckham)
Banks (1929), Chickering (1946), Nentwig & Cutler (1983)
Barro Colorado Island, Gamboa, Canal Zone

Erica eugenia Peckham & Peckham
F.O. Pickard-Cambridge (1900), Petrunkevitch (1925), Chickering (1946)
Bugaba
(Eris aurantius (Lucas) = Paraphidippus a.)
(Eris perpastus (Chickering) = Paraphidippus p.)
(Eris perpolitus (Chickering) = Paraphidippus p.)
(Eris validus (Chickering) = Paraphidippus v.)

Eustiromastix sp.
Nentwig & Cutler (1983)
Gamboa

Fluda princeps Banks
Banks (1929), Chickering (1946), Nentwig & Cutler (1983)
Barro Colorado Island, Canal Zone, El Valle, Portobelo

Freya albosignata (F.O. Pickard-Cambridge)
Chickering (1936, 1946), Nentwig & Cutler (1983)
Barro Colorado Island, Canal Zone, Gamboa, Portobelo

Freya arraijanica Chickering
Chickering (1946)
Arraijan

Freya bicavata (F.O. Pickard-Cambridge)
(= Cyrene b.)
F.O. Pickard-Cambridge (1901), Petrunkevitch (1925), Chickering (1946)
Bugaba

Freya bifida (F.O. Pickard-Cambridge)
(= Cyrene b.)
F.O. Pickard-Cambridge (1901), Petrunkevitch (1925), Chickering (1946)
Bugaba

Freya bifurcata (F.O. Pickard-Cambridge)
(= Cyrene b.)
(= Freya unicinella Chamberlin & Ivie)
F.O. Pickard-Cambridge (1901), Petrunkevitch (1925), Banks (1929), Chamberlin & Ivie (1936), Chickering (1946), Nentwig & Cutler (1983)
Bugaba, Canal Zone, Gamboa, Barro Colorado Island, El Valle, Boquete, Bella Vista, Carti Road

Freya bilobata (F.O. Pickard-Cambridge)
(= Cyrene b.)
F.O. Pickard-Cambridge (1901), Petrunkevitch (1925), Banks (1929)
Bugaba, Canal Zone, Gamboa

(Freya bulbosa (F.O. Pickard-Cambridge) = Phiale b.)

Freya demarcata Chamberlin & Ivie
Chamberlin & Ivie (1936), Chickering (1946), Nentwig & Cutler (1983)
Barro Colorado Island, Canal Zone, Gamboa, Arraijan

(Freya distincta Petrunkevitch = F. grisea (F.O. Pickard-Cambridge))

Freya flava F.O. Pickard-Cambridge
Banks (1929), Chickering (1946)
Canal Zone

Freya frontalis Banks
(= F. sagittifer Chamberlin & Ivie)
Banks (1929), Chamberlin & Ivie (1936), Chickering (1946), Nentwig & Cutler (1983)
Barro Colorado Island, Gamboa, Canal Zone, Portobelo

Freya grisea (F.O. Pickard-Cambridge)
(= Cyrene g.)
(= Freya distincta Petrunkevitch)
F.O. Pickard-Cambridge (1901), Petrunkevitch (1925), Chickering (1936, 1946)
Bugaba, Wilcox Camp, Barro Colorado Island

Freya infuscata (F.O. Pickard-Cambridge)
(= Cyrene i.)
F.O. Pickard-Cambridge (1901), Petrunkevitch (1925), Chickering (1936, 1946), Nentwig & Cutler (1983)
Bugaba, Barro Colorado Island, Canal Zone, Gamboa

Freya justina Banks
Banks (1929), Chickering (1946), Nentwig & Cutler (1983)
Barro Colorado Island, Gamboa, Canal Zone, El Valle

Freya longispina (F.O. Pickard-Cambridge)
Petrunkevitch (1925), Chickering (1936, 1946)
Barro Colorado Island, Canal Zone, Wilcox Camp

Freya minuta (F.O. Pickard-Cambridge)
(= Cyrene m.)
F.O. Pickard-Cambridge (1901), Petrunkevitch (1925), Chickering (1946)
Bugaba

(Freya nannispina Chamberlin & Ivie = Phiale dybowskii (Taczanowski))

Freya petrunkevitchi Chickering
Chickering (1946)
Barro Colorado Island, Canal Zone

Freya prominens F.O. Pickard-Cambridge
Banks (1929), Chickering (1946)
Barro Colorado Island, Canal Zone

Freya rustica Peckham & Peckham
Banks (1929)
Barro Colorado Island

(Freya sagittifer Chamberlin & Ivie = F. frontalis Banks)
(Freya serrapophysis Chamberlin & Ivie = Corythalia s.)

Freya simplicicava (F.O. Pickard-Cambridge)
(= Phiale s., new combination)
Petrunkevitch (1925), Banks (1929), Chickering (1946), Nentwig & Cutler (1983)
Barro Colorado Island, Canal Zone, Las Sabanas, Gamboa, Boquete, Panama City, Wilcox Camp

(Freya striata Chickering = Akela charlottae Peckham & Peckham)
(Freya unicinella Chamberlin & Ivie = F. bifurcata (F.O. Pickard-Cambridge))

Frigga pratensis (Peckham & Peckham)
Galiano (1979), Nentwig & Cutler (1990)
Panama, Gamboa

Fuentes pertinax Peckham & Peckham
Chickering (1936)
Barro Colorado Island

(Fuentes punctata Peckham & Peckham = Breda p.)

Gorgasella eximia Chickering
Chickering (1946)
Barro Colorado Island

Gypogyna sp.
Nentwig & Cutler (1990)
Gamboa

Habronattus banksi (Peckham & Peckham)
Nentwig & Cutler (1983)
Gamboa

Habronattus mexicanus (Peckham & Peckham)
Nentwig & Cutler (1983)
Gamboa

(Habronattus paratus (Peckham & Peckham) = Pellenes p.)
(Hamillus radians Peckham & Peckham = Pensacola r.)

Hasarius adansoni (Audouin)
Petrunkevitch (1925)
Panama

(Helorus extensus Peckham & Peckham = Cobanus e.)
(Helorus flavens Peckham & Peckham = Cobanus f.)
(Helorus mandibularis Peckham & Peckham = Cobanus m.)

Hentzia parallela (Peckham & Peckham)
(= Parahentzia insignita Chickering))
Chickering (1946), Nentwig & Cutler (1983), Richman (1989)
Barro Colorado Island, Gamboa, Canal Zone, Boquete, El Valle

Hentzia tibialis Bryant
(= Balmaceda peckhami Chickering (partim))
Chickering (1946), Richman (1989)
Barro Colorado Island

Hentzia vernalis (Peckham & Peckham)
(= Wala v.)
Chickering (1936), Richman (1989)
Barro Colorado Island

Homalattoides longipes F.O. Pickard-Cambridge
F.O. Pickard-Cambridge (1901)
Bugaba

(Homalattoides roseus F.O. Pickard-Cambridge = Dryphias maccuni Peckham & Peckham)

Hypaeus benignus (Peckham & Peckham)
(= Amycus b.)
Peckham & Peckham (1896), F.O. Pickard-Cambridge (1897), Petrunkevitch (1925), Banks (1929), Chamberlin & Ivie (1936), Chickering (1946), Galiano (1968), Nentwig & Cutler (1983)
Bugaba, Barro Colorado Island, Canal Zone, Gamboa, Portobelo, El Valle

Itata completa (Banks)
(= Lyssomanes c.)
Banks (1929), Chickering (1946), Nentwig & Cutler (1990)
Barro Colorado Island, Gamboa

(Jollas minutus Chickering = J. pompatus Peckham & Peckham)

Jollas geniculatus Simon
(= J. peritas Banks)
Banks (1929), Chickering (1946), Nentwig & Cutler (1983), Galiano (1991)
Barro Colorado Island, El Valle, Taboga Island

(Jollas peritas Banks = J. geniculatus Simon)

Jollas pompatus Peckham & Peckham
(= J. minutus Chickering)
Chickering (1946), Nentwig & Cutler (1983), Galiano (1991)
Barro Colorado Island, Canal Zone, Gamboa

(Lyssomanes banksi Chickering = L. unicolor (Taczanowski))

Lyssomanes bitaeniatus Peckham, Peckham & Wheeler
Nentwig & Cutler (1983)
Gamboa

Lyssomanes bryantae Chickering
Chickering (1946), Galiano (1980), Nentwig & Cutler (1990)
Canal Zone, Barro Colorado Island, Gamboa

(Lyssomanes completus Banks = Itata c.)

Lyssomanes consimilis Banks
(= L. montanus Chickering)
Banks (1929), Chickering (1946), Galiano (1980), Nentwig & Cutler (1983)
Barro Colorado Island, Canal Zone, El Valle, Gamboa

Lyssomanes deinognathus F.O. Pickard-Cambridge
Chickering (1946)
Canal Zone

Lyssomanes dissimilis Banks
Banks (1929), Chickering (1946), Galiano (1980)
Barro Colorado Island, Canal Zone, El Valle

Lyssomanes eatoni Chickering
Galiano (1980), Nentwig & Cutler (1983)
Cerro Punta

Lyssomanes elegans F.O. Pickard-Cambridge
(= L. zeteki Chickering)
F.O. Pickard-Cambridge (1900), Petrunkevitch (1925), Chickering (1946), Galiano (1980)
Bugaba, Barro Colorado Island

Lyssomanes jemineus Peckham & Peckham
(= L. spinifer F.O. Pickard-Cambridge)
Petrunkevitch (1925), Chickering (1936, 1946)
Barro Colorado Island, Boquete, El Valle

Lyssomanes mandibulatus F.O. Pickard-Cambridge
Banks (1929), Galiano (1980)
Barro Colorado Island

(Lyssomanes montanus Chickering = L. consimilis Banks)

Lyssomanes parki Chickering
Chickering (1946), Galiano (1980)
Barro Colorado Island

Lyssomanes patens Peckham & Peckham
(= L. trifurcatus F.O. Pickard-Cambridge)
Peckham & Peckham (1896), F.O. Pickard-Cambridge (1900), Petrunkevitch (1925), Chickering (1936, 1946), Galiano (1980)
Bugaba, Barro Colorado Island

Lyssomanes reductus Peckham & Peckham
(= L. simplicipes F.O. Pickard-Cambridge)
F.O. Pickard-Cambridge (1900), Petrunkevitch (1925), Galiano (1980)
Bugaba

Lyssomanes remotus Peckham & Peckham
Peckham & Peckham (1896), F.O. Pickard-Cambridge (1897), Petrunkevitch (1925), Galiano (1980), Nentwig & Cutler, (1983)
Gamboa

(Lyssomanes simplicipes F.O. Pickard-Cambridge = L. reductus Peckham & Peckham)
(Lyssomanes scropshirei Chickering = L. unicolor (Taczanowski))
(Lyssomanes trifurcatus F.O. Pickard-Cambridge = L. patens Peckham & Peckham)

Lyssomanes unicolor (Taczanowski)
(= L. banksi Chickering)
(= L. scropshirei Chickering)
Chickering (1946), Galiano (1980), Nentwig & Cutler (1983)
Barro Colorado Island, Canal Zone, Gamboa

Lyssomanes viridis (Walckenaer)
Petrunkevitch (1925)
Panama
(Very doubtful, this species is only known from SE USA; Cutler, pers. comm. 1987)

(Lyssomanes zeteki Chickering = L. elegans F.O. Pickard-Cambridge)
(Lyssomonea saphes Chamberlin & Ivie = Amycus benignus (Peckham & Peckham))

Mabellina prescotti Chickering
Chickering (1946)
Barro Colorado Island, El Valle

Marpissa magna (Peckham & Peckham)
(= Menemerus delus Chamberlin & Ivie)
Banks (1929), Chickering (1946), Nentwig & Cutler (1990)
Barro Colorado Island, Canal Zone, Gamboa

(Marpissa melanognatha Lucas = Menemerus bivittatus (Dufour))

Martella furva Chickering
(= Sarinda f.)
Galiano (1964), Reiskind (1977)
Barro Colorado Island

Menemerus bivittatus (Dufour)
(= Marpissa melanognatha Lucas)
F.O. Pickard-Cambridge (1901), Petrunkevitch (1925, 1939 b), Banks (1929), Chickering (1946), Nentwig & Cutler (1983)
Bugaba, Colon, Gamboa, Canal Zone, Taboga Island

(Menemerus delus Chamberlin & Ivie = Marpissa magna Peckham & Peckham)

Messua desidiosa Peckham & Peckham
(= Zygoballus d.)
Peckham & Peckham (1896), F.O. Pickard-Cambridge (1901), Petrunkevitch (1925), Chickering (1946)
Panama

Metaphidippus bicavatus F.O. Pickard-Cambridge
F.O. Pickard-Cambridge (1901), Petrunkevitch (1925), Chickering (1936)
Bugaba

Metaphidippus bispinosus F.O. Pickard-Cambridge
F.O. Pickard-Cambridge (1901), Petrunkevitch (1925), Chickering (1946)
Bugaba

Metaphidippus centralis (Peckham & Peckham)
(= Dendryphantes c.)
Peckham & Peckham (1896), F.O. Pickard-Cambridge (1901), Petrunkevitch (1925), Chickering (1946)
Panama

Metaphidippus cupreus (F.O. Pickard-Cambridge)
(= Dendryphantes c.)
F.O. Pickard-Cambridge (1901), Petrunkevitch (1925), Banks (1929), Chickering (1946)
Bugaba, Barro Colorado Island, Canal Zone, La Campana

Metaphidippus expallidatus F.O. Pickard-Cambridge
F.O. Pickard-Cambridge (1901), Petrunkevitch (1925), Chickering (1946)
Bugaba

(Metaphidippus flavolineatus F.O. Pickard-Cambridge = Beata f.)

Metaphidippus furcatus F.O.P.C
Nentwig & Cutler (1983)
Cerro Punta

Metaphidippus iridescens F.O. Pickard-Cambridge
F.O. Pickard-Cambridge (1901), Petrunkevitch (1925), Chickering (1946)
Chiriqui

Metaphidippus jacetus Chickering
Chickering (1946)
Boquete

Metaphidippus jastosus Chickering
Chickering (1946)
Barro Colorado Island, Canal Zone, Portobelo, El Valle

Metaphidippus laetabilis (Peckham & Peckham)
(= Dendryphantes l.)
Peckham & Peckham (1896), F.O. Pickard-Cambridge (1901), Petrunkevitch (1925), Banks (1929), Chickering (1946)
Barro Colorado Island, Canal Zone

Metaphidippus laetificus Chickering
Chickering (1946)
El Valle

Metaphidippus lanceolatus F.O. Pickard-Cambridge
Petrunkevitch (1925), Chickering (1946)
Boquete, Cerro Flores

Metaphidippus latus Chickering
Chickering (1946)
El Valle

Metaphidippus laxus Chickering
Chickering (1946)
Boquete

Metaphidippus levispinus (F.O. Pickard-Cambridge)
F.O. Pickard-Cambridge (1901), Petrunkevitch (1925)
Bugaba

Metaphidippus longipalpus F.O. Pickard-Cambridge
(= Dendryphantes l.)
F.O. Pickard-Cambridge (1901), Petrunkevitch (1925), Banks (1929), Chickering (1946)
Bugaba, Barro Colorado Island

Metaphidippus momus F.O. Pickard-Cambridge
(= Dendryphantes m.)
F.O. Pickard-Cambridge (1901), Petrunkevitch (1925), Banks (1929), Chickering (1946)
Bugaba, Boquete, Canal Zone, Barro Colorado Island, El Valle

Metaphidippus ochraceus F.O. Pickard-Cambridge
Chickering (1946)
Boquete

Metaphidippus octonotatus F.O. Pickard-Cambridge
Petrunkevitch (1925), Chickering (1946), Harrison (1968)
Cerro Flores, Changuinola

Metaphidippus perscitus Chickering
Chickering (1946)
El Valle

Metaphidippus quadrinotatus F.O. Pickard-Cambridge
(= Dendryphantes q.)
F.O. Pickard-Cambridge (1901), Petrunkevitch (1925), Chickering (1946)
Bugaba, Canal Zone, Barro Colorado Island

Mica longithorax Petrunkevitch
Petrunkevitch (1925), Chickering (1946)
Canal Zone

Monaga benigna Chickering
Chickering (1946), Nentwig & Cutler (1983)
Barro Colorado Island, Gamboa

Myrmarachne centralis (Peckham & Peckham)
(= Toxeus c.)
F.O. Pickard-Cambridge (1900), Petrunkevitch (1925), Banks (1929), Chickering (1946)
Bugaba, Barro Colorado Island, Taboga Island

(Myrmarachne championi (F.O. Pickard-Cambridge) = Sarinda armata (Peckham & Peckham))

Myrmarachne parallela (Fabricius)
Reiskind (1977), Nentwig & Cutler (1983)
Barro Colorado Island, Gamboa

Myrmarachne chickeringi Galiano
Galiano (1969)
Panama

Neonella sp.
Nentwig & Cutler (1983)
Gamboa

Noegus spiralifer (F.O. Pickard-Cambridge)
(= Amycus s.)
Petrunkevitch (1925), Banks (1929), Chickering (1946), Galiano (1968)
Barro Colorado Island, Canal Zone, Portobelo, Wilcox Camp

Nycerella decorata (Peckham & Peckham)
(= Phiale panamae (panamensis) Chamberlin)
Petrunkevitch (1925), Banks (1929), Chickering (1946), Galiano (1982), Nentwig & Cutler (1983)
Bella Vista, Gamboa

Nycerella delecta (Peckham & Peckham)
(= Cyrene d.)
(= Phiale d.)
(= Phiale interrupta (F.O. Pickard-Cambridge))
F.O. Pickard-Cambridge (1901), Petrunkevitch (1925), Banks (1929), Chickering (1936, 1946), Galiano (1982), Nentwig & Cutler (1983)
Bugaba, Gamboa, Remedios, Portobelo, Arraijan, Canal Zone, El Valle, Carti Road, Barro Colorado Island

Nycerella donaldi (Chickering)
(= Phiale d.)
Chickering (1946), Galiano (1982), Nentwig & Cutler (1983)
Barro Colorado Island, Canal Zone, Gamboa, El Valle

Nycerella neglecta Galiano
Galiano (1982), Nentwig & Cutler (1983)
Gamboa, Canal Zone, El Valle, Carti Road

(Nycerella panamae / panamensis (Chamberlin) = N. decorata (Peckham & Peckham))

Nycerella sanguinea (F.O. Pickard-Cambridge)
(= Freya s.)
Galiano (1982)
Bugaba

Orvilleus crassus Chickering
Chickering (1946)
Canal Zone, Barro Colorado Island

(Pachomius dybowskii (Taczanowski) = Phiale d.)

(Pachomius flavescens Peckham & Peckham = Phiale f.)

Paradamoetas formicina Peckham & Peckham
F.O. Pickard-Cambridge (1900), Petrunkevitch (1925), Banks (1929), Chickering (1946)
Barro Colorado Island, Bugaba

Paradamoetas changuinola Cutler
Cutler (1982)
Bocas del Toro

Parafluda banksi Chickering
Chickering (1946), Nentwig & Cutler (1983)
Canal Zone, El Valle

(Parahentzia insignita Chickering = Hentzia parallela (Peckham & Peckham)
(Paramarpissa insignis Banks = Carabella i.)

Paraphidippus aurantius (Lucas)
(= Eris a.)
Chickering (1946)
Boquete

Paraphidippus perpastus Chickering
(= Eris p.)
Chickering (1946)
Boquete

Paraphidippus perpolitus Chickering
(= Eris p.)
(Misidentification as *Dendryphantes dentiger* F.O. Pickard-Cambridge by Banks (1929))
Chickering (1946)
Barro Colorado Island

Paraphidippus validus Chickering
(= Eris v.)
Chickering (1946)
El Cermeno, Canal Zone, Barro Colorado Island

(Pardessus mimicus C.L.Koch = Phiale m.)

Parkella fusca Chickering
Chickering (1946)
Barro Colorado Island

Parkella venusta Chickering
Chickering (1946)
Canal Zone

Parnaenus convexus Chickering
Chickering (1946)
Barro Colorado Island

Parnaenus recurvus Chickering
Chickering (1946), Nentwig & Cutler (1983)
Barro Colorado Island, Canal Zone, Gamboa, El Valle, El Cermeno

(Partona roseus (F.O. Pickard-Cambridge) = Dryphias maccuni Peckham & Peckham)

Peckhamia prescotti Chickering
Chickering (1946)
Barro Colorado Island, Canal Zone, Arraijan, Boquete

Peckhamia variegata (F.O. Pickard-Cambridge)
(= Synageles v.)
F.O. Pickard-Cambridge (1900), Petrunkevitch (1925), Banks (1929), Chickering (1936, 1946), Nentwig & Cutler (1983)
Barro Colorado Island, Canal Zone, Las Sabanas, Taboga Island, Arraijan, El Valle, Bugaba

"*Pellenes*" *paratus* (Peckham & Peckham)
(= Habronattus p.)
Banks (1929), Chickering (1946), Harrison (1968), Nentwig & Cutler (1983), Griswold (1987)
Barro Colorado Island, Canal Zone, Gamboa, Carti Road, El Porvenir, Changuinola

Pensacola radians (Peckham & Peckham)
(= Hamilus r.)
Peckham & Peckham (1896), F.O. Pickard-Cambridge (1901), Petrunkevitch (1925)
Panama

Phiale aliceae Chickering
Chickering (1946)
Portobelo

Phiale bicuspidata (F.O. Pickard-Cambridge)
(= Cyrene b.)
F.O. Pickard-Cambridge (1901), Petrunkevitch (1925), Chickering (1936, 1946)
Bugaba, Barro Colorado Island, Canal Zone, Boquete, Wilcox Camp, Portobelo, El Valle

Phiale bilobata (F.O. Pickard-Cambridge)
Chickering (1946), Nentwig & Cutler (1983)
Canal Zone, Barro Colorado Island, El Valle, Carti Road, Gamboa

Phiale bulbosa (F.O. Pickard-Cambridge)
(= Cyrene b.)
(= Freya b.)
F.O. Pickard-Cambridge (1901), Petrunkevitch (1925), Galiano (1981 b)
Bugaba
(*P. bulbosa* of Chickering (1946) is *P. crocea* C.L. Koch; Cutler, pers. comm. 1987)

Phiale crocea C.L.Koch
(Misidentified by F.O. Pickard-Cambridge (1901), Petrunkevitch (1925), Chickering (1946) as *P. bulbosa* (F.O. Pickard-Cambridge), *P. mimica* C.L. Koch, *P. gratiosa* C.L. Koch; Galiano (1981 b), Cutler, pers. comm. 1987)
F.O. Pickard-Cambridge (1901), Petrunkevitch (1925), Chickering (1946), Galiano (1981 b), Nentwig & Cutler (1983)
Bugaba, La Campana, Barro Colorado Island, Canal Zone, Gamboa, El Valle, Arraijan

(Phiale delecta (Peckham & Peckham) = Nycerella d.)
(Phiale donaldi Chickering = Nycerella d.)

Phiale dybowskii (Taczanowski)
(= Freya nannispina Chamberlin & Ivie)
(= Pachomius d.)
(= Phiale lineola Chamberlin & Ivie)
Banks (1929), Chamberlin & Ivie (1936), Chickering (1946), Nentwig & Cutler (1983)
Canal Zone, Barro Colorado Island, Gamboa, Arraijan, Portobelo, El Valle, Carti Road

Phiale elegans (F.O. Pickard-Cambridge)
(= Cyrene e.)
F.O. Pickard-Cambridge (1901), Petrunkevitch (1925, 1939 b), Chickering (1946)
Bugaba

Phiale flavescens (Peckham & Peckham)
(= Cyrene f.)
(= Pachomius f.)
Peckham & Peckham (1896), F.O. Pickard-Cambridge (1901), Petrunkevitch (1925), Chickering (1946)
Panama

Phiale fusca (F.O. Pickard-Cambridge)
Chickering (1936)
Barro Colorado Island

Phiale geminata F.O. Pickard-Cambridge
(= Cyrene g.)
F.O. Pickard-Cambridge (1901), Petrunkevitch (1925), Banks(1929), Chickering (1946)
Bugaba, Barro Colorado Island

(Phiale gratiosa C.L. Koch = misidentification *P. crocea* C.L. Koch; Galiano 1979)

Phiale guttata (C.L.Koch)
(= P. magnifica Banks)
Banks (1929), Chickering (1946), Galiano (1979, 1981 a), Nentwig & Cutler (1983), Nentwig (1985 a)
Barro Colorado Island, Gamboa, Canal Zone

(Phiale interrupta (F.O. Pickard-Cambridge) = Nycerella delecta Peckham & Peckham)
(Phiale lineola Chamberlin & Ivie = P. dybowskii (Taczanowski))

Phiale maculosa (Chickering)
Chickering (1946)
Canal Zone, Barro Colorado Island

(Phiale magnifica Banks = P. guttata (C.L. Koch))
(Phiale mimica C.L. Koch = misidentification of *P. crocea* C.L. Koch; Galiano (1979 a))

Phiale niveo-guttata (F.O. Pickard-Cambridge)
(= Cyrene n.)
F.O. Pickard-Cambridge (1901), Petrunkevitch (1925), Chickering (1946)
Bugaba

Phiale orvillei Chickering
Chickering (1946)
El Valle
Phiale pallida (F.O. Pickard-Cambridge)
Chickering (1946)
Barro Colorado Island, Canal Zone

(Phiale panamae / panamensis Chamberlin = Nycerella decorata (Peckham & Peckham))

Phiale pratensis (Peckham & Peckham)
Chickering (1936, 1946)
Canal Zone, Barro Colorado Island, El Valle

(Phiale simplicicava (F.O. Pickard-Cambridge) = Freya s.)

Phidippus aureopilosus F.O. Pickard-Cambridge
F.O. Pickard-Cambridge (1897), Petrunkevitch (1925), Chickering (1946)
Bugaba

Plexippus paykullii (Audouin)
Petrunkevitch (1925, 1939 b), Banks (1929), Chickering (1946), Nentwig & Cutler (1983)
Canal Zone, Taboga Island, Gamboa

Psecas euoplus Chamberlin & Ivie
(misidentified by Banks (1929) as *Deza sumptuosa* (Perty))
Banks (1929), Chamberlin & Ivie (1936), Chickering (1946)
Barro Colorado Island, Canal Zone, Portobelo, El Valle

(Psecas sumptuosa Perty, see P. euoplus Chamberlin & Ivie)

Rudra geniculata Peckham & Peckham
Banks (1929)
Barro Colorado Island

(Saitis canalis Chamberlin = Corythalia c.)

Sarinda armata (Peckham & Peckham)
(= Myrmarachne championi (F.O. Pickard-Cambridge))
(= Sarinda stricta Chickering)
(= Toxeus championi)
F.O. Pickard-Cambridge (1900), Petrunkevitch (1925), Chickering (1936, 1946)
Bugaba, Portobelo, Canal Zone, Barro Colorado Island

Sarinda bicavata Chickering
Chickering (1946)
Barro Colorado Island

(Sarinda formosa (Banks) = Zuniga magna Peckham & Peckham)

(Sarinda furva (Chickering) = Martella f.)

Sarinda hentzi (Banks)
Galiano (1965), Nentwig & Cutler (1983)
Canal Zone, Gamboa, El Valle

Sarinda panamae Galiano
Galiano (1965)
Panama

Sarinda silvatica Chickering
Chickering (1946), Nentwig & Cutler (1983)
Barro Colorado Island, Gamboa, Canal Zone

(Sarinda stricta Chickering = S. armata (Peckham & Peckham))

Scopocira panamena Chamberlin & Ivie
Chamberlin & Ivie (1936), Chickering (1946), Nentwig & Cutler (1983)
Barro Colorado Island, Canal Zone, Gamboa, Portobelo, El Valle

Sebastira plana Chickering
Chickering (1946)
Barro Colorado Island, Canal Zone, El Valle, Arraijan

Semiopyla cataphracta Simon
Cutler (1971), Galiano (1985)
Canal Zone

(Shira designata Peckham & Peckham = Chira d.)
(Sidusa albicincta F.O. Pickard-Cambridge = Corythalia a.)

Sidusa marmorea F.O. Pickard-Cambridge
F.O. Pickard-Cambridge (1901)
Bugaba

(Sidusa nigriventer F.O. Pickard-Cambridge = Corythalia n.)
(Sidusa nigropicta F.O. Pickard-Cambridge = Capidava n.)
(Sidusa quadriguttata F.O. Pickard-Cambridge = Corythalia q.)

Sidusa recondita Peckham & Peckham
Peckham & Peckham (1896), F.O. Pickard-Cambridge (1901), Petrunkevitch (1925), Banks (1929), Chickering (1946)
Bugaba, Barro Colorado Island, Canal Zone, Portobelo

(Sidusa spiralis F.O. Pickard-Cambridge = Corythalia s.)
(Sidusa spirorbis F.O. Pickard-Cambridge = Corythalia s.)
(Sidusa sulphurea F.O. Pickard-Cambridge = Corythalia s.)
(Sidusa tulineba Chamberlin & Ivie = Corythalia bicincta Petrunkevitch)
(Sidusa voluta F.O. Pickard-Cambridge = Corythalia v.)
(Simonella americana Peckham & Peckham = Synemosyna a.)
(Simonella bicolor Peckham & Peckham = Synemosyna americana Peckham & Peckham)

Simprulla niger (Chickering)
(= Donatinus n.)
Chickering (1946)
Barro Colorado Island

(Synageles variegata F.O. Pickard-Cambridge = Peckhamia v.)

Synemosyna americana (Peckham & Peckham)
(= Simonella a.)
(= Simonella bicolor Peckham & Peckham)
F.O. Pickard-Cambridge (1900), Petrunkevitch (1925), Banks (1929), Chickering (1946), Galiano (1966), Reiskind (1977), Cutler (1988)
Bugaba, Barro Colorado Island, Canal Zone, Gamboa

Synemosyna edwardsi Cutler
Nentwig & Cutler (1983), Cutler (1988)
Gamboa

Thiodina crucifera (F.O. Pickard-Cambridge)
(= Colonus c.)
F.O. Pickard-Cambridge (1901), Petrunkevitch (1925), Chickering (1936)
Bugaba, Barro Colorado Island

(Thiodina pseustes Chamberlin & Ivie = T. sylvana (Hentz))
(Thiodina puerpera (Hentz) = misidentification for *T. sylvana* (Hentz), Edwards, pers. comm. 1988)

Thiodina sylvana (Hentz)
(= Colonus p.)
(= T. pseustes Chamberlin & Ivie)
F.O. Pickard-Cambridge (1901), Petrunkevitch (1925), Banks (1929), Chickering (1936, 1946), Harrison (1968)
Bugaba, Barro Colorado Island, Canal Zone, Gamboa, Bella Vista, El Valle, Arraijan, Portobelo, Boquete, El Cermeno, Changuinola

Titanattus cretatus Chickering
Chickering (1946)
Barro Colorado Island

Titanattus paganus Chickering
Chickering (1946)
Barro Colorado Island

Toloella eximia Chickering
Chickering (1946)
Barro Colorado Island

(Toxeus centralis (Peckham & Peckham) = Myrmarachne c.)
(Toxeus championi (F.O. Pickard-Cambridge) = Sarinda c.)

Uluella formosa Chickering
Chickering (1946), Reiskind (1977)
Canal Zone, Barro Colorado Island, El Valle, Boquete

(Wala vernalis (Peckham & Peckham) = Hentzia v.)
(Zuniga formosa (Banks) = Z. magna Peckham & Peckham)

(Zuniga furva Chickering = Martella f.)

Zuniga laeta (Peckham & Peckham)
Nentwig & Cutler (1983)
Canal Zone

Zuniga magna Peckham & Peckham
(= Sarinda formosa (Banks))
(= Zuniga f.)
Banks (1929), Chickering (1946), Reiskind (1977), Nentwig & Cutler (1983)
Canal Zone, Barro Colorado Island, El Valle

(Zygoballus desidiosus (Peckham & Peckham) = Messua d.)

Zygoballus electus Chickering
Chickering (1946), Nentwig & Cutler (1983)
Barro Colorado Island, Portobelo, El Valle

Zygoballus incertus (Banks)
(= Atelurius i.)
Banks (1929), Chickering (1946), Nentwig & Cutler (1983), Galiano (1987)
Canal Zone, Gamboa, Barro Colorado Island, El Valle

Zygoballus maculatipes Petrunkevitch
Petrunkevitch (1925)
Sona, Wilcox Camp

Zygoballus optatus Chickering
Chickering (1946), Nentwig & Cutler (1983)
Barro Colorado Island, Canal Zone, El Valle, El Cermeno, Gamboa

Zygoballus rufipes Peckham & Peckham
Banks (1929), Chickering (1946)
Barro Colorado Island

Zygoballus tibialis O. Pickard-Cambridge
Banks (1929)
Canal Zone

Chapter 10. Key to the families and to most genera of Panamanian spiders

This key and the figures (redrawn, often modified) are adapted and modified from Beatty (1970), Chickering (1945 a, b, 1951 c, 1957 b, 1967), Coddington (1986), Forster (1958), Gertsch (1941), Heimer (1986), Kaston (1980), Lachmuth et al. (1985), Lehtinen (1967), Levi (1962), Levi & Levi (1968), Opell (1979, 1988), Oliveira & Sazima (1984), Petrunkevitch (1925), Platnick (1974 b, 1975 b), Platnick & Forster (1982), Platnick & Shadab (1976, 1978 a, b, 1979, 1981 a, b), Raven (1981, 1985), Roth (1985), Valerio (1982). The key for Araneidae and Tetragnathidae bases on the "Key to the Neotropical genera of Araneidae" and the "Key to the Neotropical genera of Tetragnathidae" of Herbert Levi (with the help of Laura Leibensperger), second draft 1989 (used with the kind permission of the author). The key to Pisauridae / Trechaleidae has been written by Carico (pers. comm.) and the pholcid key has been contributed by Deeleman-Reinhold (pers. comm.).

Only Panamanian taxa are included. If the key does not fit satisfying the key may be wrong or the taxa under consideration is not mentioned in the key. This may be because the taxa is new for Panama or new to science. All these possibilities will certainly occur.

In all cases the key leads to the family, in most cases also to the genus. In a few families, genus identification is difficult: Dictynidae, Salticidae, Ctenidae, Heteropodidae, Thomisidae, Linyphiidae and Trechaleidae. No modern limitations of their genera exists and many species are certainly placed in the wrong genus. Therefore the key ends with a family name but further identification may be possible when considering the special literature.

The key is divided into the following subkeys:

I Infraorders, Mygalomorphae

1. Two pairs of book lungs, cheliceral fangs parallel (Figure 38) (Mygalomorphae) 2
- One pair of book lungs or none, cheliceral fangs opposing each other or oblique in some species (Araneomorphae) .. II (p. 178)

2. Claw tufts present 3
- Claw tufts absent 11

3. Anterior maxillary lobe (arrow) distinct and produced (Figure 39b) (Theraphosidae) 4
- Anterior maxillary lobe short, if present (Figure 39a) .... Barychelidae

4. Prolateral maxillae with distinct area of bacilliform setae forming lyra (Selenocosmiinae) .... *Psalmopeus*
- Otherwise 5

5. Leg tarsi clearly broader than metatarsi (Aviculariinae) .... *Avicularia*
\- Leg tarsi narrower than metatarsi (Theraphosinae) 6

6. Prolateral face of femur IV with distinct scopula 7
\- Prolateral face of femur IV without scopula 8

7. Scopula for full length of metatarsus IV .... *Xenesthis*
\- Scopula not for full length of metatarsus IV .... *Mygalarachne*

8. Scopula of tarsi II divided by setae 9
\- Scopula of tarsi II not divided by setae 10

9. Tibial spur absent .... *Crypsidromus*
\- Tibial spur present (Figure 40a) .... *Cyclosternum*

10. Lower spur of the male tibial spur widens apically *Aphonopelma*
\- Lower spur of the male tibial spur tapers apically .... *Euathlus*

11. Booklung apertures small, round, body length < 1 mm (Figure 41) .... Microstigmatidae
\- Booklung apertures normal, slitlike, body length 1 mm 12

12. Labium much longer than wide and eyes occupy at least half of head width .... Actinopodidae
\- Labium wider and/or eyes occupy less than half of head width 13

13. Apical segment of posterior lateral spinnerets short, domed, or triangular 14
\- Apical segment of posterior lateral spinnerets digitiform or longer, spinnerets very long (Figure 42) (Dipluridae) 17

14. A distinct dorsal excavation on tibia III of males and females (Figure 43b) (Ctenizidae) .... *Ummidia*
\- Otherwise 15

15. Anterior lateral eyes being set far in advance of other eyes (Idiopidae) .... *Idiops*
\- Eyes being set in a wider group 16

16. Eye group occupying more than half of the prosoma width (Figure 44) (Idiopidae) .... *Neocteniza*
\- Eye group occupying about one third of the prosoma width (Figure 45) (Cyrtaucheniidae) .... *Bolostromus*

17. Two rows of teeth on paired claws 18
- One row of teeth on paired claws 19

18. Maxillary lyra present (Figure 46) .... *Diplura*
- Maxillary lyra absent .... *Linothele*

19. One row of cheliceral teeth (Figure 47) .... *Masteria*
- Two rows of cheliceral teeth .... *Ischnothele*

II Araneomorphae

1. Spiders with two or four eyes (rarely one) 2
- Spiders with six or eight eyes 7

2. Two-eyed spiders (rarely one) 2
- Four-eyed spiders 6

3. Opisthosoma sclerotized, dorsal plate covering the whole dorsum and lateral plates (Tetrablemmidae) 4
- Opisthosoma not so sclerotized (Caponiidae) 5

4. Front eyes fused to a single eye (Figure 48) .... *Monoblemma*
- Two eyes widely separated .... *Caraimatta*

5. Maxillae robust, strongly convergent, obtusely terminated distally (Figure 49) .... *Nops*
- Maxillae narrowed distally .... *Caponina*

6. Opisthosoma globular (Figure 50) (Symphytognathidae) .... *Anapistula*
- Prosoma and opisthosoma elongated, eyes on prominent lateral tubercles (Figure 51) (Uloboridae) .... *Miagrammopes*

7. Six-eyed spiders 8
- Eight-eyed spiders .... IV (p. 181)

8. Chelicerae fused at the base, each provided distally with a lamella drawn out to a tooth, with the fang forms a kind of chela (chelate) (Figure 52a) 9
- Chelicerae of the usual type, not fused at the base, but with or without a lamella distally (not chelate) (Figure 52b) III (p. 179)

9. Prosoma elevated, much arched behind, tarsi 3-clawed, colulus pointed, twice as long as wide (Figure 53) (Scytodidae) .... *Scytodes*
- Otherwise 10

10. Tarsi 3-clawed, long legs, tarsi long and flexible, spiracle lacking, labium broader than long (*partim*) .... Pholcidae (p. 182)
- Tarsi 2-clawed, thoracic furrow conspicuous and longitudinal (Figure 54), prosoma flat above, labium pointed distally, longer than broad, sternum pointed behind (Loxoscelidae) .... *Loxosceles*

III Araneomorph, six-eyed, not chelate spiders

1. Male palp and female epigyne simple (haplogyne) 2
- Male palp complex, several sclerits visible, female epigyne a sclerotized plate with complex structure (e.g. fig. 54) (entelegyne) 14

2. Body length (5-)10-15 mm, tarsi with 3 claws (Figure 55) (Segestriidae) .... *Ariadna*
- Body length < 5 mm 3

3. Tarsi with 2 claws, without claw tufts, eyes in a ⑥ compact group, cheliceral margins not toothed, most species with dorsal and/ or ventral scuta (Oonopidae) 4
- Tarsi with 3 claws and onychium, cheliceral margins toothed, without scuta .... Ochyroceratidae

4. Both dorsal and ventral abdominal scuta present 5
- With a dorsal scutum or a ventral scutum (not both) or completely lacking scuta 8

5. Legs well supplied with spines 6
- Legs without true spines 7

6. Sternum conspicuously grooved, abdomen nearly completely covered by dorsal and ventral scuta .... *Dysderina*
- Sternum not conspicuously grooved, quite smooth and glabrous, abdomen with a less extensive dorsal scutum and a ventral scutum restricted to anterior half of venter .... *Triaeris*

7. Anterior lateral eyes well separated, posterior row of eyes slightly recurved, abdomen definitely depressed dorsoventrally .... *Opopaea*
- Anterior lateral eyes barely separated, posterior row of eyes slightly recurved .... *Scaphiella* (males)

8. Anterior lateral eyes contiguous, posterior row of eyes procurved, with numerous ventral leg spines, with a small dorsal scutum but lacking a ventral scutum .... *Ischnothyreus*
- Otherwise 9

9. Anterior lateral eyes well separated, posterior row of eyes recurved, legs without spines, all scuta lacking 10
- Otherwise 11

10. Posterior row of eyes strongly recurved, maxillae long, slender and strongly convergent .... *Oonopinus*
- Posterior row of eyes moderately recurved, maxillae also long, slender, convergent but with tips divergent .... *Oonopoides*

11. Anterior lateral eyes separated, legs without spines, with a ventral scutum which extends laterally to middorsal line, abdomen laterally compressed .... *Scaphiella* (females)
- Otherwise 12

12. Anterior lateral eyes moderately well separated, legs with numerous spines 13
- Anterior lateral eyes only slightly separated, legs without spines, scuta lacking .... *Stenoonops*

13. Palpal patella with a conspicuous prolateral bulge bearing robust spines, palpal femur and tibia also with conspicuous spines .. .. *Heteroonops*
- Palpal patella without notable distortion, palpal segments without conspicuous spines .... *Oonops*

14. Male palp and female epigyne as Figure 56 (Gnaphosidae) .... *Lygromma*
- Otherwise (Anapidae) 15

15. Coarsely punctuate prosoma, sternum and ventral abdominal scutum, male pedipalpus and epigyne as Figure 57 .... *Pseudanapis*
- Otherwise 16

16. Posterior eye row strongly procurved, medially excavated chelicerae with a distal denticulate plate (Figure 58) .... *Anapis*
- Posterior eye row straight, chelicerae without such plate (Figure 59) .... *Anapisona*

IV Araneomorph, eight-eyed spiders

1. Cribellum and calamistrum absent in both sexes .... V (p. 182)
- Cribellum present in both sexes, calamistrum present at least in females (often reduced or absent in males) 2

2. Posterior median eyes very large, projecting forward, opisthosoma 4 x as long as broad, legs very long, leg I more than twice, and leg II almost twice the length of the body, large spiders (Figure 60) (Deinopidae) .... *Deinopis*
- Otherwise 3

3. Anal tubercle large and prominent, with a fringe of long hairs, prosoma circular, small spiders about 2 mm (Figure 61) (Oecobiidae) .... *Oecobius*
- Otherwise 4

4. Labium fused to the sternum, chelicerae fused together at the base (at least 1/4 their length), each provided distally with a lamella drawn out to a tooth with which the fang forms a kind of chela, margins lacking teeth, eyes compactly arranged upon a hump, haplogyne (Figure 62) (Filistatidae) .... *Kukulcania*
- Otherwise 5

5. Tarsi scopulate, 2-clawed only on tarsi II - IV, trochanter III and IV moderately notched, large spiders .... Tengellidae
- Otherwise 6

6. A row of trichobothria on femora II - IV, a ventral row of short, stout setae from mid-metatarsus to the tip of the tarsus, metatarsus IV compressed and concave above, mostly orbweavers, some with reduced webs, small spiders (Uloboridae) 7
- Otherwise 10

7. Posterior median eyes three times as far from one another as from posterior lateral eyes, body length 2.5 mm (Figure 63) *Ariston*
- Posterior median eyes equally spaced, body length of most species > 3 mm 8

8. Prosoma with a broad, transverse thoracic groove, male prosoma oval (Figure 64a), female prosoma pear-shaped (Figure 64b), posterior eye row (especially in males) nearly straight *Philoponella*
- Otherwise 9

9. Male prosoma circular (Figure 65b), female prosoma with ± parallel sides (Figure 65a) .... *Zosis*
- Male prosoma pear-shaped (Figure 66b), female opisthosoma with a pair of dorsal tubercles in the anterior half (Figure 66a) .... *Uloborus*

10. Calamistrum in most species not appearing bipartite, nearly as long as metatarsus, tarsi without a dorsal row of trichobothria, small species .... Dictynidae
- Calamistrum usually appearing bipartite and less than 1/2 length of metatarsus, cribellum narrower than the area of spinnerets, fovea on prosoma lacking, labium long, cheliceral teeth 2 + 3, tarsi with a dorsal row of trichobothria, opisthosoma with a ± indistinct folium, medium-sized species (Titanoecidae) .... *Goeldia*

V Araneomorph, eight-eyed, ecribellate spiders

1. Chelicerae chelate, fused at base, with membranous attachment along midline for more than 1/4 - 1/2 length, mesial margin with transparent lamina apically thickened, lobe- or tooth-like, legs slender, 6 or 8 eyes (*partim*) (Pholcidae) 2
- Chelicerae not chelate 9

2. Prosoma with central deep circular depression in the middle of the thorax; head not or barely raised above thorax; brown or yellowish coloured spiders, usually in artificial environments 3
- Prosoma not with deep round depression 5

3. Opisthosoma elongate, with black median band on venter .... *Smeringopus elongatus*
- Opisthosoma globular 4

4. Large spider, > 5 mm body length, stridulating ridged on outer face of chelicerae .... *Artema atlanta*
- Spider 5 mm or smaller, stridulating tooth on thorax, stridulating plate on anterior face of opisthosoma ... *Physocyclus globosus*

5. Eyes clustered on raised head .... *Modisimus*
- Eyes not close together on raised head 6

6. Posterior eyes and anterior laterals forming semicircles with median concavity, male palpal femur with ventral process, epigyne with 2 circular openings; femur I 2 - 3 x body length .... *Blechroscelis modesta*
- Posterior eyes and anterior lateral eyes congregated in compact triads 7

7. Opisthosoma globular, with spinnerets close to epigastric fold; femur I never longer than prosoma and opisthosoma together .... *Anopsicus*
- Opisthosoma longer than wide, spinnerets on the ventral side; femur I longer than prosoma and opisthosoma together; fragile green spider with 6 eyes 8

8. Tip of opisthosoma acuminate, eye triads their diameter apart; male palpal paracymbium with several branches .... *Metagonia panama*
- Tip of opisthosoma rounded, eye triads more than their diameter apart; male palpal paracymbium conspicuously black, long and thin, S-shaped .... *"Micromerys" delicatus*

9. Tarsi 2-clawed, usually scopulate with claw tufts, sometimes recuced or lacking (e.g. in Thomisidae), spiders usually do not build a snare but construct a silken retreat which may be enlarged 10
- Tarsi 3-clawed, scopulae and claw tufts usually lacking (rarely present), web-building spiders, in many families, however, snares are reduced ....VI (p. 187)

10. Leg I strongly modified: Femur expanded dorsally to twice the height of femur II, coxa and trochanter elongated, patella enormously elongated and longer than the tibia (Figure 67) (Palpimanidae) .... *Otiothops*
- Otherwise 11

11. Anterior median eyes much the largest, eyes in 4 rows (Lyssomaninae) (Figure 68a) or in 3 rows (other salticid subfamilies) (Figure 68b, c), the posterior eyes widely separated, situated usually on lateral edges of the prosoma, often minute .... Salticidae*
- Otherwise 12

12. Eyes in 3 rows with 2 - 4 - 2 eyes as in Figure 69 (Ctenidae) . 13
- Eyes otherwise 14

13. A third claw between the claw tufts (difficult to see), leg spination as in Figure 70, anterior lateral eyes always circular (Figure 69) .... *Cupiennius*
- No third claw, leg spination and eyes otherwise .... other ctenid genera

14. Six large eyes in a recurved row or anterior row of 6 eyes, lateral eyes of the row tiny (Figure 71) (Selenopidae) .... *Selenops*
- Eyes otherwise 15

15. Endites strongly convergent, labium very long and narrow, prosoma tuberculate, some tubercles bearing spines, ant mimics (Figure 72) .... Aphantochilidae
- Otherwise 16

16. At least leg 1 and 2 laterigrade, i.e. turned so that the morphologically dorsal surface is posterior and the prolateral surface appears to be dorsal, "crab-spiders" 17
- Legs otherwise 21

17. Large spiders, usually > 10 mm in length, both cheliceral margins toothed, the tip of metatarsus I with a soft trilobate membrane (arrow in Figure 73), tarsi with heavy scopulae and claw tufts .... Heteropodidae
- Smaller spiders, usually < 10 mm in length, metatarsus without trilobate membrane 18

*(**NOTE:** Determining genera in this family is difficult. The relatively simple genitalia do not offer the wealth of characters found in some other families, i.e. Araneidae. Many generic placements of species are incorrect, other species belong in new genera, and some genera are synonyms of others. A sucessful modern key would depend on generic revisions which are accumulating slowly, but are far from completion. The only key for Panamanian salticids is that of Chickering (1946). Because of the complex nature of taxonomic identification in this family, this key is not easy to use. With practice over time, the nuances of the key can be mastered, it is probably easier to use Chickering's key if one is identifying 100 specimens than merely one or two. In addition to Chickering (1946), a copy of the salticid portions of Biologia Centrali-Americana (by F.O. and O. Pickard-Cambridge) is necessary. The papers of Banks (1929), Chamberlin & Ivie (1936), and Petrunkevitch (1925) are helpful for a few species. Finally the recent revisions on antlike salticids by Cutler (1982, 1988) and other revisions of Galiano are important for selected genera.)

18. Claw tufts lacking, scopulae present or absent, leg I and II longer than leg III and IV, cheliceral promargin without teeth .... Thomisidae
- Claw tufts and scopulae present, all legs about the same length (or only leg 2 much longer), cheliceral promargin with 1 - 2 teeth (Philodromidae) 19

19. Posterior median eyes distinctly closer to posterior lateral eyes than to each other .... *Philodromus*
- Posterior eyes nearly uniformely spaced 20

20. Prosoma longer than wide .... *Apollophanes*
- Prosoma not longer than wide .... *Thanatus*

21. Tracheal spiracle (arrow) nearer middle of abdomen or nearer epigastric furrow (Figure 74), notched trochanters, claw tufts of 5 - 10 lamelliform setae (Figure 75) (Anyphaenidae) 22*
- Tracheal spiracle adjacent to spinnerets, trochanters not notched 25

22. Tracheal spiracle much closer to epigastric furrow than to spinnerets (Figure 74b) .... *Aysha*
- Tracheal spiracle midway between epigastric furrow and spinnerets (Figure 74a) 23

23. Legs very long and thin, leg 1 greatly elongated, tibial index (width / length x 100) usually 5 or less .... *Wulfila*
- Legs stout, tibial index usually 8 or more 24

24. Prosoma with 2 dark paramedian longitudinal bands .... *Anyphaena*
- Prosoma without dark bands .... *Teudis*

25. Anterior spinnerets cylindrical, noticably separated (Figure 76a), posterior median eyes commonly elliptical or triangular, cheliceral margins weakly toothed, with one or both margins lacking teeth, with only denticles, keeled, or with only 2 promarginal teeth, endites convergent, usually with distinct oblique median indentation (Figure 77), claw tufts always present (Gnaphosidae) 26

*(**NOTE:** Platnick (1974) mentions that there are great problems in the limitations of the anyphaenid genera. This situation obviously is still existing and I am unable to give a key for all 8 genera known for Panama. The here presented key is modified from Platnick (1974) and leads only to 4 genera (with *Arachosia*, *Macrophyes*, *Pelayo* and *Sillus* lacking) and covers about 80 % of the Panamanian species.)

- Anterior spinnerets conical, contiguous (Figure 76b), eyes concolorous, posterior median eyes always circular (or oval and tibia I with 5 - 6 pairs of ventral spines), cheliceral margins strongly toothed, promargin usually with 3 teeth (or 2 teeth and 5 - 6 pairs of tibial spines), endites nearly parallel, often widened distally without median indention, claw tufts absent in some genera 31*

26. Metatarsus III and IV with distal preening comb (Figure 78a) .... *Camillina*
- Without preening comb, sometimes with preening brush (Figure 78b) 27

27. Cheliceral retromargin lobed .... *Eilica*
- Cheliceral retromargin not lobed, with or without teeth 28

28. Posterior eye row strongly procurved (Fig. 79a) 29
- Posterior eye row slightly procurved, straight or slightly recurved (Figure 79b) 30

29. Chelicerae with 3 promarginal teeth .... *Zimiromus*
- Chelicerae with a promarginal row of stiff, bent setae *Neozimiris*

30. Opisthosoma with pale transverse markins, occasionally with pale longitudinal lateral stripes, almost connected to form an H (Figure 80) .... *Gertschosa*
- Opisthosoma with 3 - 4 longitudinal black lines, dorsally on a pale background, occasionally lines broken or spotted ...... *Cesonia*

31. Sternum with deep antero-lateral grooves near margin, clypeus distinctly grooved just dorsal to margin, no scutum on opisthosoma (Liocranidae) .... *Parachemmis*
- Otherwise 32

32. Terminal segment of posterior spinnerets long and flexuose (Figure 80 c) (Miturgidae) .... *Eutichurus*
- Otherwise 33

33. Opisthosoma with complete dorsal sclerite (males) or only anterodorsal sclerite (females), epigastric area sclerotized to petiole, tibia I with only 2 - 3 pairs of ventral spines (Corinnidae, *partim*) 37**

*(**NOTE:** Clubionidae-Liocranidae-Corinnidae-Miturgidae-complex, many problems with limitations of genera and families, especially in tropical taxa.)

(NOTE**: This key covers 4 of 5 genera with *Megalostrata* lacking)

- Combination otherwise 34

34. Scutum on opisthosoma usually small or lacking, tibia I with 5 - 8 pairs of ventral spines or two dense rows, body length > 5 mm, trochanter I and II not notched (Corinnidae) .... *Corinna*
- Combination otherwise 35

35. Distal segments of posterior spinnerets cylindrical and about as long as basal segment, legs stout (Figure 81) (Miturgidae) .... *Teminius*
- Distal segments of posterior spinnerets short, hemispherical or much shorter than basal segment 36

36. Tibia I with 1 - 3 pairs of ventral spines (Clubionidae) 40
- Tibia I lacking ventral spines (Corinnidae) .... *Trachelas*

37. Prosoma elongated, constricted between the cephalic and thoracic regions with no further constrictions in the thoracic region ... . *Sphecotypus*
- Otherwise 38

38. Opisthosoma (from above) with distinct rugose petiole and pair of stout spines on base of dorsal abdominal sclerite .... *Mazax*
- Opisthosoma lacking rugose petiole, no pair of stout spines on base of dorsal abdominal sclerite 39

39. Posterior eye row slightly to moderately procurved, prosoma 1/2 - 2x as long as wide, thoracic furrow present, sometimes weak .... *Castianeira*
- Posterior eye row recurved, prosoma about 2x as long as wide, thoracic furrow absent, replaced by a shallow depression .... *Myrmecotypus*

40. Trochanter III and IV notched, longitudinal thoracic groove not present .... *Cheiracanthium*
- Trochanter III and IV not notched, sometimes only trochanter IV with shallow notch, longitudinal thoracic groove present .... *Clubiona*

VI Araneomorph, eight-eyed, ecribellate spiders with 3-clawed tarsi

1. Posterior spinnerests long, slender, tapering, with distal segment 5 x as long as basal, legs break at patella-tibia joint (Hersiliidae) .... *Tama*
- Otherwise 2

2. Anterior spinnerets long and robust, other pairs minute, endites strongly convergent, lacking serrulae (Zodariidae) 3
- Otherwise 5

3. Prosoma with 7 - 11 strong spines, projecting from head region, body length 2 - 4 mm (Figure 83) .... *Leprolochus*
- Prosoma without such spines 4

4. Clypeus height less than 4 times the diameter of an ALE, body length about 5 - 6 mm (Figure 84) .... *Tenedos*
- Clypeus height more than 5 times the diameter of an ALE, body length about 3 - 4 mm .... *Ishania*

5. Tibia and metatarsus of legs I and II armed prolaterally with series of long spines interspersed with much shorter spines (Figure 85a) (Mimetidae) 6
- Otherwise 7

6. Femur I with a retrolateral row of very short spines in the proximal half (Figure 85b) .... *Mimetus*
- Femur I without a distinct row of spines, if spines are present they are long and not in a distinct row, opisthosoma triangular .... *Gelanor*

7. Spiders press against the substrate and appear very flat spiders, legs and opisthosoma often clothed with a lateral fringe of curving hairs, usually white, eyes grouped as in Figure 86, with lateral eyes on a prominent joint tubercle (Figure 86) (Senoculidae) .... *Senoculus*
- Otherwise 8

8. Two or more dorsal tarsal trichobothria present, usually in 1 - 2 rows, hunting spiders, some of them with juvenile webs or nursery webs or other forms of permanent webs (space webs, funnel webs) 9
- Tarsal trichobothria absent or only one present, web-building spiders, only few species with reduced snares, many orbweavers 33

9. Tarsi flexible and usually bent in preservation ... "Trechaleidae"*
- Tarsi stiff 10

*(This family, created and later abandoned by Simon, is currently under redescription and revision by J.E. Carico. It includes species originally assigned to: *Trechalea*, *Syntrechalea*, *Hesydrus*, *Paradossenus*, *Enna* and *Dyrines*. A key is not yet available.)

10. Tarsal trichobothria irregular, in 2 dorsal rows, all trochanters slightly to deeply notched (arrow in Figure 87b, c) 12
- Tarsal trichobothria in single, dorsal row, trochanter I and II lacking notches (Figure 87a) (Agelenidae) 11

11. Both eye rows strongly procurved, appearing to be in 3 rows of 2 - 4 - 2 eyes, posterior median eyes distinctly posterior to posterior lateral eyes (Figure 88b) .... *Melpomene*
- Both eye rows almost straight (Figure 88a) .... *Tegenaria*

12. Cheliceral margins with one tooth or lacking teeth, all trochanters slightly notched (Figure 87c), plates on the dorsal side of the pedicel (lorum) undivided dorsally, anterior lateral eyes and posterior eyes usually arranged in a hexagon, anterior median eyes located below posterior lateral eyes (Figure 89), legs with numerous large, erect spines (Oxyopidae) 13
- Cheliceral margins strongly toothed, plates on the dorsal side of the pedicel (lorum) divided dorsally (arrow in Figure 90), all trochanters deeply notched (Figure 87b), eyes not arranged in a hexagon, anterior eyes pro- or recurved but form a single row 17

13. Posterior cheliceral margin without teeth, anterior lateral eye row obviously wider than posterior median eye row, posterior eye row only slightly procurved 15
- Posterior cheliceral margin with a single tooth on each side, anterior lateral eye row subequal to posterior median eye row or posterior median eye row much wider than anterior lateral eyes, posterior eye row strongly procurved 14

14. Distance between posterior median eyes subequal to distance between posterior median eyes and posterior lateral eyes on each side .... *Oxyopes*
- Distance between posterior median eyes much greater than distance from posterior median eyes to posterior lateral eyes on each side .... *Hamataliwa*

15. Posterior eye row straight or slightly recurved, clypeus smaller than eye area, web-building spider, space-web in form of a sheet-web, prosoma yellow-brown with two black stripes, opisthosomal black with a median brown stripe and lateral whitish dots (Figure 91) ....*Tapinillus*
- Posterior eye row recurved, clypeus equal to or higher than eye area 16

16. Praetarsus with 5 greenish spines, labium > 2x longer than wide .... *Peucetia*
- Praetarsus with 1 apical spine, labium < 2x longer than wide ... . *Schaenioscelis*

17. Posterior lateral eyes situated directly behind posterior median eyes or slightly laterally, forming rectangle or trapezoid 1,6 x wider behind, eyes 4 - 2 - 2 (Figure 92), unpaired tarsal claw with one tooth or none, lorum (plates on the dorsal side of the pedicel) as Figure 90, male palpi lacking tibial apophyses, egg cocoon attached to spinnerets, spiderlings cling to abdomen of female (Lycosidae) 19
- Otherwise 18

18. Posterior lateral eyes situated lateral to posterior median eyes and slightly behind, forming strongly recurved posterior eye row, posterior eyes forming trapezoid twice as wide behind, eyes 4 - 4 or barely 4 - 2 - 2 (Figure 93), unpaired tarsal claw with 2 - 3 teeth, lorum different from Figure 90, male palpi with tibial apophyses, egg cocoon carried under sternum and held in place by the chelicerae and palpi, spiderlings hatch into a nursery web built by the mother (Pisauridae) 25
- Both eye rows recurved, prosoma widest near rear of coxae II with elongate cephalic region about half of maximum width, chelicerae with 3 promarginal and 3 retromarginal teeth, endites rectangular, labium no longer than wide, sternum about as long as wide, opisthosoma dark brown with wide orange median longitudinal stripe with anterior spinnerets thick, conical, posterior spinnerets no longer than anteriors, colulus large, oval, body length 4 - 6 mm (see p. 141) .... *Barrisca*

19. Palea (p in Figure 94a) absent from the male palp , female prosoma with V-shaped mark anterior to thoracic furrow .... *Pirata*
- Male palp with palea (p in Figure 94a), female prosoma without V-shaped mark 20

20. (Females, go to 22) Median apophysis of the male palp (m in Figure 94a) not dorsally excavated .... *Pardosa*
- Median apophysis dorsally excavated, with embolus (e in Figure 94a) lying within excavation 21

21. Median apophysis (m in Figure 94a) of the male palp with 2 pointed processes .... *Allocosa*
- Median apophysis (m in Figure 94a) of the male palp with 1 process 22

22. (Males, go to 24) Epigyne lacking atrium (a in Figure 94c), copulatory openings located on posterior margin of epigynal plate .... *Allocosa*
- Epigyne with atrium (a in Figure 94c), copulatory opening located at sides of median septum 23

23. Eye area noticeably swollen dorsally, prosoma sides vertical anteriorly .... *Pardosa*
- Eye area not swollen dorsally, prosoma sides sloped anteriorly 24

24. Tegulum (t in Figure 94a) of male palp lacking ear-shaped lobe, terminal apophysis (t in Figure 94a) in 2 inequally sclerotized parts, epigyne lacking hood (h in Figure 94c) .... *Arctosa*
- Tegulum (t in Figure 94a) with large ear-shaped lobe upon which embolus (e) rests, terminal apophysis (t in Figure 94a, b) sickle-like or lobe-like, epigyne usually with hood (h in Figure 94c) . ... *Lycosa*

25. Anterior row of eyes strongly recurved, anterior lateral eyes nearer to posterior lateral eyes than to anterior median eyes .... *Ancylometes*
- Anterior row of eyes not strongly recurved, anterior lateral eyes nearer to anterior median eyes than to posterior lateral eyes 26

26. Males 27
- Females 30

27. Palpal organ apparently simple, without conspicuous protruding elements (Figure 95a, b) 28
- Palpal organ complex with conspicuous protruding elements (Figure 95c) 29

28. Tibial apophysis not hammer-like and not articulating (Figure 95a) .... *Architis*
- Tibial apophysis hammer-like and articulating (Figure 95b) .... *Staberius*

29. Palpal organ with a long, slender embolus looping outside the limits of the cymbium (Figure 95c) .... *Thaumasia*
- Palpal organ not as above .... *Tinus*

30. Ventral side of epigynum with a single median anterior opening (Figure 95d) .... *Staberius*
- Epigynum not as above 31

31. Ventral side of epigynum composed of a pair of prominent posterior elevations and a small, median, anterior lobe or elevation (Figure 95e) .... *Tinus*
- Epigynum not as above 32

32. Dorsal (internal) side of epigynum complex, with coiled and looped tubes (Figure 95f) .... *Thaumasia*
- Dorsal (internal) side of epigynum complex, with 2 or 4 sac-like components (Figure 95g) .... *Architis*

33. Pits (arrow in Figure 96a) on the prolateral margins of the sternum, sternum broadly truncated behind, mostly orb weavers, length 1.0 - 2.5 mm (Theridiosomatidae) 34
- Otherwise 39

34. Fourth legs longer than first or, in males, subequal .... *Ogulnius*
- First legs longer than fourth 35

35. Legs relatively short and thick (Figure 96b) 36
- Legs relatively long and slender (Figure 96c) 37

36. Epigyne with closely fitting, anteriorly hinged flap covering copulatory bursae (Figure 97a), male with straight embolus, tip of embolus blunt, rounded, sawtoothed (Figure 98b) .... *Naatlo*
- Epigyne a domed vault, hint rim appressed to venter (Figure 97b), embolus smooth, tip acute (Figure 98a) .... *Epeirotypus*

37. Posterior median eye separation 1 diameter or more .... *Wendilgarda*
- Posterior median eye separation 1/2 diameter or less 38

38. Posterior epigynal rim blunt, not pointed, not a protruding scape (Figure 99a), embolic division fragmented into bristle-like parts, not blunt lobes, median apophysis (m in Figure 100a) attenuate, slightly grooved (Figure 100a) .... *Theridiosoma*
- Epigyne triangular, medially indented plate with short scape protruding (Figure 99b), embolus division (e) not fragmented, median apophysis (m) small and grooved (Figure 100b) .... *Baalzebub*

39. The following families are often difficult to separate, so a table of the most important characters seems to be an appropriate and easy way for identification.

	Araneidae & Tetragnathidae	Linyphiidae	Nesticidae *Gaucelmus*	Theridiidae
rebordered labium	present	present	present	absent
endites	parallel	parallel	parallel	strongly convergent in many species
cheliceral margins	both margins toothed	both margins toothed	promargin 2-3 teeth, retromargin 6-12 denticles	usually without cheliceral teeth, few with 1-3 on margin
stridulation structures	absent	usually on side of chelicerae	absent	if present, not on chelicerae
tarsus IV				
- serrated bristles	absent	absent	present	rarely absent
- third claw	present	absent	absent	absent
abdominal scutum	absent	present or absent	absent	present or absent
web type	usually orbweb	irregular	irregular	irregular
autospasy between	coxa-tibia	patella-tibia	coxa-tibia	coxa-tibia

Araneidae and Tetragnathidae VII (p. 193)
Theridiidae VIII (p. 199)

VII Araneidae and Tetragnathidae

Araneidae and Tetragnathidae are two closely related families of orb-weaving spiders which are difficult to separate. Many characters which hold for palearctic genera do not hold for tropical genera. So I am unable to give criteria which allow a clear identification on family level. In many cases Tetragnathidae show the primitive character and Araneidae show the derived character. Araneidae characters: reduced tapetum in the posterior median eyes, spin prey when wrapping, the rotation of the bulb within the cymbium of the male palpus (the sclerites face submedially, the cymbium dorsally; in the left palpus the cymbium is on the left of the sclerites, *Mecynogea* and *Cyrtophora* are intermediate), shortening of the opisthosoma (except *Larinia* and most *Eustala*). Tetragnathidae characters: presence of trichobothria on the femora of the legs, the boss on the chelicerae is absent or rudimentary, large and powerful chelicerae in most species, relatively "simple" genitalia.

1. Araneidae (in some genera sexual dimorphism is pronounced and males cannot always be identified) 2
- Tetragnathidae 35

2. Posterior eye row procurved (lateral eyes anterior to medians) . 3
- Posterior eye row straight or recurved 4

3. Anterior median eyes closer to each other than to laterals; female total length 12 mm (Figure 101) .... *Argiope*
- Anterior median eyes equally spaced, or anterior median eyes closer to laterals than to each other; female total length 5 mm (Figure 102) .... *Gea*

4. Sclerotized ring around spinnerets 5
- No such sclerotized ring 6

5. A hump between genital groove and spinnerets (Figure 103) .... *Gasteracantha*
- Venter of opisthosoma flat, prosoma without modifications, fourth femora longer than first, opisthosoma with large spines, males with opisthosoma rectangular or violin-shaped (Figure 104)... .... *Micrathena*

6. Third tibia with brush of long trichobothria (Figure 105) .... *Mangora*
- Third tibia without a brush of long trichobothria 7

7. Head region of the prosoma with a median swelling (Figure 106) .... *Scoloderus*
- Otherwise 8

8. Head, eye region narrow, less than half the width of prosoma, opisthosoma oval, longer than wide, legs heavy, male palpus with club-shaped median apophysis (Figure 107) (*partim*)..... .... *Spilasma*
- Head wider, more than half width of prosoma 9

9. Opisthosoma spherical to longer than wide, oval, widest anteriorly sometimes with anterior humps on opisthosoma 24
- Female opisthosoma wider than long, cylindrical, or modified with constriction, humps, spines, or shields 10

10. Opisthosoma wider than long 11
- Opisthosoma about as wide as long or longer than wide 12

11. Abdomen laterally drawn out, epigynum with 3 lobes, male with soft tubercles on third coxa (Figure 108) *Epeiroides bahiensis*
- Opisthosoma without lateral humps or tublercles, male without tubercles on coxae, prosoma not hairy, embolus partly hidden by cymbium (Figure 109) (*partim*) .... *Bertrana*

12. Opisthosoma with turtle-like dorsal scutes (Figure 110) .... *Hypognatha*
- Opisthosoma with humps or otherwise modified 13

13. Opisthosoma cylindrical, or posterior widest, or shield-shaped. 14
- Opisthosoma with humps, spines or outgrowths 17

14. Opisthosoma cylindrical, marked with anterior dorsal black anchor; posterior dark patch and undulating lateral white line, epigynum a hood (Figure 111) .... *Mecynogea*
- Opisthosoma otherwise 15

15. Female opisthosoma longer than wide, often widest behind middle, with large dark median or paired white dorsal marks; epigynum with lateral lobes, total length more than 10 mm; male palpus with small paracymbium (Figure 112) .... *Cyrtophora*
- Opisthosoma otherwise 16

16. Posterior median eyes touching, to their radius apart, opisthosoma longer than wide, overhanging spinnerets or with humps, venter black with a pair of white spots, scape not extending beyond genital groove; male palpus wider than long (Figure 113) (*partim*) .... *Cyclosa*
- Opisthosoma longer than wide to shield-shaped, epigynum with transverse lobe or humps, opisthosoma often brightly colored (Figure 114) (*partim*) .... *Alpaida*

17. Opisthosoma subtriangular, with lateral and posterior tubercles; second tibia of male with a spur (Figure 115) .... *Verrucosa*
- Opisthosoma otherwise 18

18. Diameter of posterior median eyes twice that of others, opisthosoma soft with pairs of humps (Figure 116) ....*Pronous*
- Posterior median eyes and opisthosoma otherwise 19

19. Opisthosoma longer than wide with 2 or 3 anterior spines (Figure 114) (*partim*) .... *Alpaida*
- Opisthosoma otherwise 20

20. Female opisthosoma with pair of anterior humps or spines; posterior tail constricted at its base; total length 6 - 12 mm, males with shiny, oval, convex sclerite covering opisthosoma, less than 2 mm total length (Figure 117) .... *Witica*
- Opisthosoma otherwise 21

21. Opisthosoma with 10 - 11 tubercles, epigynum a lobe with a minute scape (Figure 118) .... *Wagneriana*
- Opisthosoma and epigynum otherwise 22

22. Opisthosoma with paired and posterior median median tubercles, sides of thorax same color as head, darkest between median and lateral eyes; epigynum having long scape originating from full width of base (base not set off) (Figure 119) (*partim*) *Parawixia*
- Epigynum slightly sclerotized; if with a scape, scape not extending beyond genital groove 23

23. Posterior median eyes touching, to about their diameter apart, venter with a pair of white spots side by side on black (Figure 113) (*partim*) .... *Cyclosa*
- Posterior median eyes their diameter or more apart, epigynum with lightly sclerotized ridge or scape; male median apophysis of palpus with 2 flagella and row of teeth (Figure 120) ...... *Kaira*

24. Opisthosoma longer than wide with shoulder humps, or shield-shaped, brownish with white (Figure 121) .... *Wixia*
- Opisthosoma otherwise 25

25. Opisthosoma oval to spherical, whitish, with short black lines highlighted by parallel light lines; embolus of male palpus partly hidden by cymbium (Figure 109) (*partim*) .... *Bertrana*
- Opisthosoma marked otherwise 26

26. Epigynum with wide lobe or median keel 27
- Epigynum with scape 28

27. Opisthosoma oval, dark with paired patches, epigynum a wide lobe; palpal median apophysis is a club-shaped projection (Figure 107) (*partim*) .... *Spilasma*
- Opisthosoma marked otherwise, epigynum with median keel (Figure 122) .... *Metazygia*

28. Scape of epigynum smooth 29
- Scape of epigynum wrinkled 30

29. Opisthosoma with a black V-mark, highlighted by a parallel white line, and enclosing a smaller V-mark (Figure 123) .... *Acacesia*
- Opisthosoma marked otherwise, spherical to oval, narrower behind; scape wide and flat; male palpal sclerites squeezed into narrow area between tegulum and transparent cymbium (Figure 124) .... *Neoscona*

30. Epigynal scape projecting anteriorly; male palpus with median apophysis a white lobe on edge of palpus pointing proximally (Figure 125) .... *Eustala*
- Scape extending posteriorly or ventrally; male palpus othewise 31

31. Opisthosoma ventrally with a median white longitudinal mark on black; scape small, indistinct, male with 2 flagella on median apophysis of pedipalpus (Figure 126) .... *Metepeira*
- Otherwise 32

32. Epigynum without separate base 33
- Epigynum with scape originating from distinct base 34

33. Scape folded on itself at origin; median apophysis long, projecting (Figure 127) .... *Eriophora*
- Scape without such fold; median apophysis large (Figure 119) (*partim*) .... *Parawixia*

34. Epigynal scape pointed at tip, total length more than 4 mm, eye region not black (Figure 128) .... *Aculepeira*
- Epigynal scape rounded at tip with a small fold or pocket 35

35. Opisthosoma elongate oval with venter black enclosing longitudinal white streak divided by a black line; scape of epigynum very short with smooth tip larger than wrinkled portion; median apophysis of male palpus with 2 prongs having a distal club (Figure 129) .... *Larinia*
- Opisthosoma and genitalia otherwise (Figure 130) .... *Araneus*

36. Anterior face of fourth femur with long, curved, feathered trichobothria 37
- Anterior face of fourth femur without such trichobothria 39

37. Epigynum with a ventrally projecting protuberance; cymbium of palpus with dorsal hook (Figure 131) .... *Leucauge argyra*
- Epigynum flat; cymbium of palpus not so modified 38

38. Posterior coxae widely separated; opisthosoma overhanging spinnerets posteriorly; mostly black with some silver patches; males minute (Figure 132) .... *Mecynometa*
- Posterior coxae at most one-third of their diameter apart; abdomen only slightly overhanging spinnerets; mostly silver coloration with some black or red on venter of abdomen (Figure 133) .... *Leucauge*

39. Projecting or enlarged chelicerae; females without epigynum (copulating pores) (except *Dolichognatha*) 40
- Chelicerae normal; adult females always with epigynum 43

40. Chelicerae extended ventrally; posterior median eyes touching; opisthosoma with two dorsal pairs of tubercles (Figure 134) (*partim*) .... *Dolichognatha*
- Chelicerae projecting or extended laterally; posterior median eyes separated 41

41. Spiracle of Opisthosoma advanced, one-third to half distance between spinnerets and genital groove; opisthosoma oval to spherical; total length less than 4 mm (Figure 135) .... *Glenognatha*
- Spiracle close to spinnerets; opisthosoma oval to cylindrical; total length usually more than 5 mm 42

42. Chelicerae extended laterally (Figure 136) .... *Cyrtognatha*
- Chelicerae projecting; opisthosoma oval to cylindrical (Figure 137) .... *Tetragnatha*

43. Posterior median eyes touching; opisthosoma with two pairs of dorsal tubercles (Figure 134) (*partim*) .... *Dolichognatha*
- Posterior median eyes separated by their diameter 44

44. Total length of body of adult females more than 18 mm; males less than 11 mm, with projecting embolus wrapped in conductor (Figure 138) .... *Nephila*

- Total length of body of adult females less than 17 mm; male palpus otherwise 45

45. Lateral eyes separated; opisthosoma higher than long (Figure 139) .... *Azilia*
- Lateral eyes touching; opisthosoma otherwise 46

46. Epigynum a transverse soft swelling with club-shaped, broken, male emboli on lateral ends; male palpus with cymbium lobed and tibia with a projection; opisthosoma subspherical (Figure 140) .... *Homalometa*
- Epigynum, palpus and opisthosoma otherwise 47

47. Opisthosoma cylindrical and silver dorsally, black ventrally; epigynum with a longer than wide depression; palpus *Leucauge*-like; sozial, above streams (Figure 141) ...... *Metabus gravidus*
- Opisthosoma short oval, often with silver spots, epigynum and palpus otherwise, not social (Figure 142) .... *Chrysometa*

VIII Theridiidae

Specimens of both sexes are necessary to key a theridiid genus because in some genera no characters are known to separate the females (e.g. *Theridion*) or the males (e.g. *Dipoena*).

1. Colulus (arrow) always present, at least half as long as its setae (Figure 143c, d) 2
- Colulus absent or if present, then shorter than 1/4 length of its setae (Figure 143a, b) 6

2. Chelicerae without teeth; lateral eyes always separated; embolus and connecting ducts of female coiled; seminal receptacles dumbbell-shaped (Figure 144) .... *Latrodectus*
- Chelicerae with one or more teeth or denticles; lateral eyes sometimes separated; embolus, connecting ducts rarely coiled; seminal receptacles not dumbbell-shaped 3

3. With denticles or teeth on posterior margin of chelicerae, opisthosoma of both sexes usually modified with humps, or narrowly elongate to worm-shaped, or higher than long; eye region of male usually with projections, male clypeus usually with projections, grooves or bulging (Figure 145) (*partim*) .... *Argyrodes*
- Posterior margin of chelicerae lacking teeth or denticles; opisthosoma otherwise 4

4. Opisthosoma suboval in outline, longer than wide or high (Figure 146) .... *Steatoda*
- Opisthosoma higher than long, or with tubercles, or narrowly elongate to worm-shaped extending beyond spinnerets 5

5. Male palpal patella without a spur; male clypeus and eye region usually modified with projections or grooves (Figure 145) (*partim*) .... *Argyrodes*
- Male palpal patella with a spur; male clypeus and eye region not so modified (Figure 147) .... *Synotaxus*

6. Colulus absent (Figure 143 a) 7
- Colulus present or colulus replaced by two setae (Figure 143 b) 18

7. Eye region of prosoma projecting above clypeus; opisthosoma heavily sclerotized, usually with folds or spines; sclerotized ring around spinnerets hiding the two colulus setae (Figure 148) (*partim*) .... *Phoroncidia*
- Eye region of prosoma not projecting, or if projecting, opisthosoma not heavily sclerotized with folds or spines 8

8. Prosoma modified, flat oval or with posterior stalk; opisthosoma extended beyond spinnerets and a sclerotized ring around pedicel; male embolus and female connecting ducts very long and threadlike (Figure 149) .... *Helvibis*
- Prosoma and opisthosoma otherwise 9

9. Six eyes; or, with eight eyes but anterior medians with maximum diameter equal to radius of posterior medians, and palpal bulb twisted with embolus facing outside (ectal) (Figure 150) (*partim*) .... *Styposis*
- Eight eyes; if anterior medians small, palpal bulb not facing ectal side 10

10. Females with four seminal receptacles, opisthosoma subtriangular; palpus with duct looping through length of median apophysis, which is broadly attached to tegulum, radix absent (Figure 151) (*partim*) .... *Euryopis*
- Females with two seminal receptacles, opisthosoma sometimes subtriangular, palpus with duct not looping through length of median apophysis when broadly attached to tegulum; radix absent or present 11

11. Opisthosoma not sclerotized in either sex, with a single posterior, dorsal tip overhanging spinnerets, usually longer than high, often with grooves on sides, males always with radix and separate median apophysis (Figure 152) .... *Chrysso*

\- Opisthosoma otherwise; or, if with posterior dorsal tip overhanging spinnerets, male with anterior sclerotized ring on opisthosoma (*Coleosoma*); or lacking radix and having median apophysis broadly attached to tegulum or embolus (*Achaearanea*) 12

12. One tooth on the posterior margin of chelicerae, palpus very simple without conductor, median apophysis, radix, opisthosoma suboval (Figure 153) .... *Paratheridula*

\- Posterior margin of chelicerae without teeth, or if with a tooth, opisthosoma with a posterior dorsal extension (*Coleosoma*) 13

13. Opisthosoma of male with a sclerotized ring around pedicel covering anterior half of venter, and opisthosoma sometimes constricted in middle, male embolus with a broad basal portion and a long tube (Figure 154) .... *Coleosoma*

\- Opisthosoma of male without sclerotized ring, or if with sclerotized ring venter not covered, and opisthosoma not constricted . 14

14. Male with only one very large palpus; female opisthosoma higher than long, with a white line from dorsal tip to anal tubercle; epigynum with a pair of openings, one on each side of a cone-shaped knob; male less than 1/4 length of female ....*Tidarren*

\- Male always with two palpi; female opisthosoma, if higher than long and with white line, then epigynum not a cone-shaped knob; male 1/2 length or more than female 15

15. Palpus without conductor; no median apophysis; tegulum attached to cymbium on both ends; female opisthosoma wider than long (Figure 155) .... *Theridula*

\- Palpus with conductor; median apophysis a separate sclerite or a lobe of the tegulum; tegulum always attached to cymbium at proximal end only; female opisthosoma sometimes wider than long 16

16. Median apophysis broadly attached to tegulum or to embolus, never a separate sclerite; palpus without radix; opisthosoma usually higher than long, often with dark patches on sides, sometimes with a posterior dorsal tubercle (Figure 156) .... *Achaearanea*

\- Median apophysis always a separate sclerite; opisthosoma usually longer than high or suboval, never with a tubercle, palpus almost always with radix 17

17. Less than 2.5 mm in total length, uniformly colored (except for discrete black spots or gray marks), first patella and tibia usually less than 1.5 x length of prosoma. Eye region of males often with grotesque projection, elevated, sometimes clypeus with a transverse groove, opisthosoma sometimes heavily sclerotized (Figure 157) .... *Thymoites*

\- Usually larger spiders, usually with pattern or band on opisthosoma containing white and black pigment, first patella and tibia usually longer; male prosoma never grotesquely modified, opisthosoma never sclerotized (except for narrow sclerotized ring around pedicel in males of some species) (Figure 158) .... *Theridion*

18. Eye region of prosoma projecting above clypeus; opisthosoma heavily sclerotized, often folded or with spines; female with two seminal receptacles; paracymbium usually a hook on margin of cymbium of male palpus (Figure 148) (*partim*) ..... *Phoroncidia*

\- Eye region not projecting (or if projecting, then four seminal receptacles and paracymbial hook on back of cymbium); opisthosoma not heavily sclerotized (or if sclerotized, oval in outline) 19

19. Opisthosoma of both sexes extended, longer behind spinnerets than in front or worm-shaped (Figure 145) (*partim*) *Argyrodes*

\- Opisthosoma barely overhanging spinnerets behind 20

20. Two pairs of round or oval seminal receptacles in female; male prosoma often with dorsal grooves; male palpus with duct usually looping through median apophysis; embolus, if long, pointing counterclockwise in left palpus 21

\- One pair of round or oval seminal receptacles (sometimes thin-walled asymmetrical pockets present); male prosoma never with dorsal grooves; male with duct rarely looping through length of median apophysis (*Spintharus*); embolus generally pointing clockwise in left palpus 22

21. Palpus without radix; median apophysis broadly attached to tegulum; opisthosoma subtriangular; male prosoma rarely different in shape from that of female (Figure 151) (*partim*) .... *Euryopis*

\- Palpus with radix; or if not, median apophysis a separate sclerite; opisthosoma usually subspherical, wider than long or with humps, rarely subtriangular; prosoma of male often high, cylindrical, with grooves on dorsum (Figure 159) ...... *Dipoena*

22. Six eyes; or if eight eyes, then maximum diameter of anterior medians equal to radius of posterior medians and palpal bulb twisted with embolus facing lateral (Figure 150) (*partim*) .... *Styposis*

- Eight eyes; maximum diameter of anterior medians more than radius of posteriors; or if maximum diameter of anterior medians equal to radius of posteriors, palpal embolus facing venter 23

23. Posterior median eyes farther from each other than laterals or equally spaced, and not more than 1/3 their diameter from laterals (but usually touching) and prosoma black in eye region (Figure 160) .... *Stemmops*

- Posterior eyes otherwise, or if similar, eye region of prosoma not black 24

24. Chelicerae with denticles or teeth on posterior margin; opisthosoma usually ovoid, not sclerotized, often with a longitudinal, dorsal dark band; paracymbium hook usually in alveolus of cymbium (Figure 161) .... *Anelosimus*

- Chelicerae without denticles or teeth on posterior margin; opisthosoma usually with humps, subtriangular or wider than long 25

25. Embolus long, with tip supported by cymbium; conductor absent or minute; epigynum an indistinct depression with connecting ducts coiled on each side and seminal receptacles anterior; opisthosoma black or dark on venter above pedicel (Figure 162) .... *Chrosiothes*

- Embolus never supported by cymbium; conductor very large; epigynum otherwise; opisthosoma colored otherwise 26

26. Opisthosoma widest at middle or posterior to middle, never higher than wide; without silvery plates; lateral eyes often separated; eyes often on tubercles; usually tubercles or horns between anterior and posterior median eyes (Figure 163) ...... *Episinus*

- Opisthosoma widest anterior to middle, or higher than wide, sometimes with silvery plates; lateral eyes touching; eyes never on tubercles; no tubercles between eyes 27

27. Posterior median eyes about their diameter or less apart; opisthosoma covered by silver plates (Figure 164) .... *Thwaitesia*
- Posterior median eyes two to three diameter apart; opisthosoma not covered by silvery plates (Figure 165) .... *Spintharus*

Chapter 11. References

Archer, A. F. 1953. Studies in the orbweaving spiders (Argiopidae). Am. Mus. Nov. 1622: 13 - 19.

Ausserer, A. 1875. Zweiter Beitrag zur Kenntnis der Arachniden-Familie der Territelariae Thorell (Mygalidae Autor.). Verh. Zool. Bot. Ges. Wien. 25: 125 - 206.

Banks N. 1929. Spiders from Panama. Bull. Mus. Comp. Zool. 69: 53 - 95.

Beatty, J. A. 1970. The spider genus *Ariadna* in the Americas (Araneae, Dysderidae). Bull. Mus. Comp. Zool. 139: 433 - 518.

Bennett, C. F. 1968. Human influences on the zoogeography of Panama. Ibero-Americana 51: 1 - 112.

Berman, J. D. and H. W. Levi. 1971. The orb weaving genus *Neoscona* in North America (Araneae: Araneidae). Bull. Mus. Comp. Zool. 141: 465 - 500.

Brady, A. R. 1964. The lynx spiders of North America, North of Mexico (Araneae: Oxyopidae). Bull. Mus. Comp. Zool. 132: 429 - 518.

Brady, A. R. 1970. The lynx spider genus *Hamataliwa* in Mexico and Central America (Araneae: Oxyopidae). Bull. Mus. Comp. Zool. 140: 75 - 128.

Brady, A. R. 1975. The lynx spider genus *Oxyopes* in Mexico and Central America (Araneae: Oxyopidae). Psyche 82: 189 - 243.

Brignoli, P. M. 1983. A catalogue of the Araneae described between 1940 and 1981. Manchester Univ. Press, Manchester.

Brillouin, L. 1956. Science and information theory. Academic Press, New York.

Bryant, E. B. 1940. Cuban spiders in the Museum of Comparative Zoology. Bull. Mus. Comp. Zool. 86: 247 - 554 .

Carico, J. E. 1976. The spider genus *Tinus* (Pisauridae). Psyche 83: 63-78.

Carico, J. E. 1981. The neotropical spider genera *Architis* and *Staberius* (Pisauridae). Bull. Am. Mus. Nat. Hist. 170: 140 - 153.

Censos nationales de 1960. 1961. Vol. 1: Lugares poblados de la Republica. Direccion de estadistica y censo, Panama.

Chamberlin, R. V. 1925. Diagnoses of new American Arachnida. Bull. Mus. Comp. Zool. 67: 211 - 234.

Chamberlin, R. V, Ivie W. 1936. New spiders from Mexico and Panama. Bull. Univ. Utah 27: 1 - 103.

Chamberlin, R. V, Ivie W. 1942. Agelenidae of the genera *Hololena*, *Novolena*, *Rualena*, and *Melpomene*. Ann. Ent. Soc. Am. 35: 203 - 241.

Chickering, A. M. 1936. Additions to the list of known species of spiders from Barro Colorado Island, Panama. Trans. Am. Microscop. Soc. 55: 449 - 456.

Chickering, A. M. 1937 a. Anyphaenidae of Barro Colorado Island, Panama Canal Zone. Pap. Michigan Acad. Sci. Arts Lett 22: 541 - 561.

Chickering, A. M. 1937 b. The Clubionidae of Barro Colorado Island, Panama. Trans. Am. Microscop. Soc. 56: 1 - 47.

Chickering, A. M. 1940 a. New Anyphaenidae from Panama with notes on known species. Trans. Am. Miscroscop Soc. 59: 78 - 122.

Chickering, A. M. 1940 b. The Senoculidae of Panama. Pap. Michigan Acad. Sci. Arts Lett. 26: 195 - 218.

Chickering, A. M. 1941. The Palpimanidae of Panama. Pap. Michigan Acad. Sci. Arts Lett 27: 235 - 239.

Chickering, A. M. 1942. Female allotype of *Landana spinosa* Petrunkevitch. Pap. Michigan Acad. Sci. Arts Lett. 28: 317 - 321.

Chickering, A. M. 1943. Twenty-one new species of *Dipoena* (Theridiidae) from Panama. Trans. Am. Microscop. Soc. 62: 329 - 378.

Chickering, A. M. 1945 a. Hypotypes of *Accola spinosa* Petrunkevitch (Dipluridae) from Panama. Trans. Connecticut Acad. Arts Sci. 36: 159 - 167.

Chickering, A. M. 1945 b. The male allotype and female hypotype of *Storena barroana* (Chamberlin). Pap. Michigan Acad. Sci. Arts Lett. 31: 133 - 140.

Chickering, A. M. 1946. The Salticidae (spiders) of Panama. Bull. Mus. Comp. Zool. 97: 1 - 474.

Chickering, A. M. 1947. The Mimetidae (Araneae) of Panama. Trans. Am. Microscop. Soc. 66: 221 - 248.

Chickering, A. M. 1948. Four new species of *Dipoena* (Araneae; Theridiidae) from Panama. Trans. Am. Microscop. Soc. 67: 331 - 340.

Chickering, A. M. 1949. The Gnaphosidae (Araneae) of Panama. Trans. Am. Microscop. Soc. 68: 314 - 331.

Chickering, A. M. 1950 a. Three new dictynids (Araneae) from Pana ma. Pap Michigan Acad. Sci. Arts Lett. 34: 85 - 96.

Chickering, A. M. 1950 b. The spider genus *Tmarus* (Thomisidae) in Panama. Bull. Mus. Comp. Zool. 103: 213 - 255.

Chickering, A. M. 1951 a. Operation Arachne. Turdox News 29: 94 - 97.

Chickering, A. M. 1951 b. Two new species of Dictynidae (Araneae) from Panama. Psyche 58: 149 - 157.

Chickering, A. M. 1951 c. The Oonopidae of Panama. Bull. Mus. Comp. Zool. 106: 207 - 245.

Chickering, A. M. 1953 a. A new species of *Hypognatha* from Panama. Breviora 23: 1 - 8.

Chickering, A. M. 1953 b. Two new species of *Senoculus* (Araneae, Senoculidae) from Panama. Trans. Am. Microscop. Soc. 72: 281 - 287.

Chickering, A. M. 1954. The spider genus *Mangora* (Argiopidae) in Panama. Bull. Mus. Comp. Zool. 111: 195 - 215.

Chickering, A. M. 1955 a. The genus *Eustala* (Araneae, Argiopidae) in Central America. Bull. Mus. Comp. Zool. 112: 391 - 518.

Chickering, A. M. 1955 b. Spider hunting in Panama. Bios 26: 69 - 74.

Chickering, A. M. 1956 a. Three new species of Mimetidae (Araneae) from Panama. Breviora 57: 1 - 14.

Chickering, A. M. 1956 b. A new species of *Ariognatha* (Araneae, Argiopidae) from Panama. Breviora 65: 1 - 7.

Chickering, A. M. 1957 a. Notes on certain species of *Tetragnatha* (Araneae, Argiopidae) in Central America and Mexico. Breviora 67: 1 - 4.

Chickering, A. M. 1957 b. A new zodariid spider from Panama. Bre viora 69: 1 - 7.

Chickering, A. M. 1957 c. The genus *Tetragnatha* (Araneae, Argiopi dae) in Panama. Bull. Mus. Comp. Zool. 116: 301 - 354.

Chickering, A. M. 1960. The female of *Lucarachne beebei* Gertsch (Araneae: Symphytognathidae). Psyche 67: 95 - 97.

Chickering, A. M. 1961. The genus *Micrathena* (Araneae, Argiopidae) in Central America. Bull. Mus. Comp. Zool. 125: 391 - 470.

Chickering, A. M. 1963 a. The female of *Bertrana hieroglyphica* Petrunkevitch (Araneae, Argiopidae). Psyche 70: 129 - 132.

Chickering, A. M. 1963 b. The male of *Mecynometa globosa* (O. P.-Cambridge) (Araneae, Argiopidae). Psyche 70: 180 - 183.

Chickering, A. M. 1963 c. A description of *Dinopis longipes* F. P.-Cambridge, 1902 (Araneae, Dinopidae). Breviora 192: 1 - 6.

Chickering, A. M. 1963 d. Three new species of *Mangora* (Araneae, Argiopidae) from Central America. Breviora 191: 1 - 11.

Chickering, A. M. 1966. Three new species of *Accola* (Araneae, Diplur- idae) from Costa Rica and Trinidad, W.I. Psyche 73: 157 - 164.

Chickering, A. M. 1967. The genus *Nops* (Araneae, Caponiidae) in Panama and the West Indies. Breviora 274: 1 - 19.

Chickering, A. M. 1968 a. The genus *Dysderina* (Araneae, Oonopidae) in Central America and the West Indies. Breviora 296: 1 - 37.

Chickering, A. M. 1968 b. The genus *Miagrammopes* (Araneae, Ulobori dae) in Panama and the West Indies. Breviora 289: 1 - 28.

Chickering, A. M. 1968 c. The genus *Scaphiella* (Araneae, Oonopidae) in Central America and the West Indies. Psyche 75: 135 - 156.

Chickering, A. M. 1968 d. The genus *Ischnothyreus* (Araneae, Oonopi dae) in Central America and the West Indies. Psyche 75: 77 - 86.

Chickering, A. M. 1968 e. The genus *Triaeris* Simon (Araneae, Oono pidae) in Central America and the West Indies. Psyche 75: 351 - 359.

Chickering, A. M. 1969 a. Description of *Pocobletus coroniger* Simon (Araneae, Linyphiidae). Psyche 76: 54 - 57.

Chickering, A. M. 1969 b. The genus *Stenoonops* (Araneae, Oonopidae) in Panama and the West Indies. Breviora 339: 1 - 35.

Chickering, A. M. 1970. New species of *Grammonota* (Araneae, Li nyphiidae) from Panama and Costa Rica. Psyche 77: 347 - 360.

Chickering, A. M. 1970 b. New species of *Erigone* (Araneae, Li nyphiidae) from Panama and the West Indies. Psyche 77: 131 - 145.

Chickering, A. M. 1970 c. The genus *Oonops* (Araneae, Oonopidae) in Panama and the West Indies. Part 1. Psyche 77: 487 - 512.

Chickering, A. M. 1972. The genus *Corinna* (Araneae, Clubionidae) in Panama. Psyche 79: 365 - 378.

Chickering, A. M. 1973. Notes on *Heteroonops* and *Triaeris* (Araneae, Oonopidae). Psyche 80: 227 - 229.

Christenson, T. E. 1984. Behaviour of colonial and solitary spiders of the theridiid species *Anelosimus eximius*. Anim. Behav. 32: 725 - 734.

Coddington, J. A. 1986. The genera of the spider family Theridiosomatidae. Smiths. Contr. Zool. 422: 1 - 96.

Croat, T. B. 1978. Flora of Barro Colorado Island. Stanford University Press, Stanford, California.

Cutler, B. 1971. The spider genus *Semiopyla* (Araneae: Salticidae). Entomol. News 82: 141-146.

Cutler, B. 1982. Description of a new species of *Paradamoetas* (Araneae: Salticidae), with a revised key to the genus. Great Lakes Entomol. 15: 219 - 222.

Cutler, B. 1988. Middle American *Synemosyna* (Araneae: Salticidae), a key and description of a new species. Stud. Neotrop. Fauna Environ. 23: 197 - 202.

Decae, A. E. 1987. Dispersal: Ballooning and other mechanisms. pp 348 - 356. In Nentwig W (ed.) Ecophysiology of spiders. Springer, Berlin.

Dietrich, W. E., Windsor, D. M., and T. Dunne. 1982. Geology, Climate, and Hydrology of Barro Colorado Island. pp 21 - 46 In: Leigh, E. G., Rand, A. S., Windsor, D. M. (eds.) The ecology of a tropical forest. Smithsonian, Washington DC.

Dondale, C. D., and J. H. Redner. 1969. The *infuscatus* and *dispar* groups of the spider genus *Philodromus* in North and Central America and the West Indies (Araneida: Thomisidae). Can. Entomol. 101: 921 - 954.

Dondale, C. D., and and J. H. Redner. 1975. Revision of the spider genus *Apollophanes* (Araneida: Thomisidae). Can. Entomol. 107: 1175 - 1192.

Dondale, C. D., and J. H. Redner. 1983 a. The wolf spider genus *Allocosa* in North and Central America (Araneae: Lycosidae). Can. Entomol. 115: 933 - 964.

Dondale, C. D., and J. H. Redner. 1983 b. Revision of the wolf spiders of the genus *Arctosa* C. L. Koch in North and Central America (Araneae: Lycosidae). J. Arachnol. 11: 1 - 30.

Dondale, C. D., and J. H. Redner. 1984. Revision of the *melvina*-group of the wolf spider genus *Pardosa*. Psyche 91: 67 - 117.

Eberhard, W. G. 1986. Subsocial behavior of the spitting spider *Scytodes intricata* (Araneae, Scytodidae). Rev. Arachnol. 7: 35 - 40.

Estadistica Panamena. 1980. Situacion fisica, seccion 121, meteorologia: anos 1978 y 1979. Direccion de estadistica y censo, Panama.

Exline, H., and H. W. Levi.S 1962. American spiders of the genus *Argyrodes* (Araneae: Theridiidae). Bull. Mus. Comp. Zool. 127: 75 - 204.

Exline, H., and H. W. Levi. 1965. The spider genus *Synotaxus* (Araneae, Theridiidae). Trans. Am. Microscop. Soc. 84: 177 - 184.

Forster, R. R., and N. I. Platnick. 1977. A revision of the spider family Symphytognathidae (Arachnida, Araneae). Am. Mus. Nov. 2619: 1 - 29.

Forster, R. R., Platnick, N. I., and J. A. Coddington. 1990. A proposal and review of the spider family Synotaxidae (Araneae, Araneoidea), with notes on theridiid interrelationships. Bull. Am. Mus. Nat. Hist. 193: 1 - 116.

Galiano, M. E. 1961. Revision del genero *Chira* Peckham, 1896 (Araneae, Salticidae). Rev. Mus. Arg. Cien. Nat. Zool. 3: 159 - 188.

Galiano, M. E. 1964. Salticidae (Araneae) formiciformes. I. *Martella* Peckham, 1892. Physis, Buenos Aires 24 (68): 353-363.

Galiano, M. E. 1965. Salticidae (Araneae) Formiciformes IV. Revi sion del genero *Sarinda* Peckham, 1892. Rev. Mus. Arg. Cien. Nat. Entomol. 1: 267 - 311.

Galiano, M. E. 1966. Salticidae (Araneae) formiciformes. V. *Synemosyna* Hentz, 1846. Rev. Mus. Arg. Cien. Nat. Entomol. 1: 339-380.

Galiano, M. E. 1968. Revision de los generos *Acragas*, *Amycus*, *Encolpius*, *Hypaeus*, *Mago* y *Noegus* (Salticidae, Araneae). Rev. Mus. Arg. Cien. Nat. Entomol. 2: 267 - 360.

Galiano, M. E. 1969. Salticidae (Araneae) Formiciformes VII. El genero "*Myrmarachne*" MacLeay, 1839, en America. Rev. Mus. Arg. Cien. Natur. Entomol. 3: 107 - 148.

Galiano, M. E. 1979 a. Revision of the genus *Phiale* C.L.Koch, 1846 (Araneae, Salticidae). II. *Phiale guttata* (C.L.Koch), new combination. Bull. Brit. Arachnol. Soc. 4: 345 - 348.

Galiano, M. E. 1979 b. Revision del genero *Frigga* C.L.Koch, 1891. (Araneae, Salticidae). Acta Zool. Lilloana 33: 113 - 135.

Galiano, M. E. 1980. Revision del genero *Lyssomanes* Hentz, 1845 (Araneae, Salticidae). Opera Lilloana (Tucuman, Argentina) 5 - 104.

Galiano, M. E. 1981a. Revision of the genus *Phiale* C.L.Koch 1846 (Araneae, Salticidae). IV. The polymorphic species of the *gratiosa* group. Bull. Br. Arachnol. Soc. 5: 205-216.

Galiano, M. E. 1981b. Revision del genero *Phiale* C.L.Koch, 1846 (Araneae, Salticidae). III Las especies polimorficas del grupo *mimica*. J. Arachnol. 9: 61 - 85.

Galiano, M. E. 1982. Revision del genero *Nycerella* (Araneae, Salti cidae). Physis (Buenos Aires), Sec C: 53 - 63.

Galiano, M. E. 1985. Two new species of *Semiopyla* with notes on *S. cataphracta* (Araneae, Salticidae). Rev. Suisse Zool. 92: 281 - 290.

Galiano, M. E. 1987. Revision de los generos del grupo Hurieae (Araneae, Salticidae). J. Arachnol. 15: 285 - 301.

Galiano, M. E. 1991. Revision des genero *Jollas* (Araneae, Salticidae). Physis, Secc. C 47: 15-29.

George, W. 1962. Animal geography. Heinemann, London.

Gertsch, W. J. 1939. A new genus in the Pholcidae. Am. Mus. Nov. 1033: 1 - 4.

Gertsch, W. J. 1941. Report on some arachnids from Barro Colorado Island, Canal Zone. Am. Mus. Nov. 1146: 1 - 14.

Gertsch, W. J. 1946. Notes on American spiders of the family Dictynidae. Am. Mus. Nov. 1319: 1 - 21.

Gertsch, W. J. 1958. The spider genus *Loxosceles* in North America, Central America, and the West Indies. Am. Mus. Nov. 1907: 31 - 39.

Gertsch, W. J. 1973. A report on cave spiders from Mexico and Central America. Assoc. Mex. Cave Stud. Bull. 5: 141 - 163.

Gertsch, W. J. 1982. The spider genera *Pholcophora* and *Anopsicus* (Araneae, Pholcidae) in North America, Central America and the West Indies. Texas Meml. Mus. Bull. 28: 95 - 144.

Gertsch, W. J. 1984. The spider family Nesticidae (Araneae) in North America, Central America, and the West Indies. Tex. Meml. Mus. Bull. 31: 1 - 91.

Gertsch, W. J. 1986. The spider genus *Metagonia* (Araneae: Pholci dae) in North America, Central America, and the West Indies. Texas Meml. Mus. Speol. Monogr. 1: 39 - 62.

Gertsch, W. J., and L. I. Davis. 1946. Report on a collection of spiders from Mexico. V. Am. Mus. Nov. 1313: 1 - 11.

Gertsch, W. J., Ennik 1983. The spider genus *Loxosceles* in North America, Central America, and the West Indies (Araneae, Loxoscelidae). Bull. Am. Mus. Nat. Hist. 175: 264 - 360.

Gertsch, W. J., and H. K. Wallace. 1937. New American Lycosidae with notes on other species. Am. Mus. Nov. 919: 1-22.

Graf, B., and W. Nentwig. 1987. Ontogenetische Änderungen der Körperfär bung und des Verhaltens von *Eriophora fuliginea* (Araneae: Araneidae). Verh. Dt. Zool. Ges. 294 - 295.

Greenberg, R., and J. Gradwohl. 1980. Leaf surface specialisation of birds and arthropods in a Panamanian forest. Oecologia 46: 115 - 124.

Griswold, C. E. 1983. *Tapinillus longipes* (Taczanowski), a web- building lynx spider from the American tropics (Araneae: Oxyopidae). J. Nat. Hist. 17: 979 - 985.

Griswold, C. E. 1987. A revision of the jumping spider genus *Habronattus* F. O. P.-Cambridge (Araneae: Salticidae) with phenetic and cladistic analyses. Univ. California Publ. Entomol. 107: 1 - 344.

Gursky, H.-J. 1987. Los complejos basales de Centroamerica meridional - hechos y problemas. Actas Fac. Cienc. Tierra UANL Linares 2: 243 - 249.

Gursky, H.-J. 1989. Presencia y origen de rocas sedimentanias en el basamento de Costa Rica. Rev. Geol. Amer. Centr., San Jose.

Halffter, G. 1987. Biogeography of the montane entomofauna of Mexico and Central America. Ann. Rev. Entomol. 32: 95 - 114.

Harrison, J. O. 1968. Some spiders associated with banana plants in Panama. Ann. Ent. Soc. Am. 61: 878 - 884.

Harrod, J. C., Levi, H. W., and L. B. Leibensperger. 1990. The neotropical orbweavers of the genus *Larinia* (Araneae: Araneidae). Psyche 97: 241-265.

Heimer, St. 1986. Notes on the spider family Mimetidae with descriptions of a new genus from Australia. Entomol. Abh. (Dresden) 49: 113 - 137.

Hoffmaster, D. K. 1982. Predator avoidance behaviors of five species of Panamanian orb-weaving spiders (Araneae; Araneidae, Uloboridae). J. Arachnol. 10: 69 - 73.

Hormiga, G., and H. G. Döbel. 1990. A new *Glenognatha* (Araneae, Tetragnathidae) from New Jersey, with redescriptions of *G. centralis* and *G. minuta*. J. Arachnol. 18: 195-204.

Hünicken, M. A. 1981. A giant fossil spider (*Megarachne servinei*) from Bajo de Veliz, Upper Carboniferous, Argentina. Bol Acad. Nat. Cien, Cordoba, Argentina 53: 317 - 326.

Jantschke, B., and W. Nentwig. 1987. Brutfürsorge bei *Ischnothele guyanensis*, einer tropischen Vogelspinne (Araneae: Dipluridae). Verh. Dt. Ges. Zool. 315.

Jocqué, R. 1988. An updating of the genus *Leprolochus* (Araneae: Zodariidae). Stud. Neotrop. Fauna Environ. 23: 77 - 87.

Jocqué, R. 1991. A generic revision of the spider family Zodariidae (Araneae). Bull. Am. Mus. Nat. Hist. 201: 1-160.

Jocqué, R., and D. Ubick. 1991. A revision of the genus *Ishania* Chamberlin (Araneae, Zodariidae). Bull. Br. Arachnol. Soc. 8:241-245.

Karsch, F. 1880. Zur Arachnidengattung *Theraphosa* Walckenaer. Z. Ges. Nat. Wiss. 53: 843 - 846.

Kaston, B. J. 1980. How to know the spiders. Brown Company Publishers, Dubuque, Iowa.

Keyserling, E. 1881. Neue Spinnen aus Amerika. II. Folge. Verh. Zool. Bot. Ges. Wien 30: 547 - .

Keyserling, E. 1884. Neue Spinnen aus Amerika. VI. Folge. Verh. Zool. Bot. Ges. Wien 34: 489 - .

Lachmuth, U., Grasshoff, M., and F. G. Barth. 1984. Taxonomische Revision der Gattung *Cupiennius* Simon 1891 (Arachnida: Araneae: Ctenidae). Senckenberg Biol. 65: 329 - 372.

Leech, R. 1972. A revision of the nearctic Amaurobiidae (Arachnida: Araneida). Mem. Ent. Soc. Can. 84: 1 - 182.

Leppik, E. E. 1955. *Dichromena ciliata*, a noteworthy entomophilous plant among Cyperaceae. Am. J. Bot. 42: 455 - 458.

Levi, H. W. 1953. New and rare *Dipoena* from Mexico and Central America (Araneae, Theridiidae). Am. Mus. Nov. 1639: 1 - 11.

Levi, H. W. 1954 a. Spiders of the genus *Euryopis* from North and Central America (Araneae, Theridiidae). Am. Mus. Nov. 1666: 16 - 39.

Levi, H. W. 1954 b. Spiders of the new genus *Theridiotis* (Araneae: Theridiidae). Trans. Am. Microscop. Soc. 73: 117 - 189.

Levi, H. W. 1954 c. The spider genus *Theridula* in North and Central America and the West Indies (Araneae: Theridiidae). Trans. Am. Microscop. Soc. 73: 331 - 343.

Levi, H. W. 1955 a. The spider genera *Chrysso* and *Tidarren* in America. (Araneae: Theridiidae). J. N. Y. Entomol. Soc. 63: 59 - 81.

Levi, H. W. 1955 b. The spider genera *Coressa* and *Achaearanea* in America North of Mexico (Araneae, Theridiidae). Am. Mus. Nov 1718: 1 - 33.

Levi, H. W. 1956 a. The spider genera *Neottiura* and *Anelosimus* in America (Araneae: Theridiidae). Trans. Am. Microscop. Soc. 75: 407 - 422.

Levi, H. W. 1956 b. The spider genus *Mysmena* in the Americas (Araneae, Theridiidae). Am. Mus. Nov. 1801: 1 - 13.

Levi, H. W. 1957 a. The North American spider genera *Paratheridula*, *Tekellina*, *Pholcomma* and *Archerius* (Araneae: Theridiidae). Trans. Am. Microscop. Soc. 76: 105 - 115.

Levi, H. W. 1957 b. The spider genera *Crustulina* and *Steatoda* in North America, Central America, and the West Indies (Araneae, Theridiidae). Bull. Mus. Comp. Zool. 117: 367 - 425.

Levi, H. W. 1959 a. The spider genus *Coleosoma* (Araneae, Theridiidae). Breviora 110: 1 - 8.

Levi, H. W. 1959 b. The spider genera *Achaearanea*, *Theridion* and *Sphyrotinus* from Mexico, Central America and the West Indies (Araneae, Theridiidae). Bull. Mus. Comp. Zool. 121: 57 - 163.

Levi, H. W. 1960. The spider genus *Styposis* (Araneae, Theridiidae). Psyche 67: 13 - 19.

Levi, H. W. 1962. More American spiders of the genus *Chrysso* (Araneae, Theridiidae). Psyche 69: 209 - 237.

Levi, H. W. 1963 a. The American spiders of the genus *Anelosimus* (Araneae, Theridiidae). Trans. Am. Miscroscop Soc. 82: 30 - 48.

Levi, H. W. 1963 b. American spiders of the genera *Audifia*, *Euryo pis* and *Dipoena* (Araneae: Theridiidae). Bull. Mus. Comp. Zool. 129: 121 - 185.

Levi, H. W. 1963 c. American spiders of the genus *Theridion* (Ara neae, Theridiidae). Bull. Mus. Comp. Zool. 129: 481 - 589.

Levi, H. W. 1963 d. American spiders of the genus *Achaearanea* and the new genus *Echinotheridion* (Araneae, Theridiidae). Bull. Mus. Comp. Zool. 129: 187 - 240.

Levi, H. W. 1963 e. The American spider genera *Spintharus* and *Thwaitesia* (Araneae: Theridiidae). Psyche 70: 223 - 234.

Levi, H. W. 1964 a. American spiders of the genus *Episinus* (Araneae: Theridiidae). Bull. Mus. Comp. Zool. 131: 1 - 25.

Levi, H. W. 1964 b. The spider genus *Helvibis* (Araneae, Theridiidae) (Araneae, Theridiidae). Trans. Am. Microscop. Soc. 83: 133 - 142.

Levi, H. W. 1964 c. The spider genus *Thymoites* in America (Araneae: Theridiidae). Bull. Mus. Comp. Zool. 130: 445 - 471.

Levi, H. W. 1964 d. The spider genera *Stemmops*, *Chrosiothes*, and the new genus *Cabello* from America. Psyche 71: 73 - 92.

Levi, H. W. 1966. American spider genera *Theridula* and *Paratheridula* (Araneae: Theridiidae). Psyche 73: 123 - 130.

Levi, H. W. 1967. Habitat observations, records, and new South American theridiid spiders (Araneae, Theridiidae). Bull. Mus. Comp. Zool. 136: 21 - 38.

Levi, H. W. 1968. The spider genera *Gea* and *Argiope* in America (Araneae: Araneidae). Bull. Mus. Comp. Zool. 136: 319 - 352.

Levi, H. W. 1970. The ravilla group of the orbweaver genus *Eriophora* in North America (Araneae: Araneidae). Psyche 77: 280 - 302.

Levi, H. W. 1973. Small orb-weavers of the genus *Araneus* North of Mexico (Araneae: Araneidae). Bull. Mus. Comp. Zool. 145: 473 - 552.

Levi, H. W. 1974. The orb-weaver genera *Araniella* and *Nuctenea* (Araneae: Araneidae). Bull. Mus. Comp. Zool. 146: 291 - 316.

Levi, H. W. 1975. The American orb-weaver genera *Larinia*, *Cercidia* and *Mangora* North of Mexico (Araneae, Araneidae). Bull. Mus. Comp. Zool. 147: 101 - 135.

Levi, H. W. 1976. The orb-weaver genera *Verrucosa*, *Acanthepeira*, *Wagneriana*, *Acacesia*, *Wixia*, *Scoloderus* and *Alpaida* North of Mexico (Araneae: Araneidae). Bull. Mus. Comp. Zool. 147: 351 - 391.

Levi, H. W. 1977 a. The orb-weaver genera *Metepeira*, *Kaira* and *Aculepeira* in America North of Mexico (Araneae, Araneidae). Bull. Mus. Comp. Zool. 148: 185 - 238.

Levi, H. W. 1977 b. The American orb-weaver genera *Cyclosa*, *Metazygia* and *Eustala* North of Mexico (Araneae, Araneidae). Bull. Mus. Comp. Zool. 148: 61 - 127.

Levi, H. W. 1978. The American orb-wever genera *Colphepeira*, *Micrathena* and *Gasteracantha* North of Mexico (Araneae, Araneidae). Bull. Mus. Comp. Zool. 148: 417 - 442.

Levi, H. W. 1980. The orb-weaver genus *Mecynogea*, the subfamily Metinae and the genera *Pachygnatha*, *Glenognatha* and *Azilia* of the subfamily Tetragnathinae North of Mexico (Araneae: Araneidae). Bull. Mus. Comp. Zool. 149: 1 - 74.

Levi, H. W. 1981. The American orb-weaver genera *Dolichognatha* and *Tetragnatha* North of Mexico (Araneae: Araneidae, Tetragnathidae). Bull. Mus. Comp. Zool. 149: 271 - 318.

Levi, H. W. 1985. The spiny orb-weaver genera *Micrathena* and *Chaetacis* (Araneae: Araneidae). Bull. Mus. Comp. Zool. 150: 429 - 618.

Levi, H. W. 1986 a. The neotropical orb-weaver genera *Chrysometa* and *Homalometa* (Araneae: Tetragnathidae). Bull. Mus. Comp. Zool. 151: 91 - 215.

Levi, H. W. 1986 b. The orb-weaver genus *Witica* (Araneae: Araneidae). Psyche 93: 35 - 46.

Levi, H. W. 1988. The neotropical orb-weaving spiders of the genus *Alpaida* (Araneae: Araneidae). Bull. Mus. Comp. Zool. 151: 365 - 487.

Levi, H. W. 1989. The neotropical orb-weaver genera *Epeiroides*, *Bertrana* and *Amazonepeira* (Araneae: Araneidae). Psyche 96: 75 - 99.

Levi, H. W. 1991 a. The neotropical orb-weaver genera *Edricus* and *Wagneriana* (Araneae: Araneidae). Bull. Mus. Comp. Zool. 152: 363-415.

Levi, H. W. 1991 b. The neotropical and Mexican species of the orb-weaver genera *Araneus*, *Dubiepeira*, and *Aculepeira* (Araneae: Araneidae). Bull. Mus. Comp. Zool. 152: 167-315.

Levi, H. W. 1991 c. The American species of the orb-weaving genus *Carepalxis* and the new genus *Rubrepeira* (Araneae: Araneidae). Psyche 98: 251-264.

Levi, H.W. 1992. Spiders of the orb-weaver genus *Parawixia* in America (Araneae: Araneidae). Bull. Mus. Comp. Zool. 153: 1-46.

Levi, H. W., and L. R. Levi. 1968. A guide to spiders and their kin. Golden Press, New York.

Levi, H. W., and V. R. D. von Eickstedt. 1989. The Nephilinae spiders of the neotropics (Araneae: Tetragnathidae). Mem. Inst. Butantan 51: 43 - 56.

Lise, A. A. 1981. Tomisideos neotropicais V: Revisao do genero *Onocolus* Simon, 1895 (Araneae, Thomisidae, Stephanopsinae). Iheringia (Zool.) 57: 3 - 97.

Lloyd, M., Zar, J. H., and J. R. Karr. 1966. On the calculation of information-theoretical measures of diversity. Am. Midl Nat. 79: 257 - 272.

Lubin, Y. D. 1978. Seasonal abundance and diversity of web-building spiders in relation to habitat structure on Barro Colorado Island, Panama. J. Arachnol. 6: 31 - 51.

Lubin, Y. D. 1980. Population studies of two colonial orb-weaving spiders. Zool. J. Linn. Soc. 70: 265 - 287.

Lubin, Y. D., Eberhard, W. G., and G. G. Montgomery. 1978. Webs of *Miagrammopes* (Araneae: Uloboridae) in the neotropics. Psyche 85: 1 - 23.

Manhart, C. in prep. Spiders on bark in a seasonal tropical rain forest (Panguana, Peru).

Mello-Leitao, CH. de. 1929. Oxyopideos do Brasil. Rev. Mus. Paul (Sao Paulo) 16: 1-48.

Millidge, A. F. 1985. Some linyphiid spiders from South America (Araneae, Linyphiidae). Am. Mus. Nov. 2836: 1 - 78.

Millidge, A. F. 1987. The erigonine spiders of North America. Part 8. The genus *Eperigone* Crosby and Bishop (Araneae, Linyphiidae). Am. Mus. Nov. 2885: 1 - 75.

Millidge, A. F. 1991. Further linyphiid spiders (Araneae) from South America. Bull. Amer. Mus. Nat. Hist. 205: 1-199.

Muma, M. H. 1953. A study of the spider family Selenopidae in North America, Central America, and the West Indies. Am. Mus. Nov. 1619: 1 - 55.

Myers, N. 1980. Conversion of tropical moist forests. Nat. Acad. Sci, Washington, D.C.

Nentwig, W. 1985 a. A mimicry complex between mutillid wasps (Hymenoptera: Mutillidae) and spiders (Araneae). Stud. Neotrop. Fauna Environ. 20: 113 - 116.

Nentwig, W. 1985 b. Feeding ecology of the tropical spitting spider *Scytodes longipes* (Araneae, Scytodidae). Oecologia 65: 284 - 288.

Nentwig, W. 1985 c. Social spiders catch larger prey: a study of *Anelosimus eximius* (Araneae: Theridiidae). Behav. Ecol. Sociobiol. 17: 79 - 85.

Nentwig, W. 1985 d. *Architis nitidopilosa*, a neotropical pisaurid with a permanent catching web (Araneae, Pisauridae). Bull. Br. Arachnol. Soc. 6: 297 - 303.

Nentwig, W. 1985 e. Prey analysis of four species of tropical orb weaving spiders (Araneae: Araneidae) and a comparison with araneids of the temperate zone. Oecologia 66: 580 - 594.

Nentwig, W. 1990. Collections of spiders from Panama. Unpublished.

Nentwig, W., and T. E. Christenson. 1986. Natural history of the non-solitary sheetweaving spider *Anelosimus jucundus* (Araneae: Theridiidae). Zool. J. Linn. Soc. 87: 27 - 35.

Nentwig, W., and B. Cutler. 1983. Collections of spiders from Panama. (Coll. W. N., det. B. C.) Unpublished.

Nentwig, W., and B. Cutler. 1990. Collections of spiders from Panama. (Coll. W. N., det. B. C.) Unpublished.

Nentwig, W., and St. Heimer. 1983. Collections of spiders from Panama. (Coll. W. N., det. St. H.) Unpublished.

Nitzsche, R. O. M. 1988. "Brautgeschenk" und Umspinnen der Beute bei *Pisaura mirabilis*, *Dolomedes fimbriatus* und *Thaumasia uncata* (Arachnida, Araneida, Pisauridae). Verh. Naturwiss. Ver. Hamburg 30: 353 - 393.

Oliveira, P., and I. Sazima. 1984. The adaptive bases of ant-mimicry in a neotropical aphantochilid spider (Araneae: Aphantochilidae). Biol. J. Linn. Soc. 22: 145 - 155.

Opell, B. D. 1979. Revision of the genera and tropical American species of the family Uloboridae. Bull. Mus. Comp. Zool. 148: 443 - 549.

Opell, B. D. 1981. New Central and South American Uloboridae (Arachnida, Araneae). Bull. Am. Mus. Nat. Hist. 170: 219 - 221.

Opell, B. D. 1983. Checklist of American Uloboridae (Arachnida: Araneae). Great Lakes Entomol. 16: 61 - 66.

Opell, B. 1988. Ocular changes accompanying eye loss in the spider family Uloboridae. J. Morphology 196: 119 - 126.

Panama Canal Company. 1972. Climatological data. Canal Zone and Panama. Annual 1971, vol. 63. Balboa Heights, Canal Zone.

Paz, N. 1988. Ecologia y aspectos del comportamiento en *Linothele* sp. (Araneae, Dipluridae). J. Arachnol. 16: 5 - 22.

Peckham, G. W, Peckham E G. 1896. Spiders of the family Attidae from Central America and Mexico. Occ. Pap. Nat. Hist. Soc. Wisconsin 3: 1 - 101.

Petrunkevitch, A. 1925. Arachnida from Panama. Trans. Connecticut Acad. Arts Sci. 27: 51 - 248.

Petrunkevitch, A. 1939 a. The status of the family Archaeidae and the genus *Landana*. Ann. Entomol. Soc. Am. 32: 479 - 501.

Petrunkevitch, A. 1939 b. The spider fauna of Panama and its Central American affiliation. Am. Nat. 63: 455 - 469.

Petrunkevitch, A. 1955. Arachnida. pp 42 - 162. In Moore, C. (ed.) Treatise on invertebrate paleontology. part P, Arthropoda 2, Chelicerata.

Pickard-Cambridge, F. O. 1897 - 1904. Arachnida. Araneida. Biol. Centr. Amer. Zool. 2: 1 - 610 (1897, 1 - 40; 1899, 41 - 88; 1900, 89 - 192; 1901, 193 - 312; 1902, 313 - 424; 1903, 425 - 464; 1904, 465 - 610).

Pickard-Cambridge, O. 1889 - 1902. Arachnida. Araneida. Biol. Centr. Amer. Zool. 1: 1 - 316 (1889, 1 - 56; 1890, 57 - 72; 1891, 73 - 88; 1892 89 - 104; 1893, 105 - 120; 1894, 121 - 144; 1895, 145 - 160; 1896, 161 - 224; 1897, 225 - 232; 1898, 233 - 288; 1899, 289 - 304; 1902, 305 - 316).

Pielou, E. C. 1966. The measurement of diversity in different types of biological collections. J. Theor. Biol. 13: 131 - 144.

Platnick, N. I. 1974. The spider family Anyphaenidae in America North of Mexico. Bull. Mus. Comp. Zool. 146: 205 - 266.

Platnick, N. I. 1975 a. A revision of the spider genus *Eilica* (Araneae, Gnaphosidae). Am. Mus. Nov. 2578: 1 - 19.

Platnick, N. I. 1975 b. A revision of the palpimanid spiders of the new subfamily Otiothopinae (Araneae, Palpimanidae). Am. Mus. Nov. 2562: 1 - 32.

Platnick, N. I. 1978. A revision of the spider genus *Barrisca* (Araneae, Rhoicininae). J. Arachnol. 6: 213 - 217.

Platnick, N. I. 1981. Spider biogeography: past, present, and future. Rev. Arachnol. 3: 85 - 96.

Platnick, N. I. 1989. Advances in spider taxonomy 1981 - 1987. Manchester University Press, Manchester.

Platnick, N. I., and R. R. Forster. 1982. On the Micromygalinae, a new subfamily of mygalomorph spiders (Araneae, Microstigmatidae). Am. Mus. Nov. 2734: 1 - 13.

Platnick, N. I., and A. Lau. 1975. A revision of the *celer* group of the spider genus *Anyphaena* (Araneae, Anyphaenidae) in Mexico and Central America. Am. Mus. Nov. 2575: 1 - 36.

Platnick, N. I., and M. U. Shadab. 1974 a. A revision of the *transquillus* and *speciosus* groups of the spider genus *Trachelas* (Araneae, Clubionidae) in North and Central America. Am. Mus. Nov. 2553: 1 - 34.

Platnick, N. I., and M. U. Shadab. 1974 b. A revision of the *bispinosus* and *bicolor* groups of the spider genus *Trachelas* (Araneae, Clubionidae) in North America and Central America and the West Indies. Am. Mus. Nov. 2560: 1 - 34.

Platnick, N. I., and M. U. Shadab. 1976 a. A revision of the spider genera *Lygromma* and *Neozimiris* (Araneae, Gnaphosidae). Am. Mus. Nov. 2598: 1 - 23.

Platnick, N. I., and M. U. Shadab. 1976 b. A revision of the spider genera *Drassodes* and *Tivodrassus* (Araneae, Gnaphosidae) in North America. Am. Mus. Nov. 2593: 1 - 29.

Platnick, N. I., and M. U. Shadab. 1976 c. A revision of the Neotropical spider genus *Zimiromus*, with notes on *Echemus* (Araneae, Gna phosidae). Am. Mus. Nov. 2609: 1 - 24.

Platnick, N. I., and M. U. Shadab. 1978 a. A review of the spider genus *Mysmenopsis* (Araneae, Mysmenidae). Am. Mus. Nov. 2661: 1 - 71.

Platnick, N. I., and M. U. Shadab. 1978 b. A review of the spider genus *Anapis* (Araneae, Anapidae), with a dual cladistic analysis. Am. Mus. Nov. 2663: 1 - 23.

Platnick, N. I., and M. U. Shadab. 1979. A review of the spider genera *Anapisona* and *Pseudanapis* (Araneae, Anapidae). Am. Mus. Nov. 2672: 1 - 20.

Platnick, N. I., and M. U. Shadab. 1980. A revision of the spider genus *Cesonia* (Araneae, Gnaphosidae). Bull. Am. Mus. Nat. Hist. 165: 335 - 386.

Platnick, N. I., and M. U. Shadab. 1981 a. A new genus of the spider family Gnaphosidae (Arachnida, Araneae). Bull. Am. Mus. Nat. Hist. 170: 176 - 180.

Platnick, N. I., and M. U. Shadab. 1981 b. Two new species of the mygalomorph spider genus *Neocteniza* (Araneae, Actinopodidae). Bull. Am. Mus. Nat. Hist. 170: 76 - 79.

Platnick, N. I., and M. U. Shadab. 1981 c. New species and records of neotropical Gnaphosidae (Arachnida, Araneae). Bull. Am. Mus. Nat. Hist. 170: 189 - 196.

Platnick, N. I., and M. U. Shadab. 1982. A revision of the American spiders of the genus *Camillina* (Araneae, Gnaphosidae). Am. Mus. Nov 2748: 1 - 38.

Platnick, N. I., and M. U. Shadab. 1989. A revision of the spider genus *Teminius* (Araneae, Miturgidae). Am. Mus. Nov. 2963: 1 - 12.

Porter, D. M. 1973. The vegetation of Panama: a review. pp 167 - 201. In Graham A (ed.) Vegetation and vegetational history of Northern Latin America. Elsevier, Amsterdam.

Raven, R. J. 1985. The spider infraorder Mygalomorphae (Araneae): Cladistics and systematics. Bull. Am. Mus. Nat. Hist. 182: 1 - 180.

Raven, R. J., and N. I. Platnick. 1981. A revision of the American spiders of the family Microstigmatidae (Araneae, Mygalomorphae). Am. Mus. Nov. 2707: 1 - 20.

Reimoser, E. 1939. Wissenschaftliche Ergebnisse der österreichischen biologischen Expedition nach Costa Rica. Ann. Naturhist. Mus. Wien 50: 328 - 386.

Reiskind, J. 1969. The spider subfamily Castianeirinae of North and Central America (Araneae, Clubionidae). Bull. Mus. Comp. Zool. 138: 163 - 325.

Reiskind, J. 1977. Ant-mimicry in Panamanian clubionid and salticid spiders (Araneae: Clubionidae, Salticidae). Biotropica 9: 1 - 8.

Rich, P. V., and T. H. Rich. 1983. The Central American dispersal route: biotic history and paleogeography. pp 12 - 34. In Janzen D H: Costa Rican natural history. University of Chicago Press, Chicago.

Richman, D.A. 1989. A revision of the genus *Hentzia* (Araneae, Salticidae). J. Arachnol. 17: 285-344.

Ridgely, R. S. 1976. Birds of Panama. Princeton University Press, Princeton.

Roberts, M. J. 1979. A study of the spiders *Dipoena alta* Keyserling, *D. lineatipes* Bryant and a new species *D. jamesi* (Araneae: Theridiidae). J. Arachnol. 7: 199 - 222.

Robinson, M. H. 1969. The predatory behaviour of *Argiope argentata* (Fabricius). Am. Zool. 9: 161 - 173.

Robinson, M. H., and B. Robinson. 1970 a. The stabilimentum of the orb-web spider, *Argiope argentata*: an improbable defense against predators. Can. Entomol. 102: 641 - 655.

Robinson, M. H., and B. Robinson. 1970 b. Prey caught by a sample population of the spider *Argiope argentata* (Araneae: Araneidae) in Panama: a year's census data. Zool. J. Linn. Soc. 49: 345 - 357.

Robinson, M. H., and H. Mirik. 1971. The predatory behavior of the gold en-web spider *Nephila clavipes* (Araneae: Araneidae). Psyche 78: 123 - 139.

Robinson, M. H., and J. Olazarri. 1971. Units of behavior and complex sequences in the predatory behavior of *Argiope argentata* (Fabricius) (Araneae: Araneidae). Smith. Contr. Zool. 65: 1 - 36.

Robinson, M. H., and B. Robinson. 1971. The predatory behavior of the ogre-faced spider *Dinopis longipes* F. Cambridge (Araneae: Dinopidae). Am. Midl. Nat. 85: 85 - 96.

Robinson, M. H., Robinson, B., and W. Graney. 1971. The predatory behavior of the nocturnal orb-web spider *Eriophora fuliginea* (C. L. Koch) (Araneae: Araneidae). Rev. Per. Entomol. 14: 304 - 315.

Robinson, M. H., and B. Robinson. 1973. The stabilimenta of *Nephila clavipes* and the origins of stabilimentum-building in araneids. Psyche 80: 277 - 288.

Robinson, M. H., and B. Robinson. 1974. Adaptive complexity: the thermoregulatory postures of the golden-web spider *Nephila clavipes* at low latitudes. Am. Midl. Nat. 92: 386 . 396.

Robinson, M. H. 1978. Symbioses between insects and spiders; an association between lepidopteran larvae and the social spider *Anelosimus eximius* (Araneae: Theridiidae). Psyche 84: 225 - 232.

Robinson, M. H., and B. Robinson. 1980. Comparative studies of the courtship and mating behaviour of tropical araneid spiders. Pac. Ins. Monogr. 36: 1 - 217.

Roth, V. D. 1985. Spider genera of North America. Distributed on behalf of the Am. Arachnol. Soc.

Schmidt, G. 1990. Zur Kenntnis der Gattung *Mygalarachne* (Araneida: Theraphosidae). Arachnol. Anz. 9:8-12.

Schmidt, K. P. 1954. Faunal realms, regions, and provinces. Q. Rev. Biol 29: 322 - 331 .

Scioscia, C. L. 1988. Determinacion de *Bryantella speciosa* y *B. smaragdus*, nueva combination, mediante la aplication de tecnicas numericas (Araneae, Salticidae). J. Arachnol. 16: 177 - 191.

Shear, W. A. 1970. The spider family Oecobiidae in North America, Mexico, and the West Indies. Bull. Mus. Comp. Zool. 140: 129 - 164.

Shear, W. A. 1984. Early land animals in North America: Evidence from Devonian Age arthropods from Gilboa, New York. Science 224: 492 - 494.

Shelly, T. 1983. Prey selection by the neotropical spider *Alpaida tuonaba* with notes on web-site tenacity. Psyche 90: 123 - 133.

Simon, E. 1880. Etude sur le genre *Labdacus* Cambr. Bull. Soc. Zool. France 5: 152 - 156.

Simon, E. 1891 a. Liste des Aviculariides qui habitent le Mexique et l'Amerique centrale. Act. Soc. 44: 327 - 339.

Simon, E. 1891 b. Descriptions d'espces et de genres nouveaux de la famille des Aviculariidae. Ann. Soc. Ent. Fr. 60: 300 - 312.

Simon, E. 1898. Descriptions d'arachnides nouveaux des familles des Agelenidae, Pisauridae, Lycosidae et Oxyopidae. Ann. Soc. Ent. Belg. 42: 5 - 34.

Simpson, G. G. 1950. History of the fauna of Latin America. Am. Sci. 361 - 389.

Smith, A. G., Briden J. C., and Drewry, G. E. 1973. Phanerozoic world maps. pp 1 - 42. In Hughes N. F. (ed.) Organisms and continents through time. Spec. Pap. Palaeont. Palaeontol. Ass., London.

Ubick, D. 1988. Spider studies in Costa Rica. Unpublished report.

Udvardy, M. D. F. 1969. Dynamic zoogeography. Van Nostrand Reinhold Company, New York.

Udvardy, M. D. F. 1975. A classification on the biogeographical provinces of the world. IUCN Occ. Pap. 18: 1 - 43.

Valerio, C. E. 1981. Spitting spiders (Araneae, Scytodidae, *Scytodes*) from Central America. Bull. Am. Mus. Nat. Hist. 170: 80 - 89.

Valerio, C. E. 1982. Aranas terafosidas de Costa Rica (Araneae, Theraphosidae). IV. Generos *Metriopelma* y *Cyclosternum*, incluyendo especies de Panama. Brenesia 19/20: 407 - 423.

Vega, C. M Z. 1980. Lista anotada de especies de aranas de Costa Rica. Brenesia 18: 301 - 352.

Vollrath, F. 1976. Konkurrenzvermeidung bei tropischen kleptoparasitischen Haubennetzspinnen der Gattung *Argyrodes*. Entomol. Germ. 3: 104 - 108.

Vollrath, F. 1978. A close relationship between two spiders: *Curimagua bayano* synecious on a *Diplura* species. Psyche 85: 347 - 353.

Vollrath, F. 1979. Behaviour of the kleptoparasitic spider *Argyrodes elevatus*. Anim. Behav. 27: 515 - 521.

Vollrath, F. 1980. Male body size and fitness in the web-building spider *Nephila clavipes*. Z. Tierpsych. 53: 61 - 78

Vollrath, F. 1982. Colony foundation in a social spider. Z. Tierpsych. 60: 313 - 324.

Vollrath, F. 1985. Web spider's dilemma: a risky move or site dependent growth. Oecologia 68: 69 - 72.

Vollrath, F. 1986. Eusociality and extraordinary sex ratio in the spider *Anelosimus eximius* (Araneae: Theridiidae). Behav. Ecol. Sociobiol. 18: 283 - 287.

Wallace, H. K., and H. Exline. 1978. Spiders of the genus *Pirata* in North America, Central America and the West Indies (Araneae: Lycosidae). J. Arachnol. 5: 1 - 112.

West, R., Dondale, C. D., and R. A. Ring. 1984. A revised checklist of the spiders (Araneae) of British Columbia. J. Ent. Soc. Brit. Columbia 81: 80 - 98.

West, R. C, Dondale, C. D., and R. A. Ring. 1988. Additions to the revised checklist of the spiders (Araneae) of British Columbia. J. Ent. Soc. Brit. Columbia 85: 77 - 86.

White, A. 1856. Description of *Mygale Emilia*, a spider from Panama, hithero apparently unrecorded. Proc. Zool. Soc. London 24: 183 - 185.

Williams, E. 1941. An ecological study of the floor fauna of the Panama rain forest. Bull. Chicago Acad. Sci. 6: 63 - 123.

Wunderlich, J. 1986. Spinnenfauna gestern und heute. Verlag J. Wunderlich, Straubenhardt.

Wunderlich, J. 1988. Die fossilen Spinnen im dominikanischen Bernstein. Verlag J. Wunderlich, Straubenhardt.

Young, O. P., and T. C. Lockley. 1989. Spiders of Spanish moss in the delta of the Mississippi. J. Arachnol. 17: 143-148.

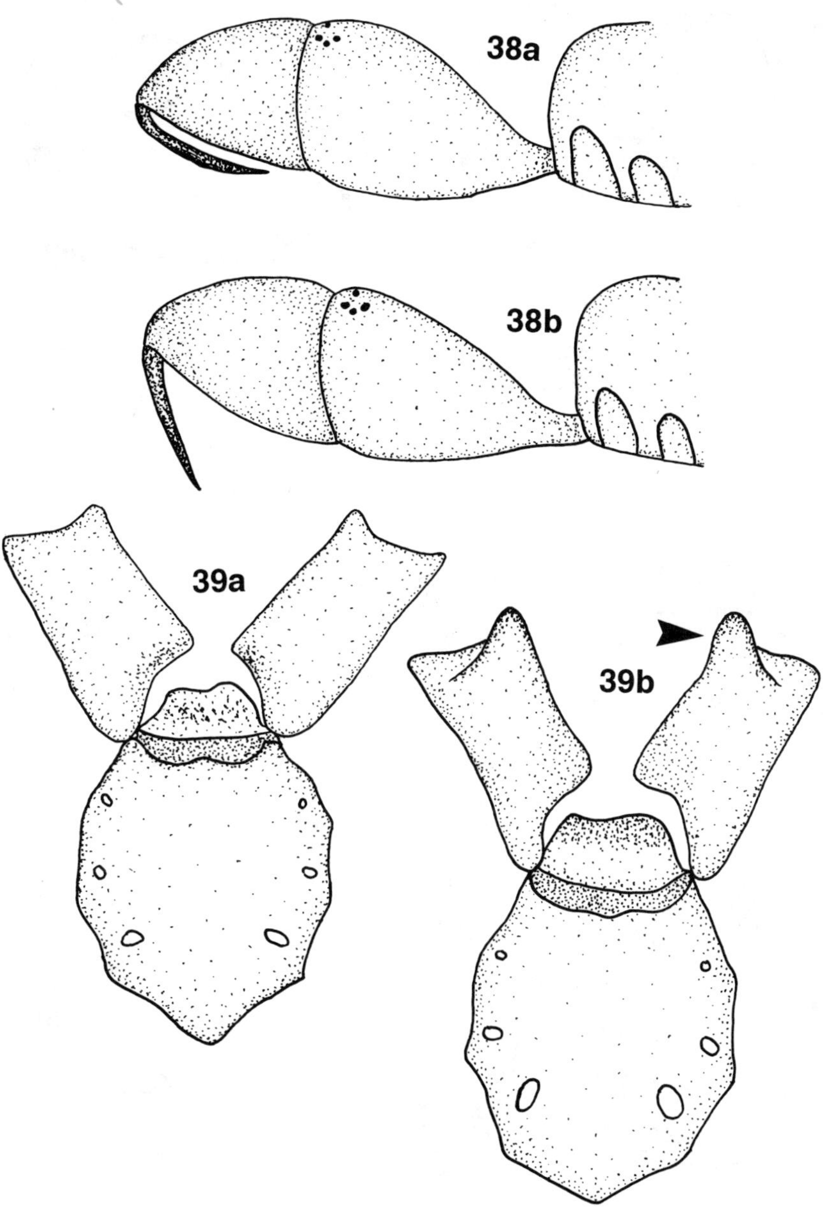

Figs. 38-39. Fig. 38. Fang action in orthognath spiders with closed fang (above) and open fang (below). Fig. 39: Sternum, maxillae and labium. a. Barychelidae, b. Theraphosidae.

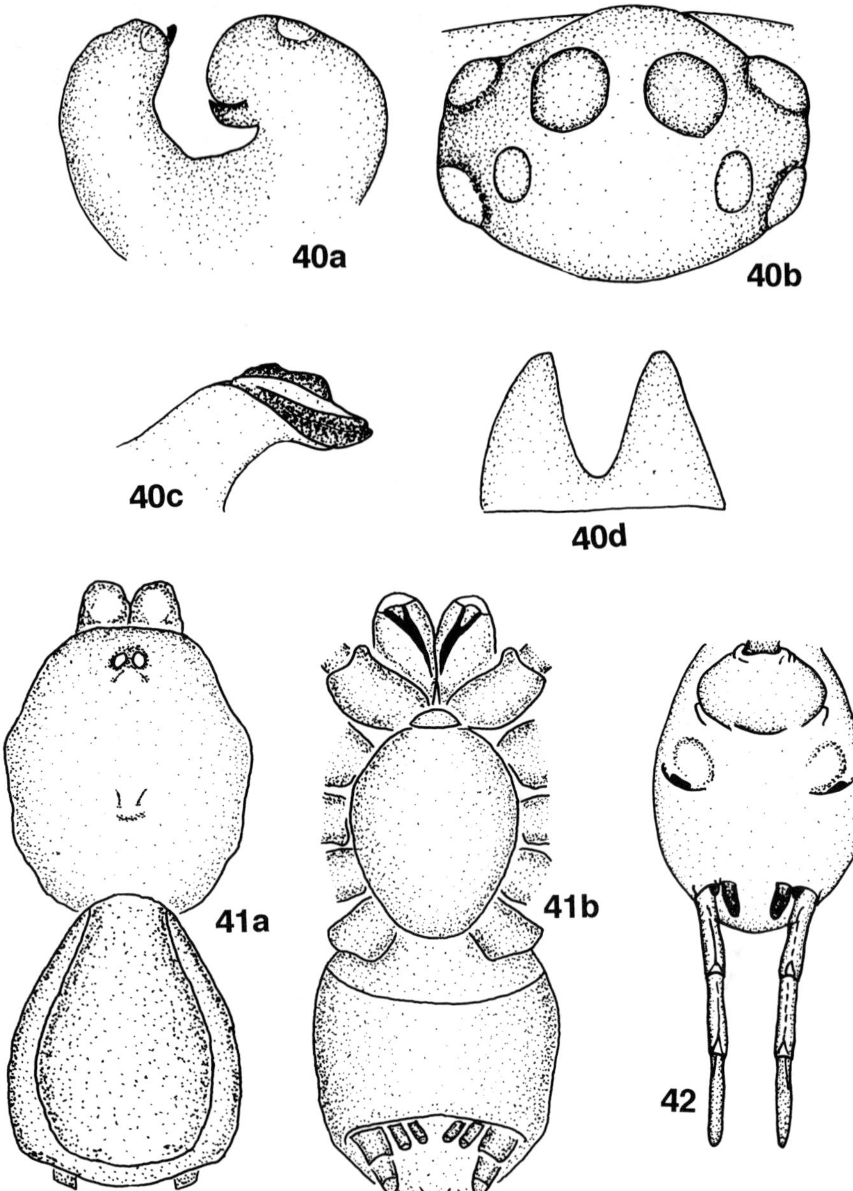

Figs. 40-42. Fig. 40. *Cyclosternum stylipus*. a. tibial spurs, right leg, b. ocular area, indicating anterior edge of prosoma, c. bulb, distal end, dorsal view, d. spermathecae, dorsal view. Fig. 41: *Micromygale diblemma*. a. dorsal view, b. ventral view. Fig. 42: Dipluridae, opisthosoma from below.

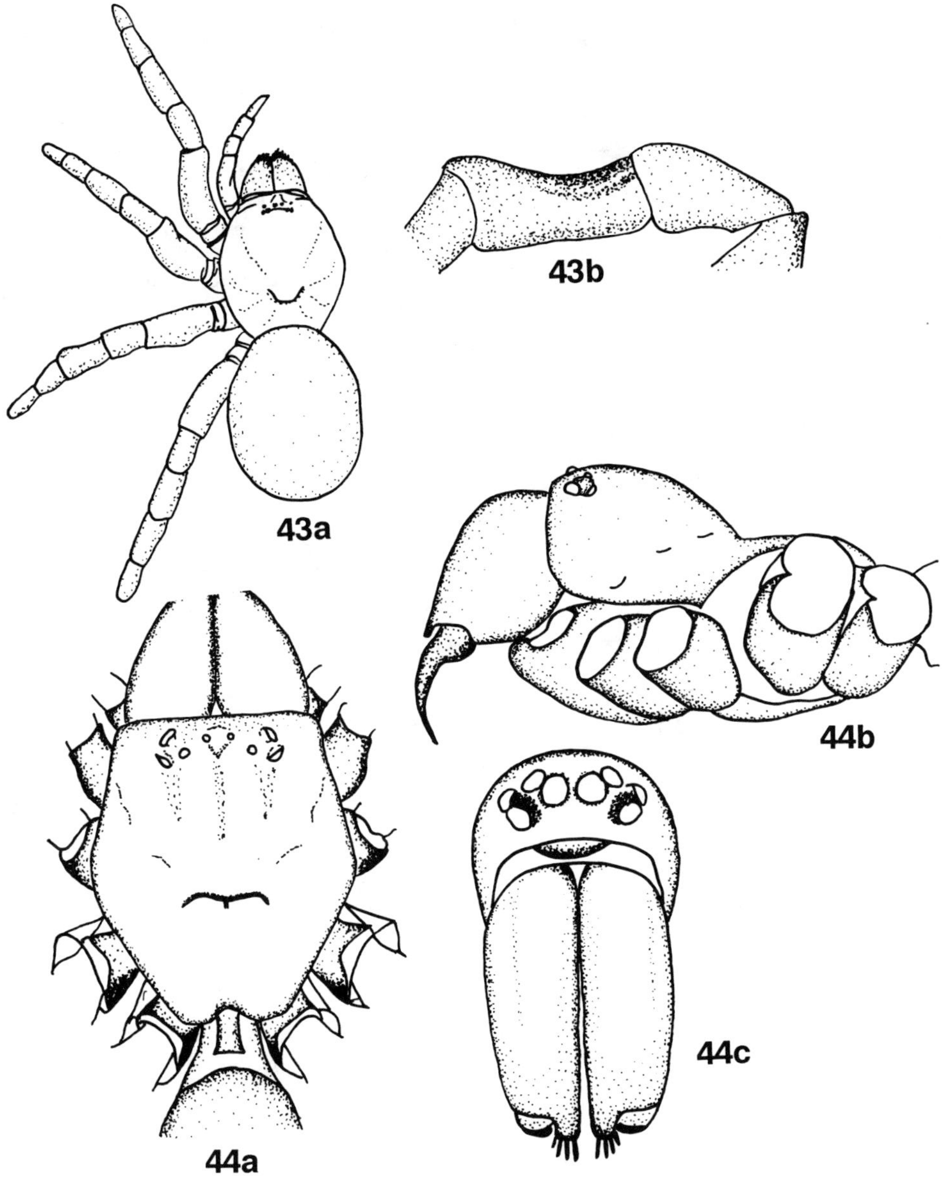

Figs. 43-44. Fig. 43: *Ummidia*. a. dorsal view of female, b. tibia III with depression. Fig. 44: *Neocteniza*. a. prosoma, dorsal view, b. prosoma, lateral view, c. prosoma and chelicerae, anterior view.

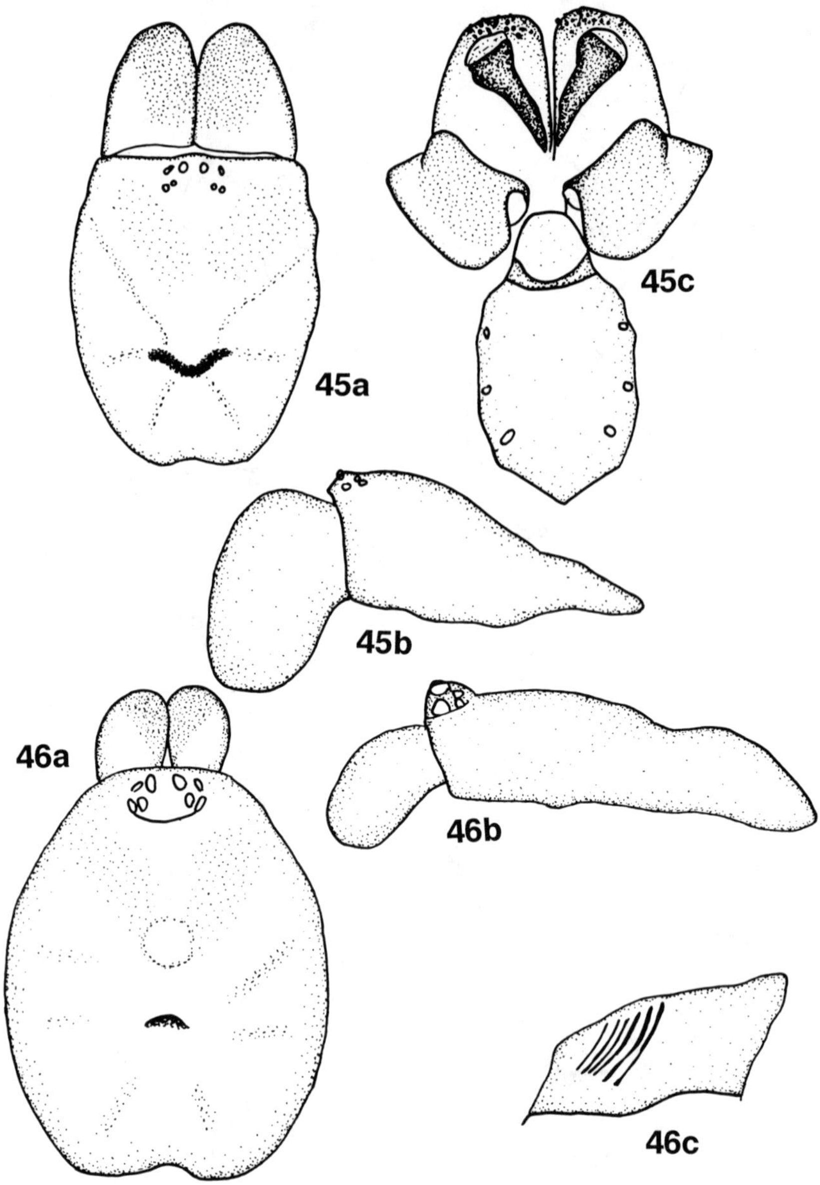

Figs. 45-46. Fig. 45: *Bolostromus*. a. prosoma, dorsal view, b. prosoma, sternum, maxillae and labium, ventral view, c. prosoma, lateral view. Fig. 46: *Diplura*. a. prosoma dorsal view, b. prosoma, lateral view, c. right maxilla, oblique view of anterior face showing lyra.

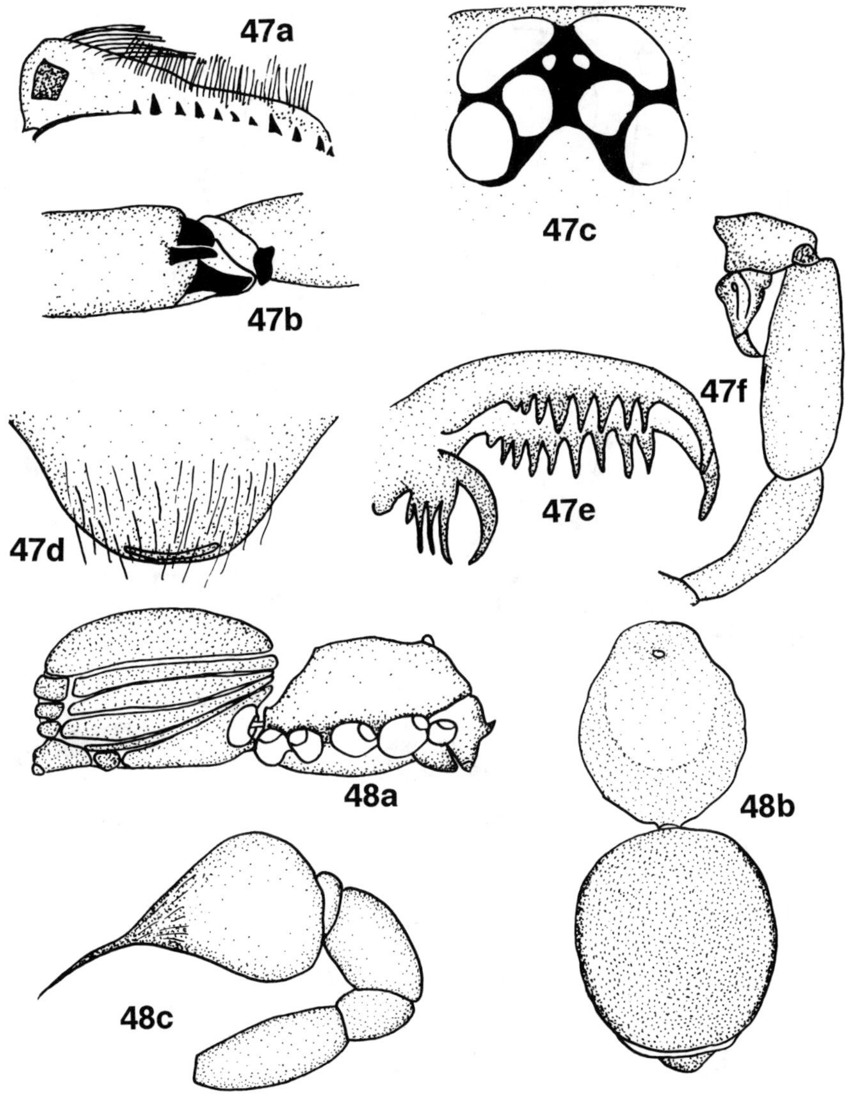

Figs. 47-48. Fig. 47: *Masteria spinosa*. a. left fang groove, prolateral side, b. spines at distal end of first tibia and proximal end of first metatarsus, prolateral view, c. eyes from above, d. epigyne, e. three tarsal claws, f. male palp, retrolateral view. Fig. 48: *Monoblemma unica*. a. lateral view, b. dorsal view, c. left male palpus, retrolateral view.

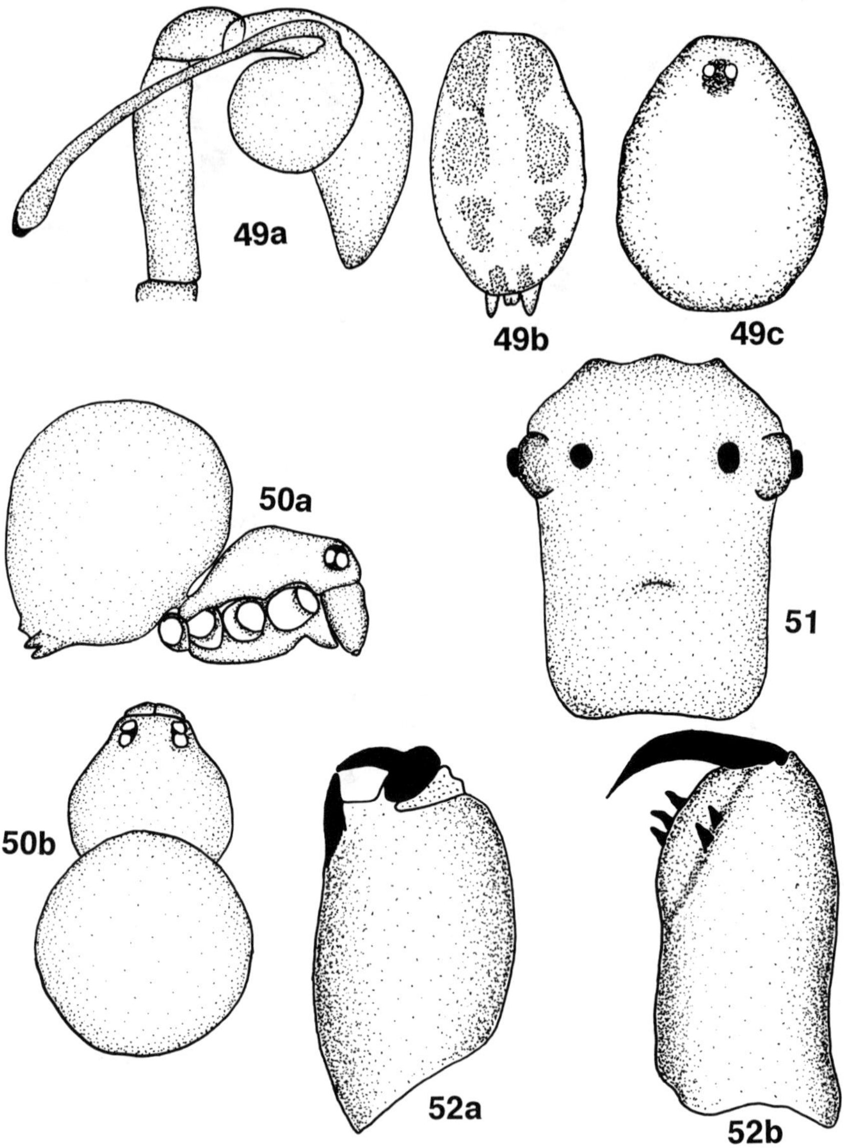

Figs. 49-52. Fig. 49: *Nops simla*. a. left male palp, prolateral view, b. opisthosoma, dorsal view, c. prosoma, dorsal view. Fig. 50: *Anapistula secreta*. a. lateral view, b. dorsal view. Fig. 51 *Miagrammopes*, prosoma, dorsal view. Fig. 52: Chelicerae types. a. chelate, b. not chelate.

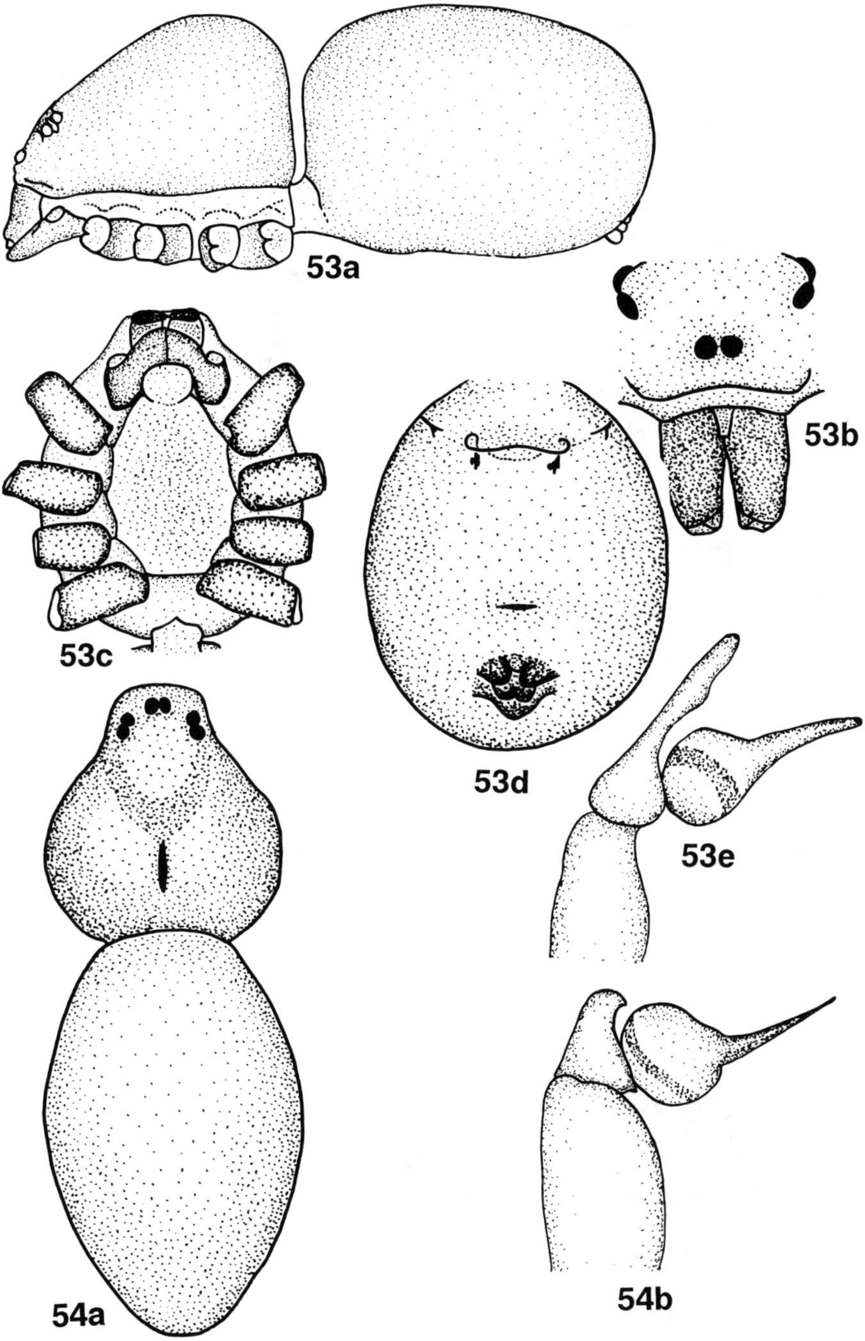

53a 53b 53c 53d 53e 54a 54b

Figs. 53-54. Fig. 53: *Scytodes*. a. lateral aspect, b. eyes and chelicerae, c. prosoma, ventral view, d. opisthosoma, ventral view, e. male palp. Fig. 54: *Loxosceles*. a. dorsal view, b. male palp.

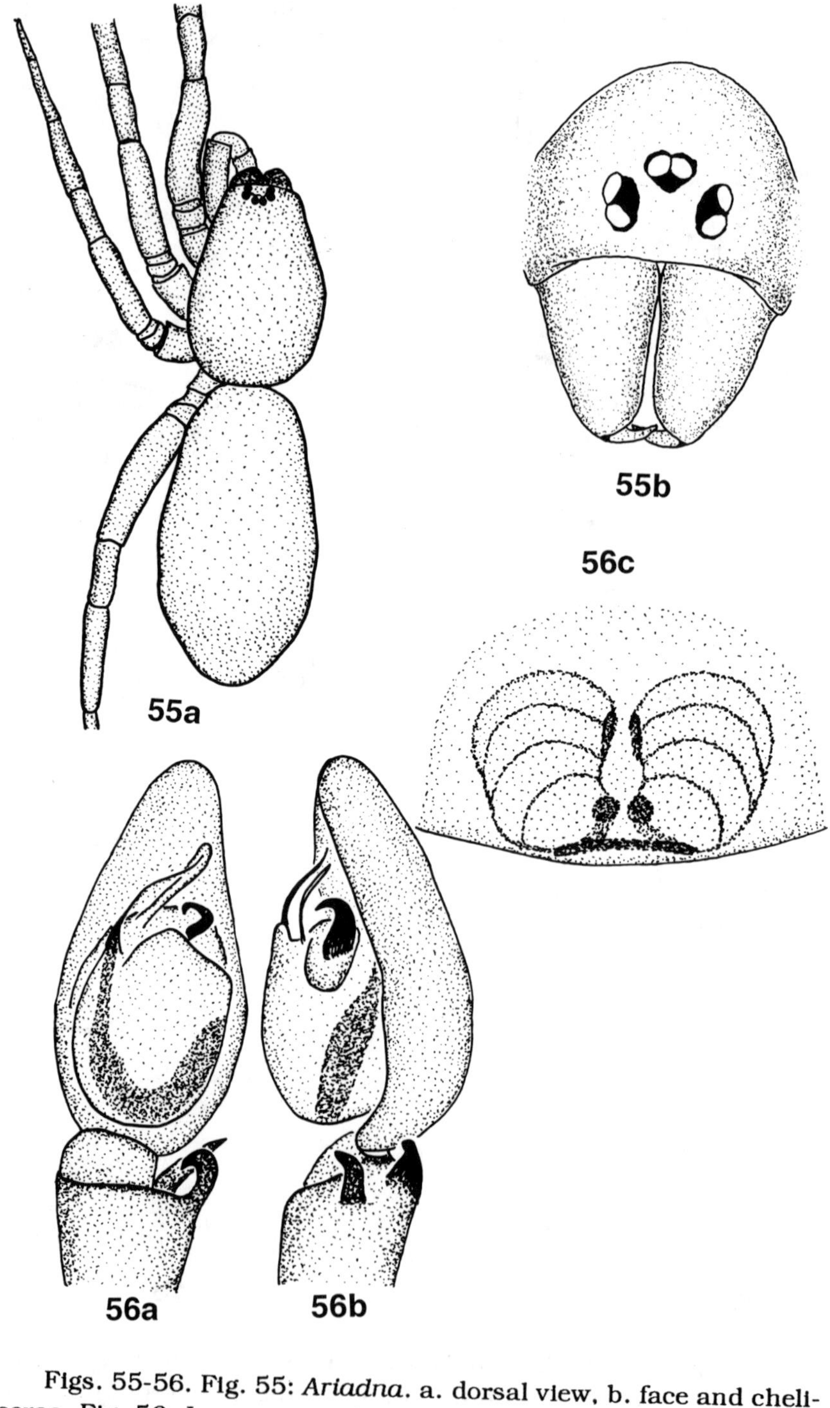

Figs. 55-56. Fig. 55: *Ariadna*. a. dorsal view, b. face and chelicerae. Fig. 56: *Lygromma volcan*. a. male palpus, ventral view, b. palpus retrolateral view, c. epigyne, ventral view.

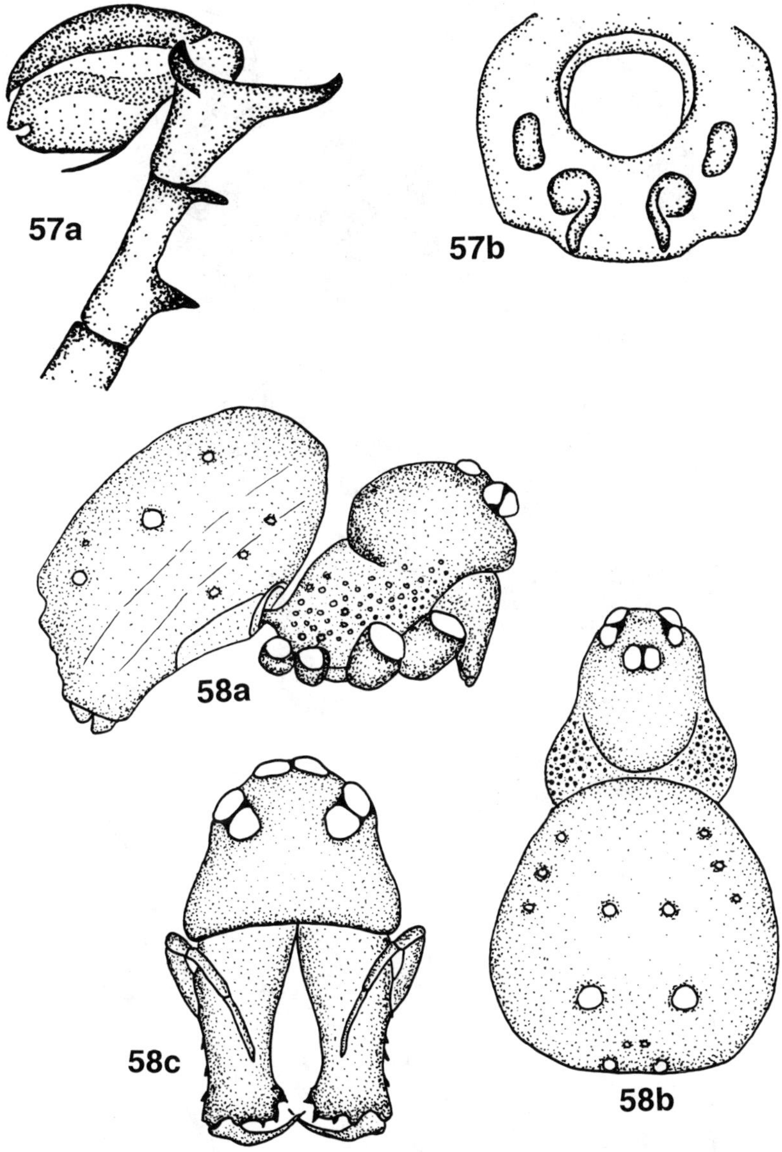

Figs. 57-58. Fig. 57: *Pseudanapis gertschi*. a. male palpus, b. epigyne. Fig. 58: *Anapis keyserlingi*. a. lateral view, b. dorsal view, c. face and chelicerae.

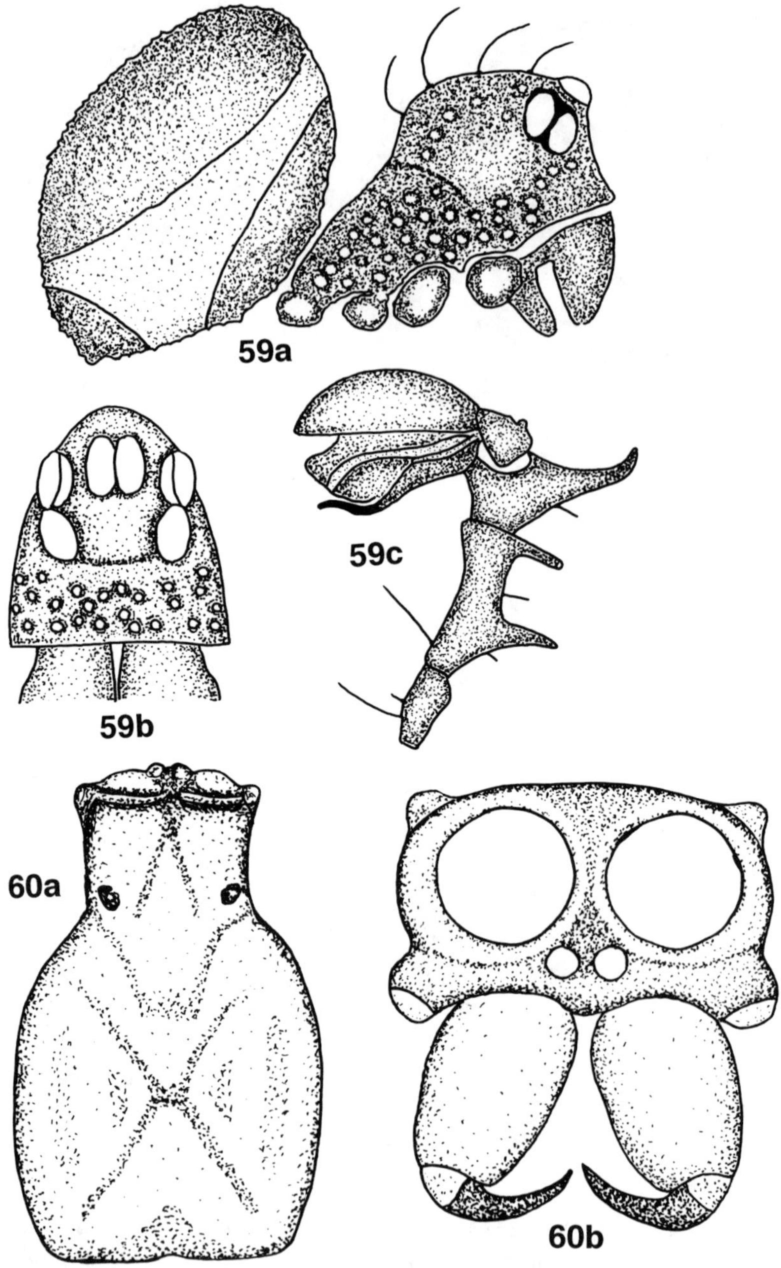

Figs. 59-60. Fig. 59: *Anapisona gertschi*. a. lateral view, b. front view, c. male palp, prolateral view. Fig. 60: *Deinopis*. a. prosoma, dorsal view, b. face.

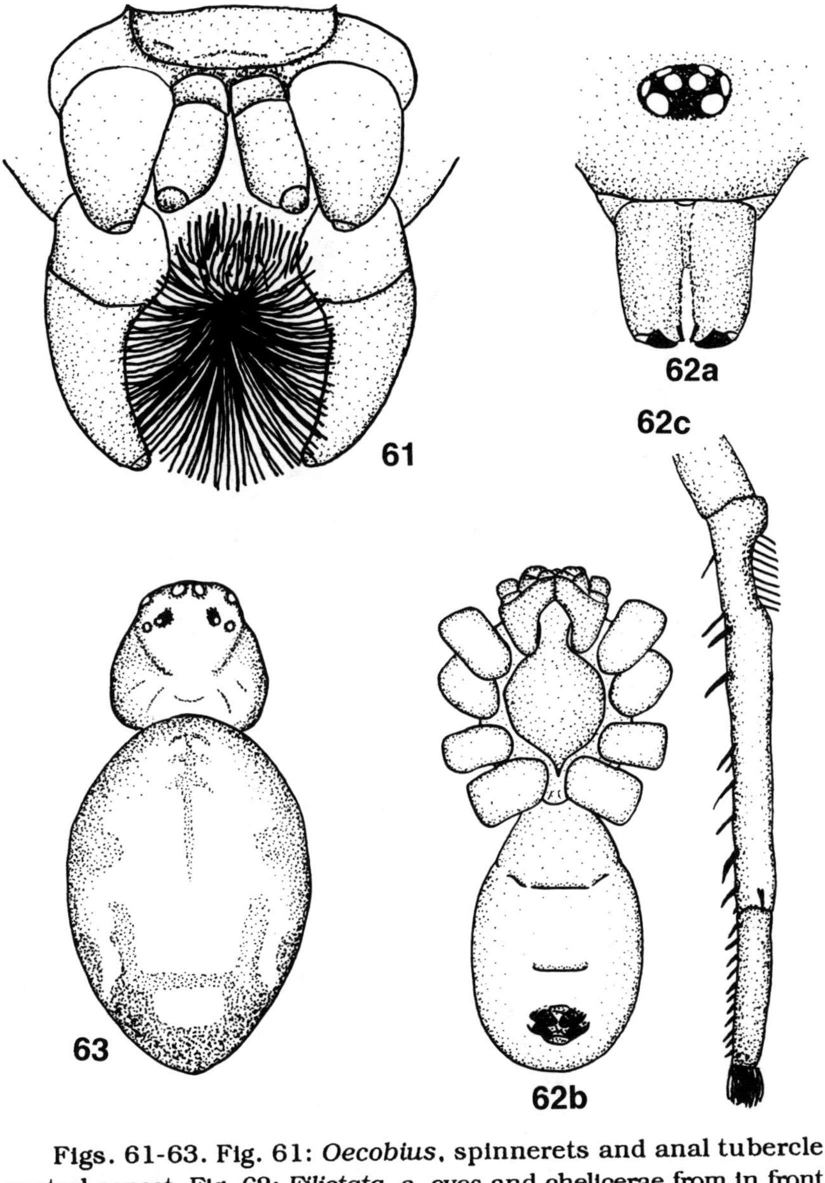

Figs. 61-63. Fig. 61: *Oecobius*, spinnerets and anal tubercle, ventral aspect. Fig. 62: *Filistata*. a. eyes and chelicerae from in front, b. ventral aspect, c. leg IV with calamistrum. Fig. 63: *Ariston aristus*, female, dorsal aspect.

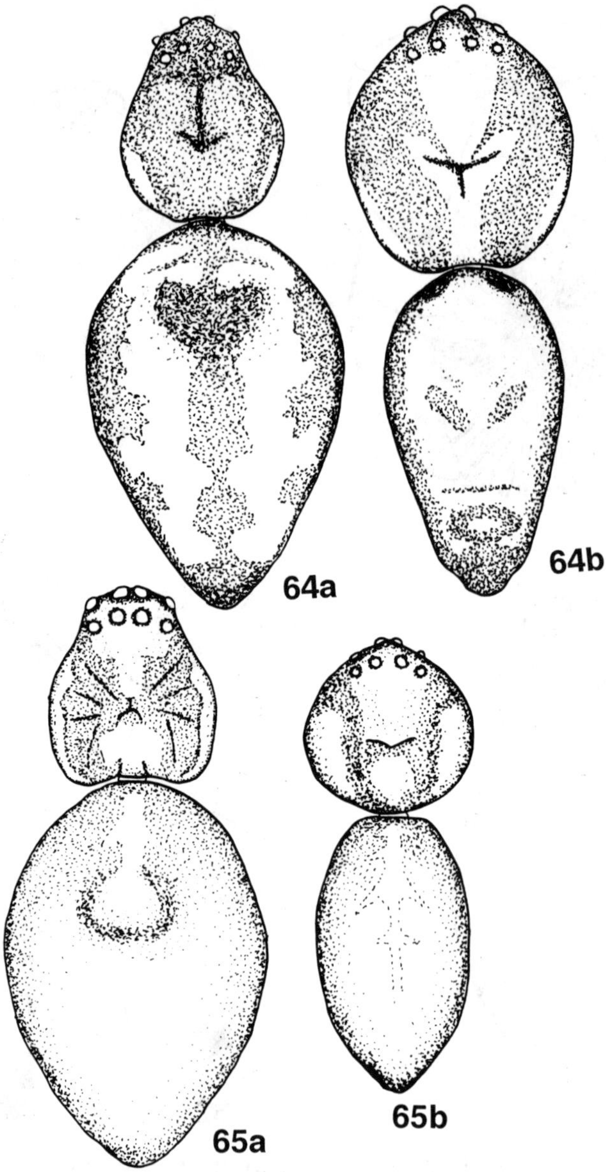

Figs. 64-65. Fig. 64: *Philoponella divisa*, dorsal aspect. a. female, b. male. Fig. 65: *Zosis geniculatus*, dorsal aspect. a. female, b. male.

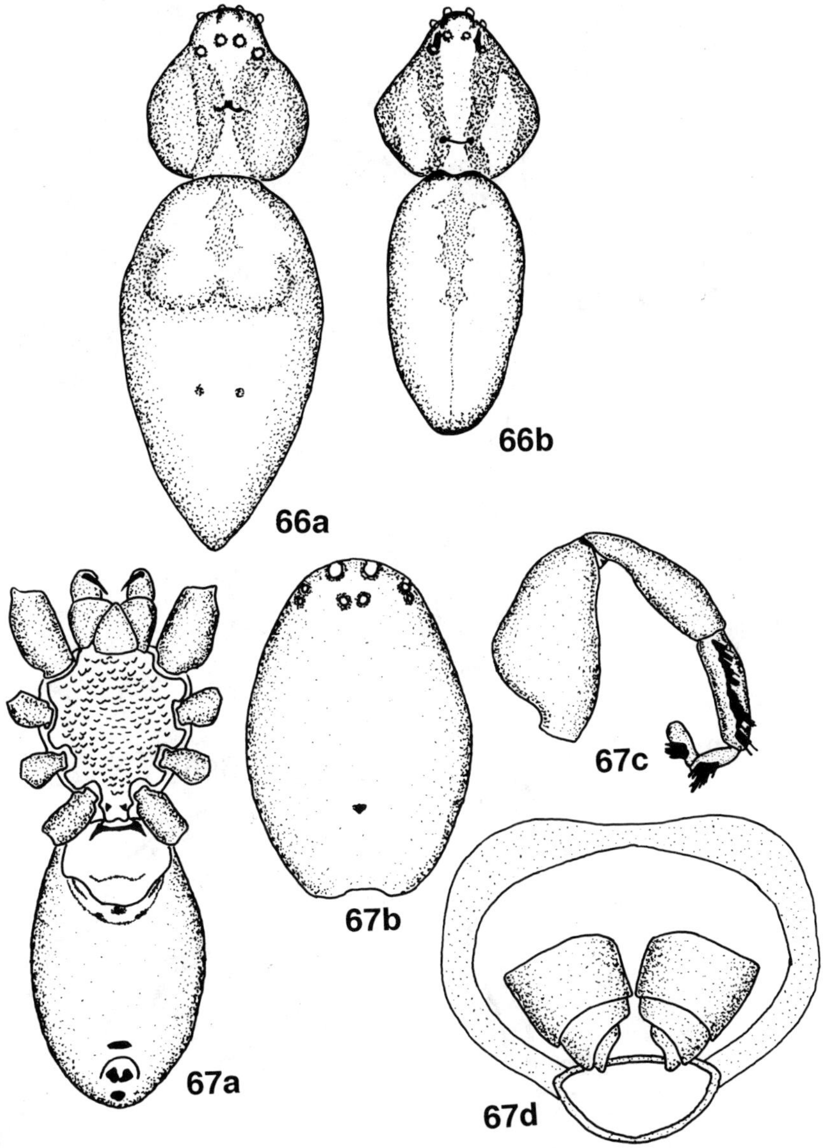

Figs. 66-67. Fig. 66: *Uloborus penicillatus*, dorsal aspect. a. female, b. male. Fig. 67: *Otiothops macleayi*. a. ventral view, b. prosoma, dorsal view, c. leg I, protateral view, d. spinnerets and anal tubercle, ventral view.

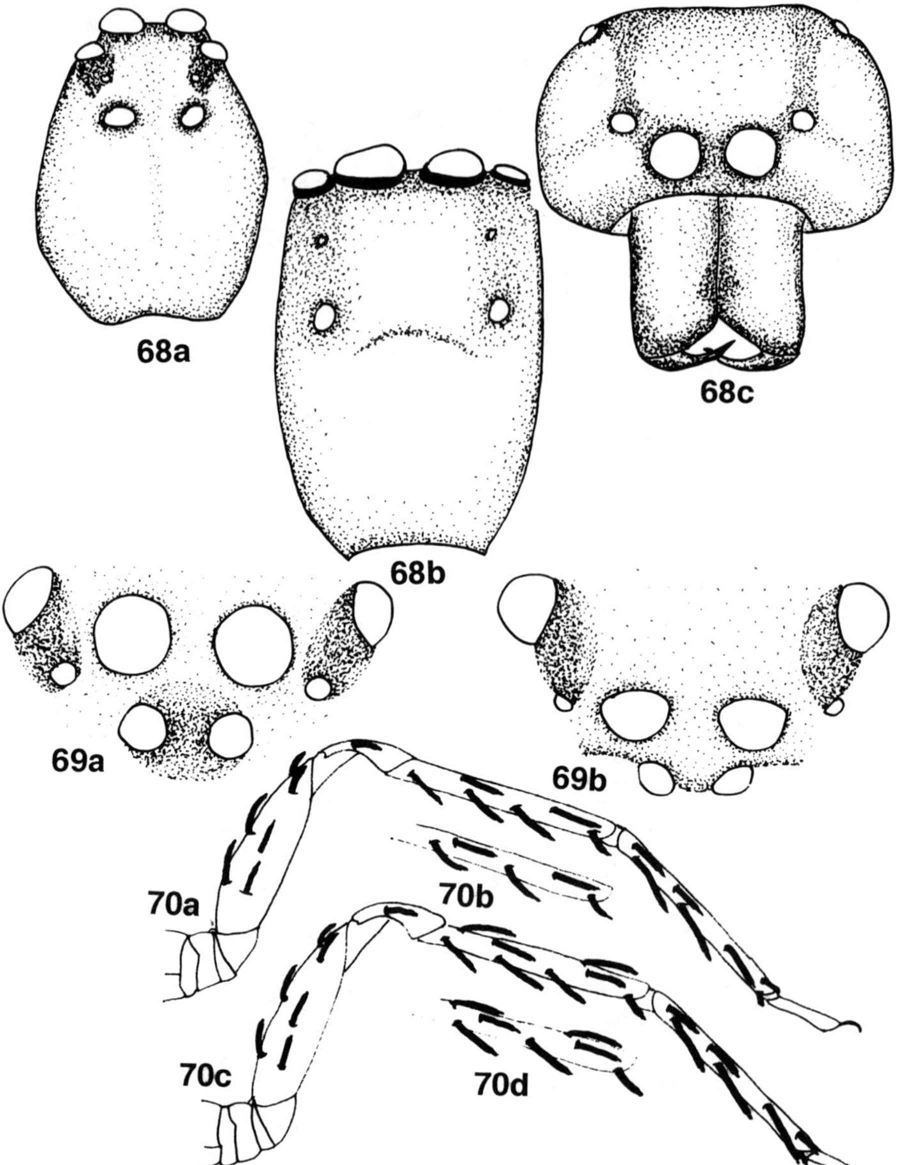

Figs. 68-70. Fig. 68: Salticidae. a. subfamily Lyssomaninae, prosoma dorsal view, b. prosoma, dorsal view, c. face and chelicerae from in front. Fig. 69: *Cupiennius*, eyes. a. from in front, b. dorsal view. Fig. 70: *Cupiennius*, spination of legs. a. female, leg I, II, b. female, tibia IV, c. male, leg I, II, d. male, tibia III, IV.

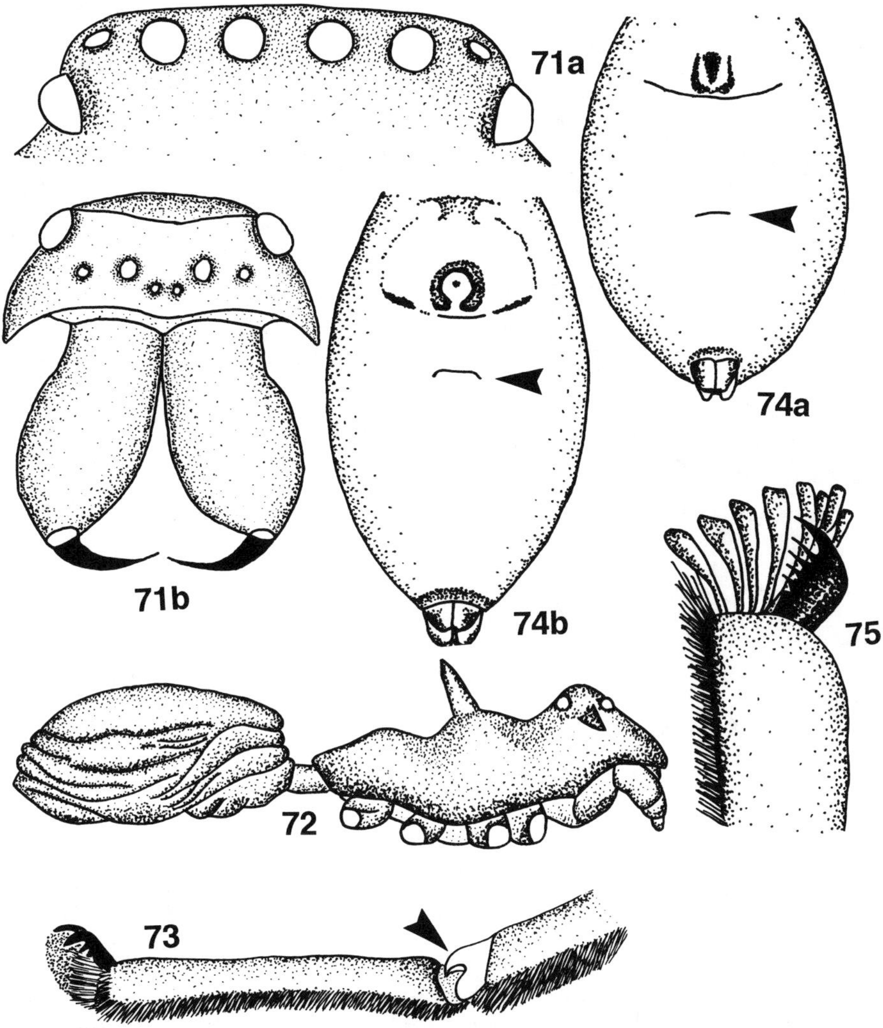

Figs. 71-75. Fig. 71: *Selenops*. a. eye group, dorsal view, b. face and chelicerae from in front. Fig. 72: *Aphantochilus*, female, lateral view. Fig. 73: *Heteropoda*, leg I with trilobate membrane (arrow), scopulae and claw tufts. Fig. 74: Anyphaenidae, opisthosoma, ventral view, showing tracheal spiracle (arrow), a. *Anyphaena*, b. *Aysha*. Fig. 75: Anyphaenidae, claw tufts consisting of lamelliform sete.

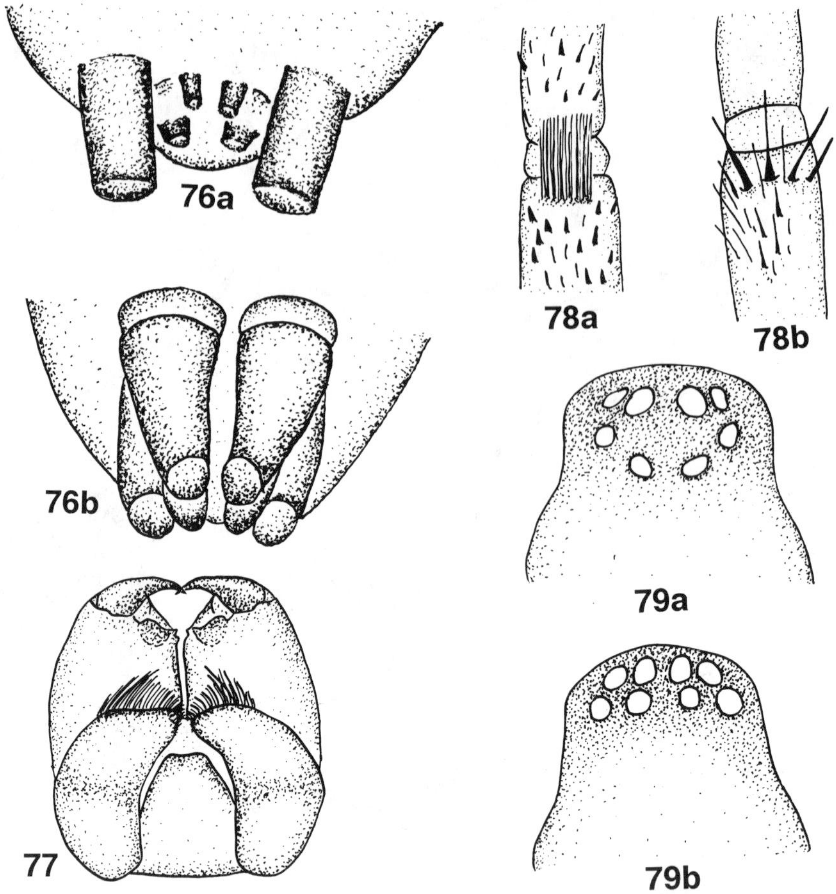

Figs. 76-79. Fig. 76: Spinnerets, ventral view. a. Gnaphosidae, b. Clubionidae s.l. Fig. 77: *Gnaphosa*, chelicerae, endites and labium, ventral aspect. Fig. 78: Gnaphosidae, tarsus and metatarsus III or IV. a. with preening comb, b. without preening comb. Fig. 79: Gnaphosidae, eye arrangement, dorsal view. a. posterior eye row strongly procurved, b. posterior eye row slightly recurved.

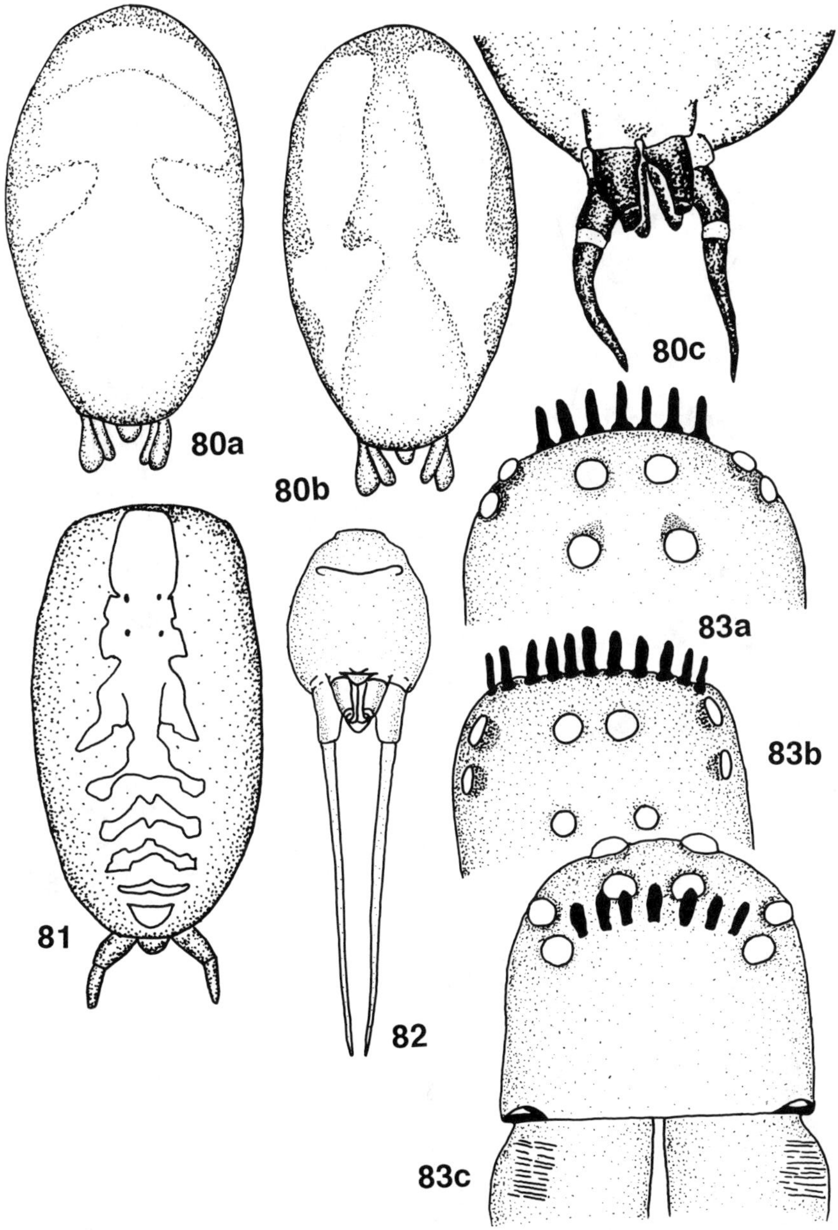

Figs. 80-83. Fig. 80: a., b.: *Gertschosa*, opisthosoma, dorsal view; c: *Eutichurus*, spinnerets, ventral view. Fig. 81: *Teminius*, opisthosoma, ventral view. Fig. 82: Hersiliidae, opisthosoma and spinnerets, ventral view. Fig. 83: *Leprolochus spinifrons*. a. cephalic region of male from above, b. cephalic region of female from above, c. face and chelicerae of male from in front.

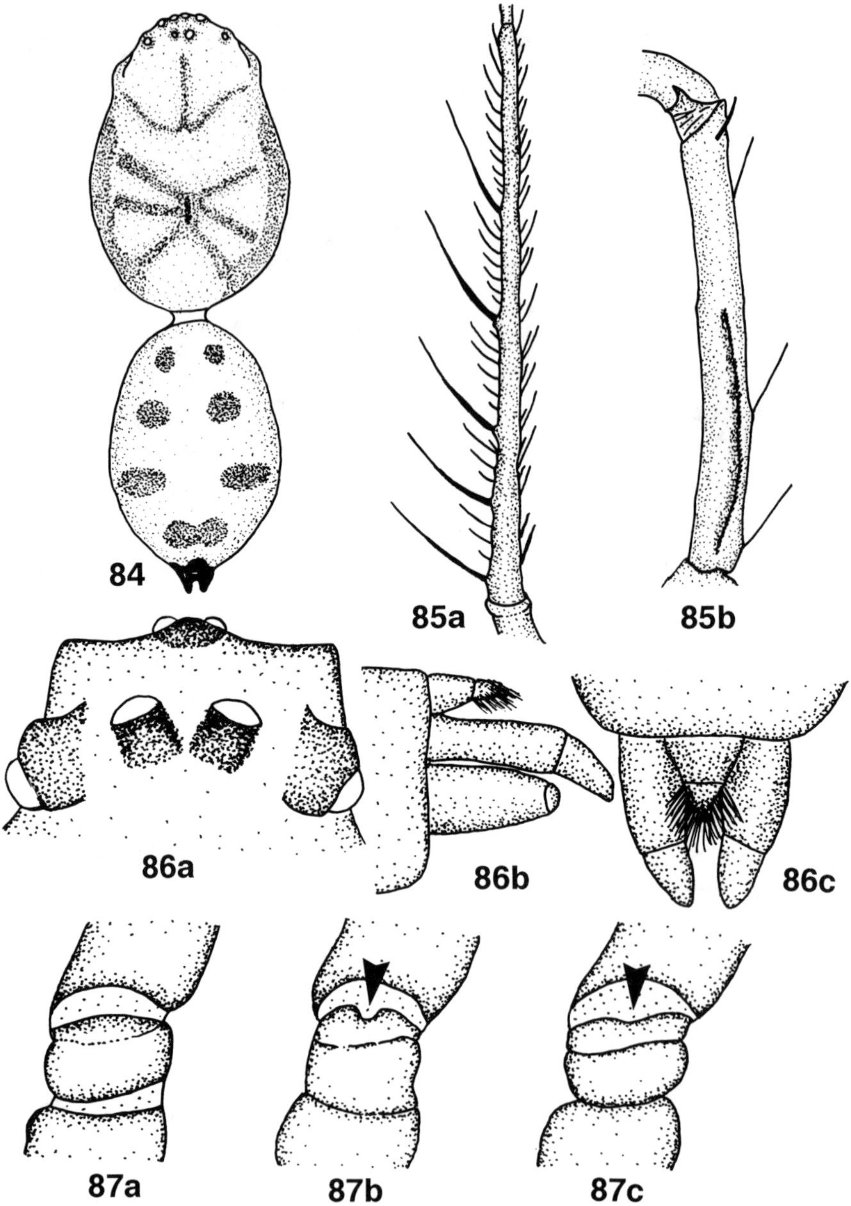

Figs. 84-87. Fig. 84: *Naibena barroana*, dorsal view. Fig. 85: *Mimetus*. a. metatarsus I showing spination, b. femur I, retrolateral view. Fig. 86: *Senoculus*. a. eye arrangement, dorsal view, b. spinnerets, lateral view, c. spinnerets, dorsal view. Fig. 87: Femur, trochanter and coxa with notches indicated by an arrow: a. Agelenidae, b. Lycosidae, Pisauridae, c. Oxyopidae.

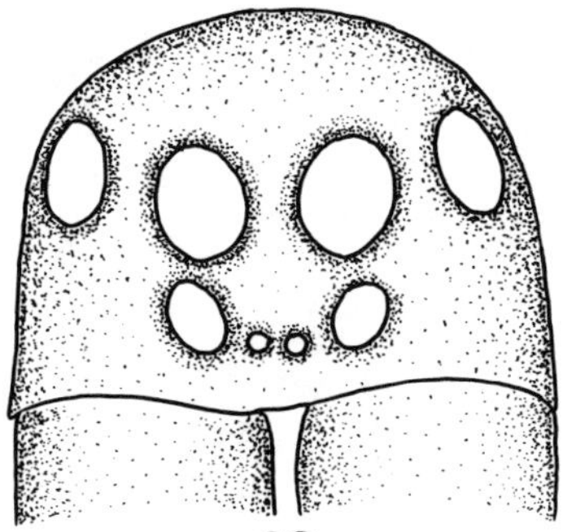

88a

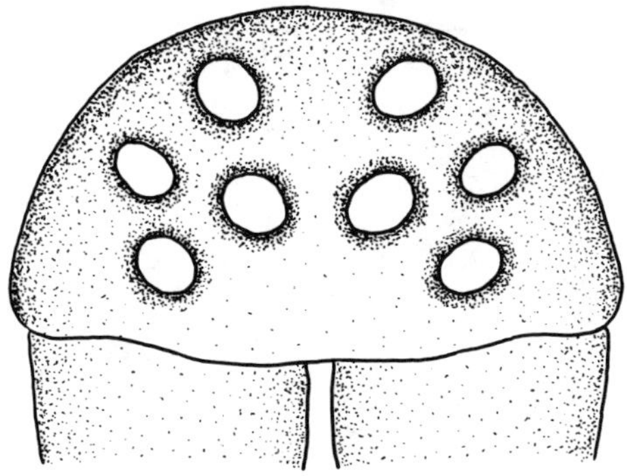

88b

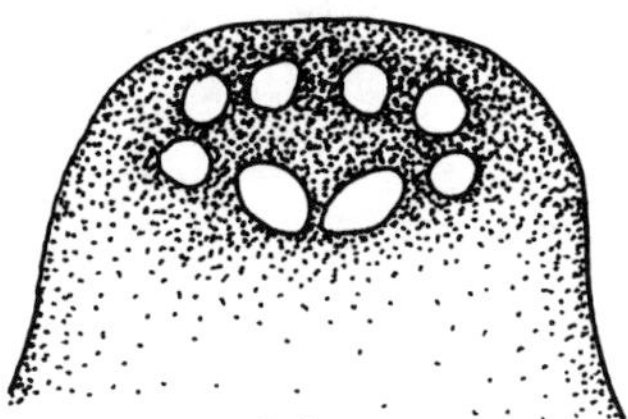

89a

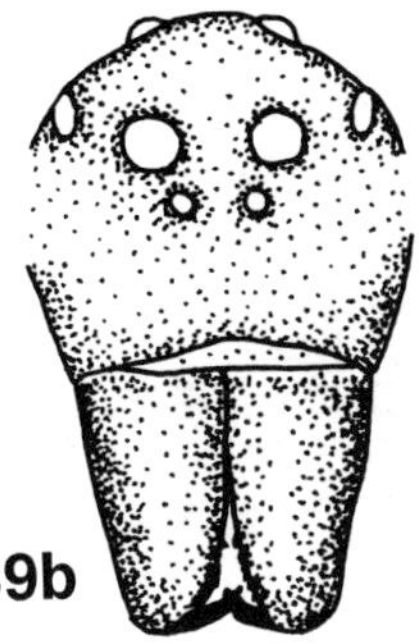

89b

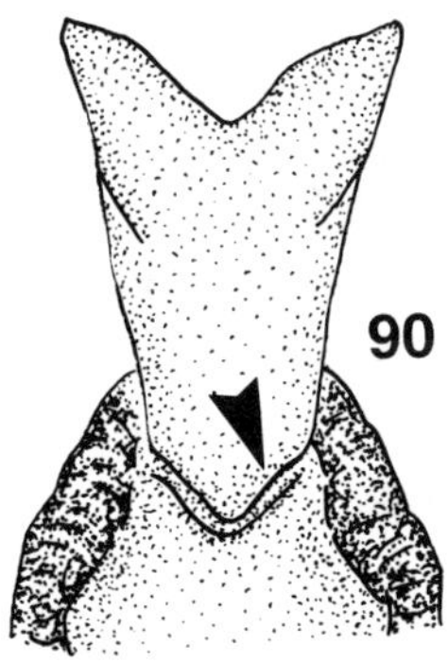

90

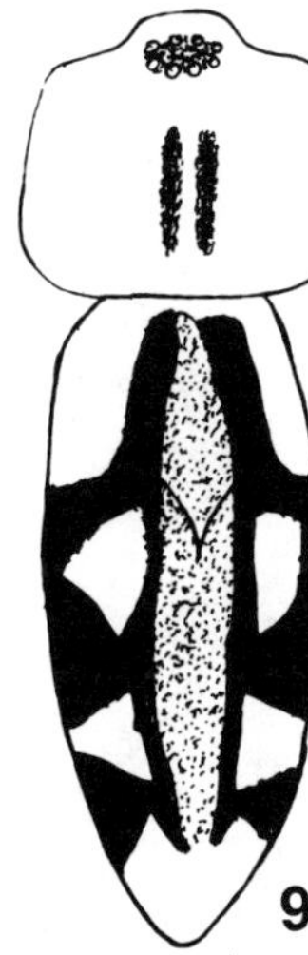

91

Figs. 88-91. Fig. 88: Eye arrangement, from in front. a. *Tegenaria*, b. *Melpomene*. Fig. 89: Oxyopidae, eye arrangement. a. dorsal view, b. face and chelicerae. Fig. 90: Lycosidae, plates on the dorsal side of the pedicel (lorum), an arrow indicates their division. Fig. 91: *Tapinillus*, colour pattern, dorsal view.

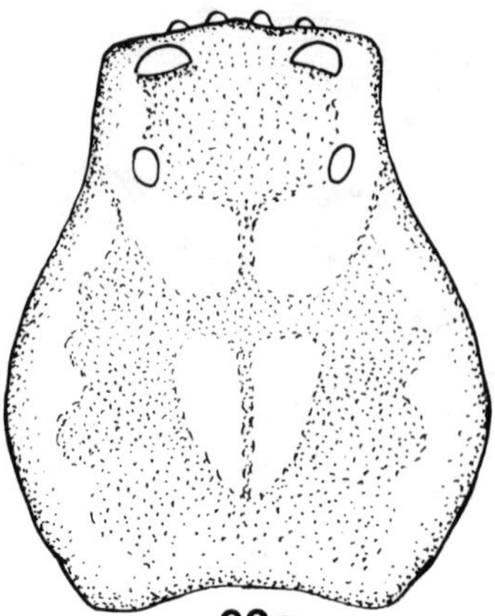

92a

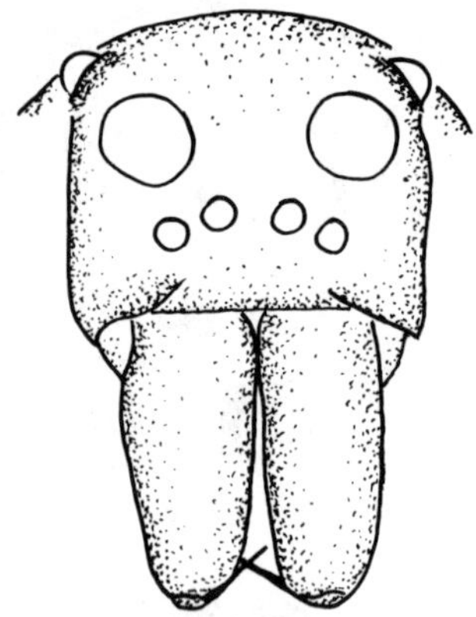

92b

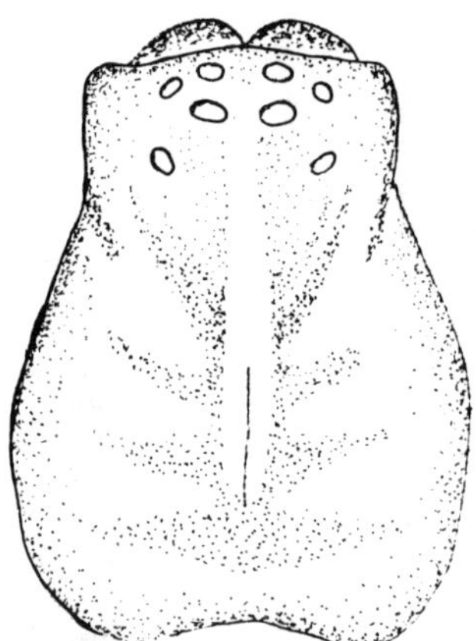

93a

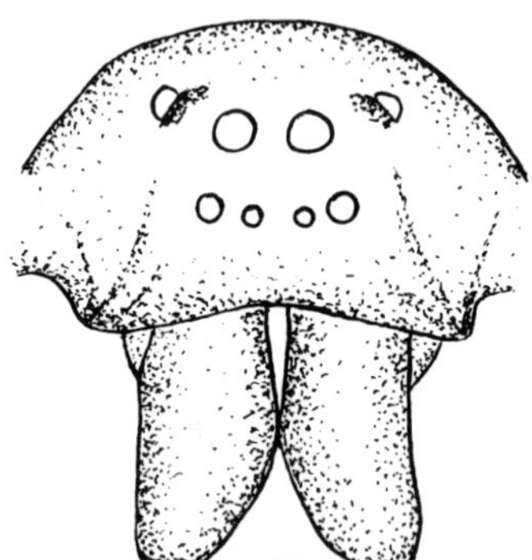

93b

Figs. 92-93. Fig. 92: Lycosidae. a. prosoma, dorsal view, b. face and chelicerae from in front. Fig. 93: Pisauridae. a. prosoma, dorsal view, b. face and chelicerae from in front.

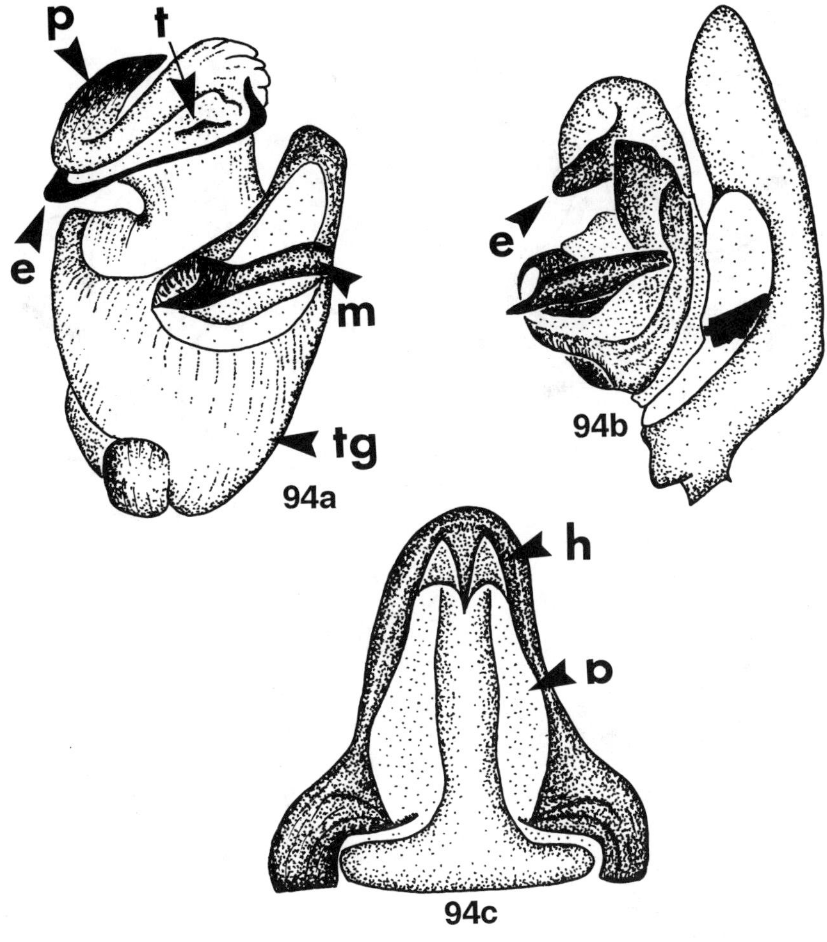

Fig. 94: Lycosidae. a., b., male pedipalpus, e embolus, ma median apophysis, p palea, t terminal apophysis, tg tegulum, c. epigyne, a atrium, h hood.

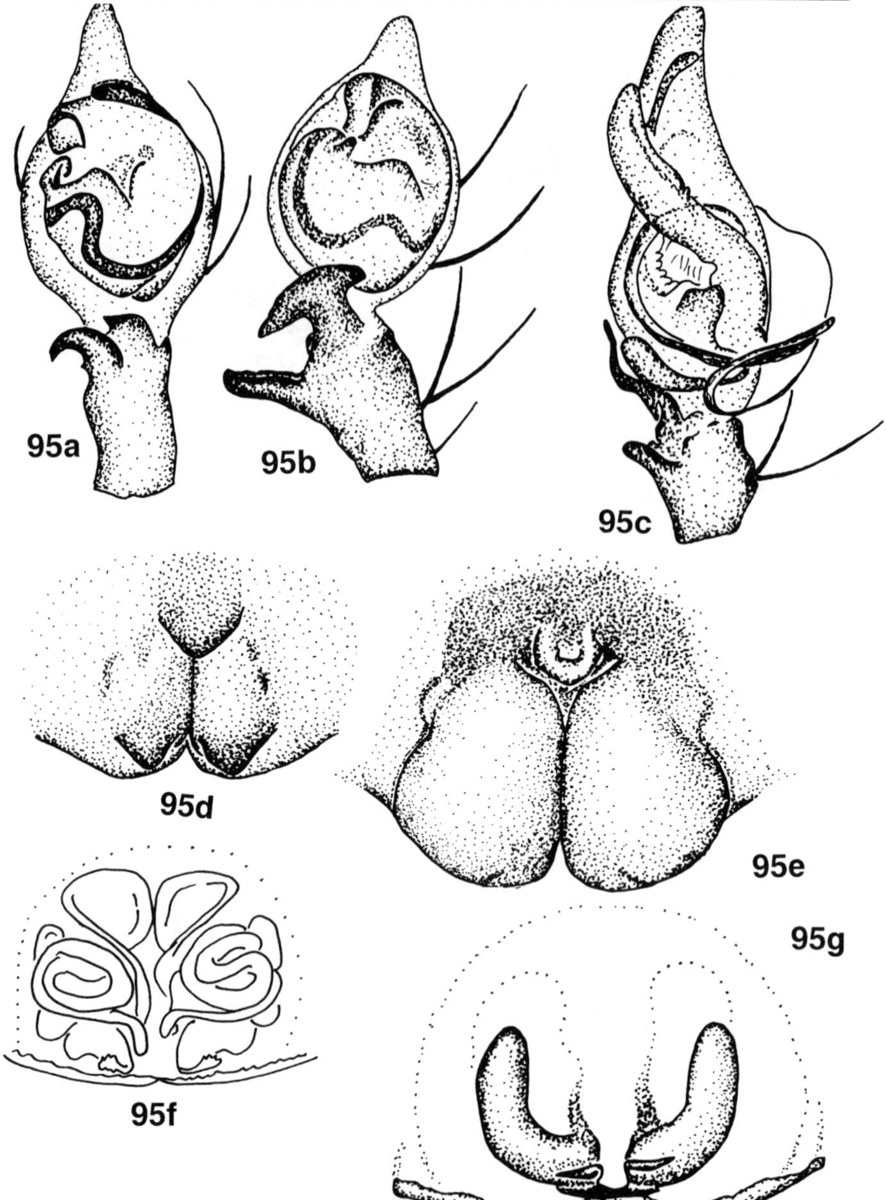

Fig. 95: Pisauridae, a.-c. male pedipalpus, a. *Architis*, b. *Staberius*, c. *Thaumasia*, d. *Staberius*, epigynum, ventral side, e. *Tinus*, epigynum, ventral side, f. *Thaumasia*, dorsal (internal) side of epigynum, g. *Architis*, dorsal (internal) side of epigynum.

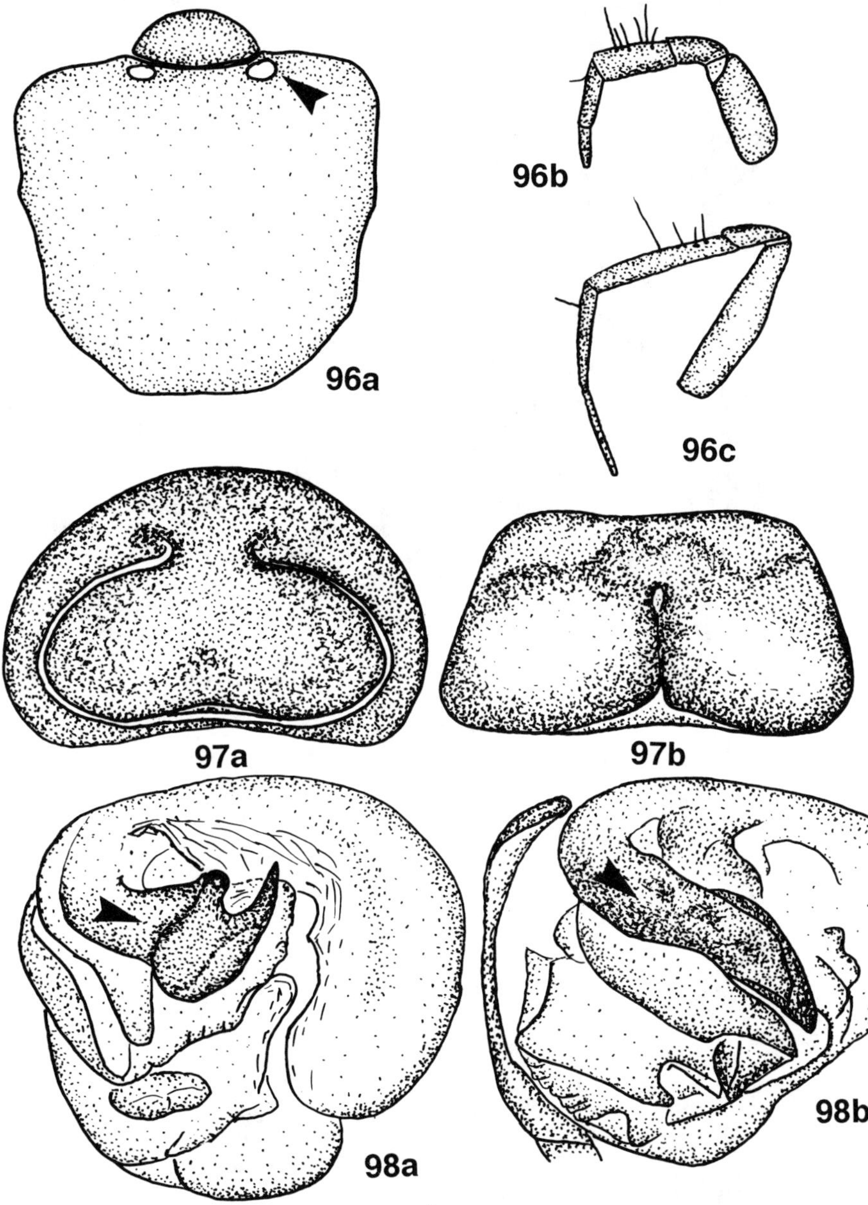

Figs. 96-98. Fig. 96 a.: *Theridiosoma*, sternum, labium and sternal pits (arrow), b., c. Theridiosomatidae, leg I. b. relatively short and thick, c. relatively long and slender. Fig. 97: Epigyne, ventral view. a. *Naatlo*, b. *Epeirotypus*. Fig. 98: Male pedipalpus, embolus indivated by arrows. a. *Naatlo*, subapical view, b. *Epeirotypus*, submesal view.

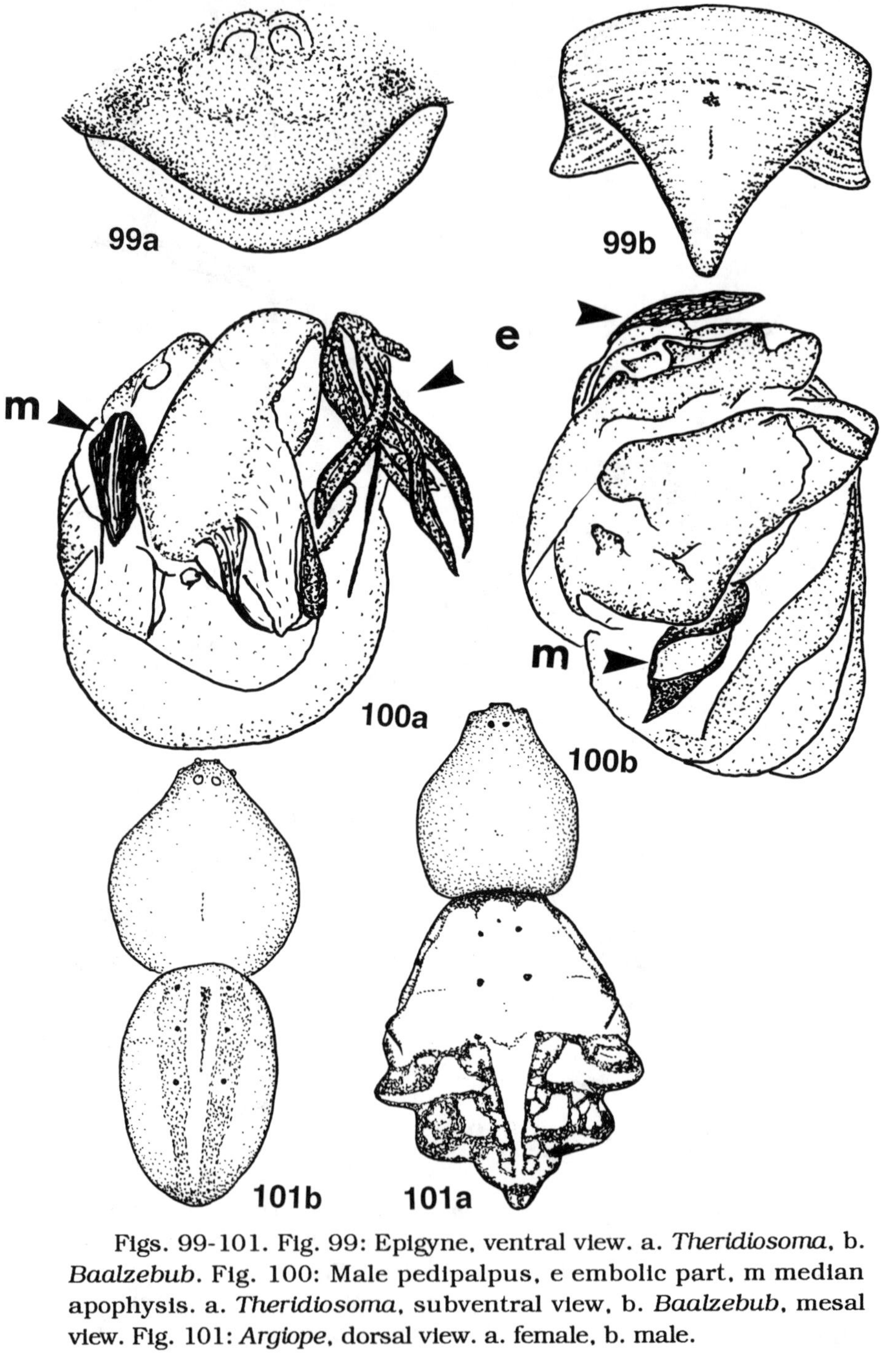

Figs. 99-101. Fig. 99: Epigyne, ventral view. a. *Theridiosoma*, b. *Baalzebub*. Fig. 100: Male pedipalpus, e embolic part, m median apophysis. a. *Theridiosoma*, subventral view, b. *Baalzebub*, mesal view. Fig. 101: *Argiope*, dorsal view. a. female, b. male.

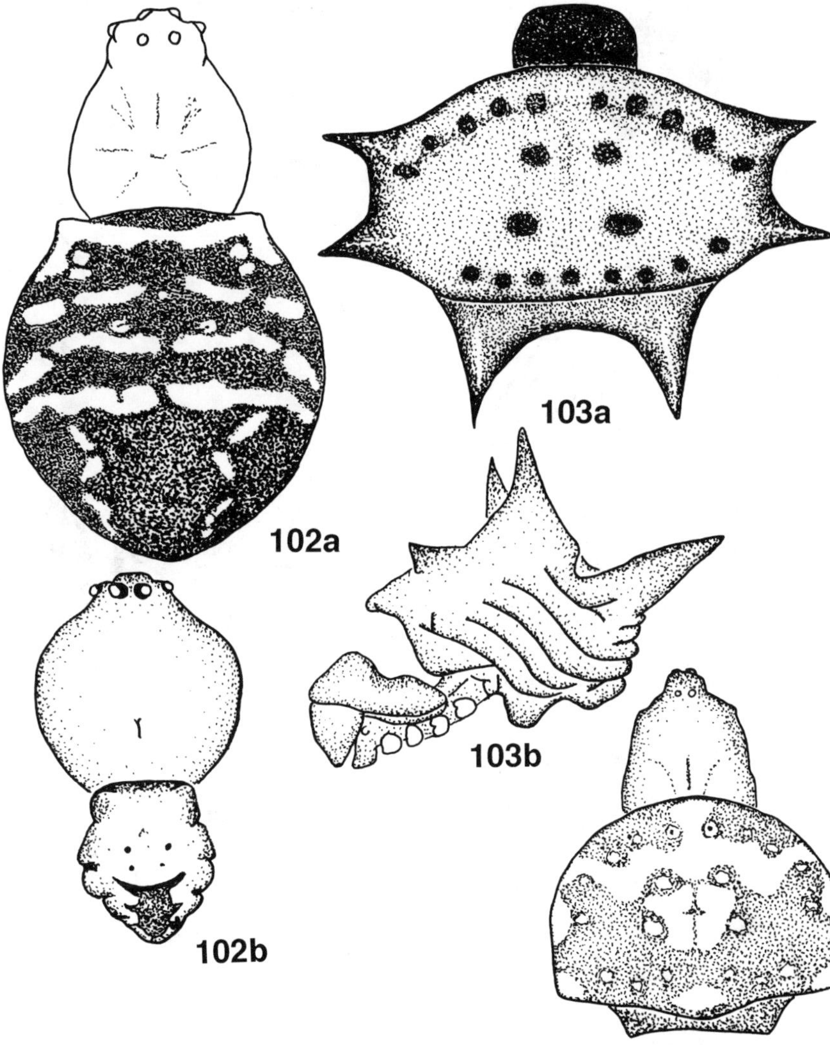

Figs. 102-103. Fig. 102: *Gea heptagon*, ventral view. a. female, b. male. Fig. 103: *Gasteracantha cancriformis*. a. female, ventral view, b. female, lateral view, c. male, dorsal view.

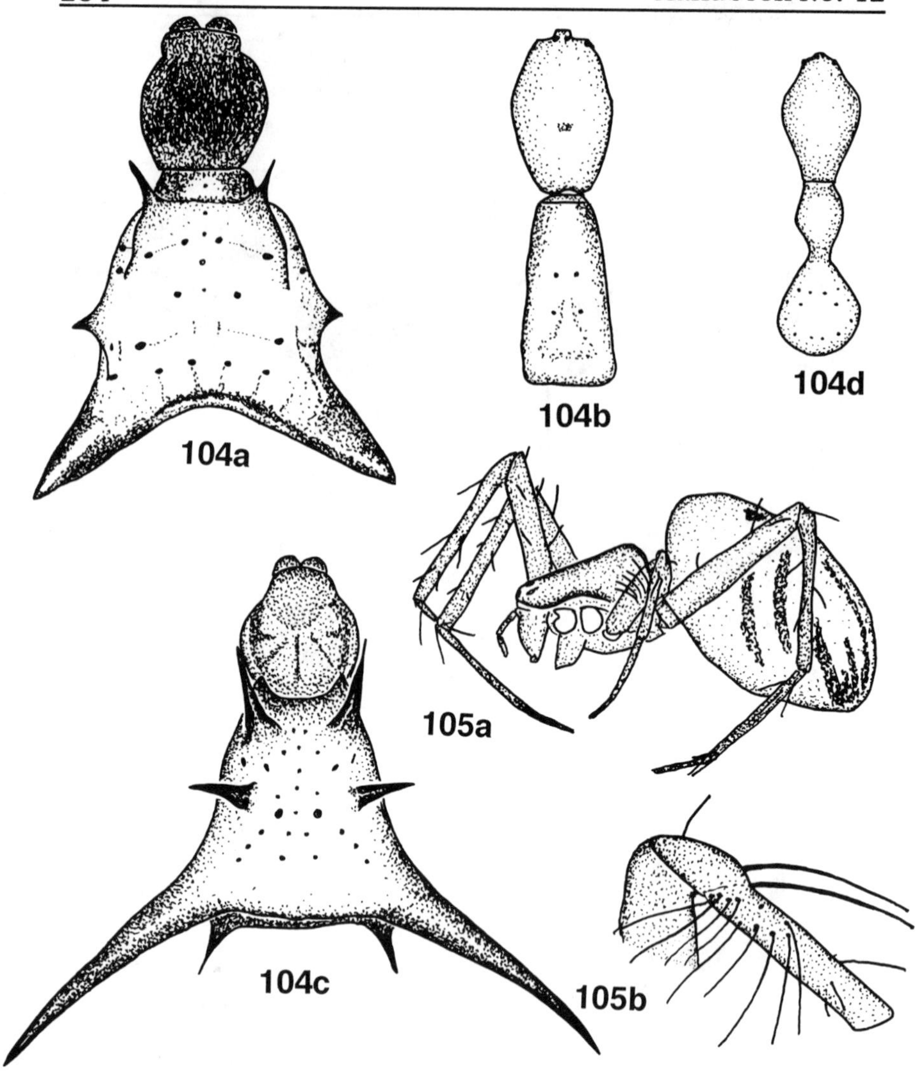

Figs. 104-105. Fig. 104: *Micrathena*, dorsal view. a., c. female, b., d. male. Fig. 105: *Mangora*. a. female, lateral view, b. tibia III showing brush of long trichobothria.

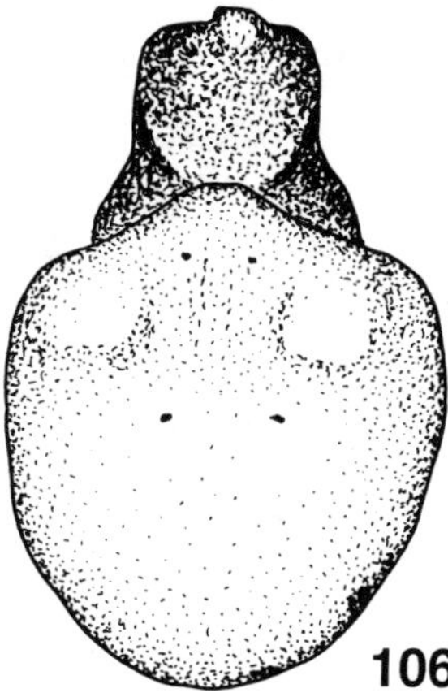

106a

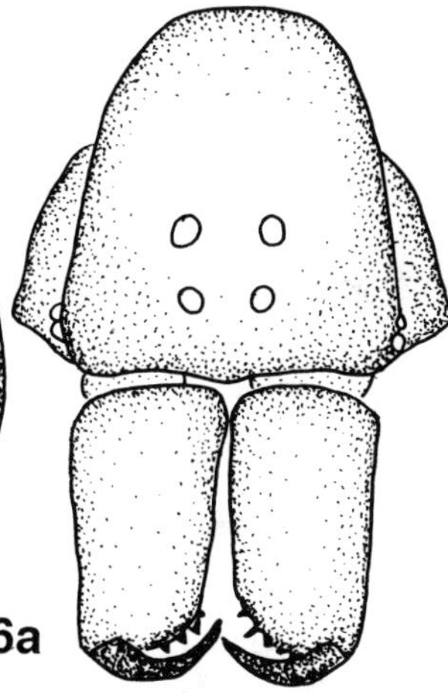

106b

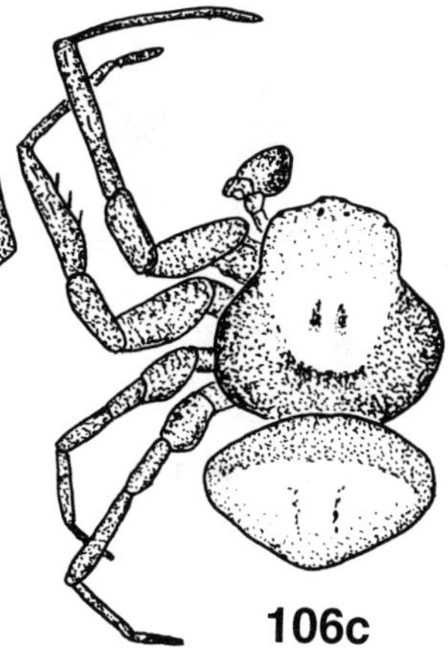

106c

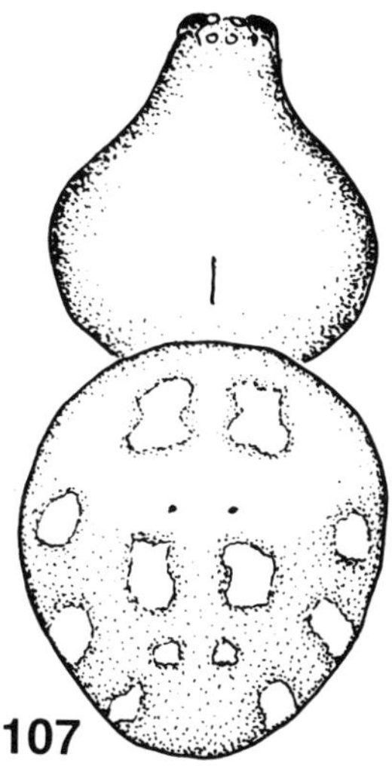

107

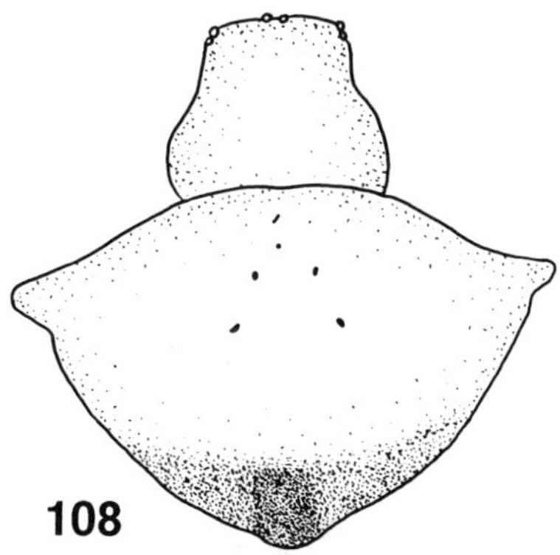

108

Figs. 106-108. Fig. 106: *Scoloderus*. a. female, dorsal view, b. female, face and chelicerae from in front, c. male, dorsal view. Fig. 107: *Spilasma*, female, dorsal view. Fig. 108: *Epeiroides bahiensis*, female, dorsal view.

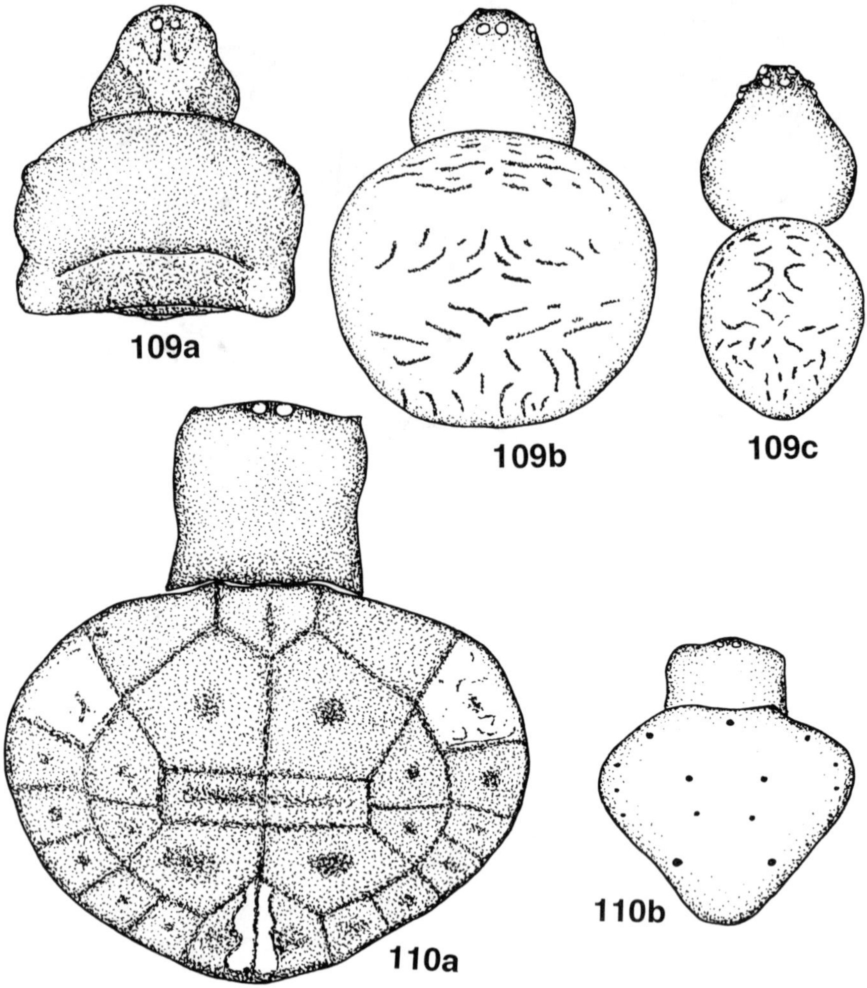

Figs. 109-110. Fig. 109: *Bertrana*, dorsal view. a., b. female, c. male. Fig. 110: *Hypognatha*, dorsal view. a. female, b. male.

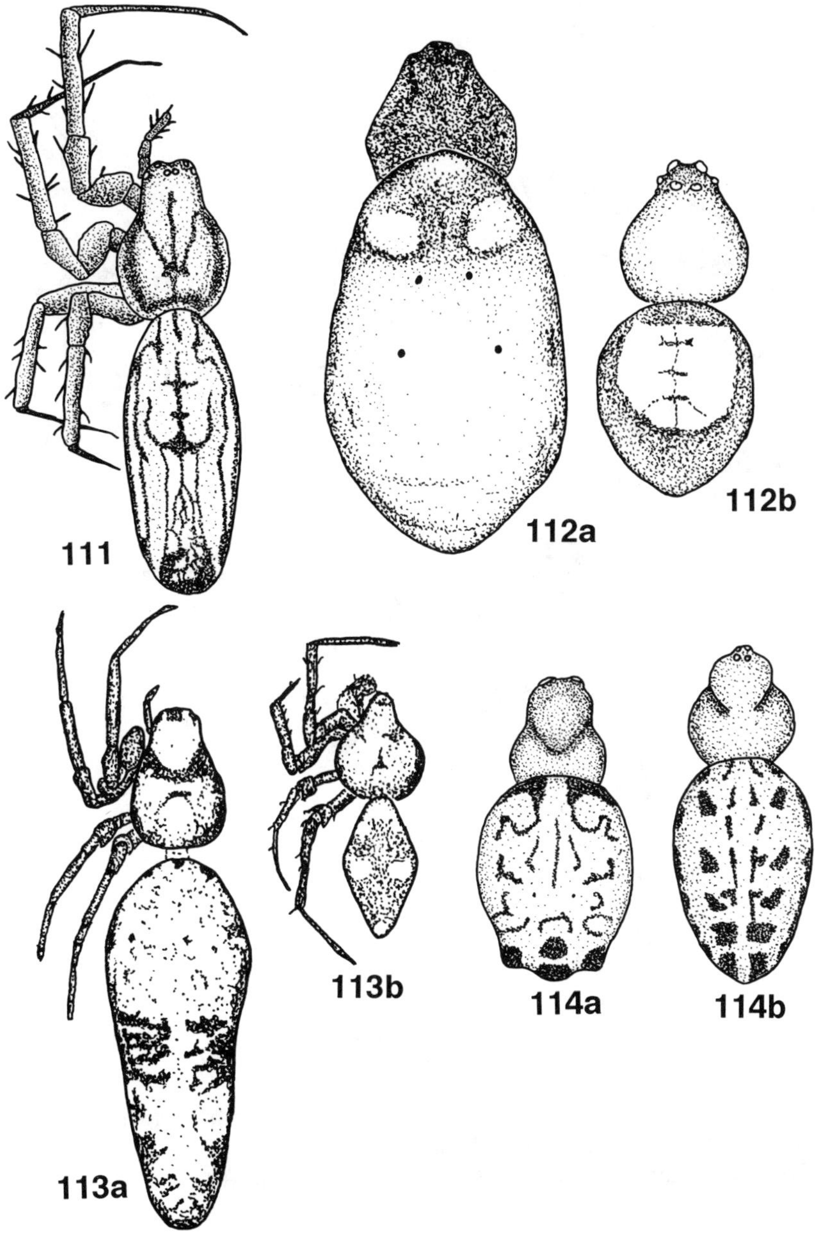

Figs. 111-114. Fig. 111: *Mecynogea*, female, dorsal view. Fig. 112: *Cyrtophora*, dorsal view. a. female, b. male. Fig. 113: *Cyclosa*, dorsal view. a. female, b. male. Fig. 114 a., b.: *Alpaida*, females, dorsal view.

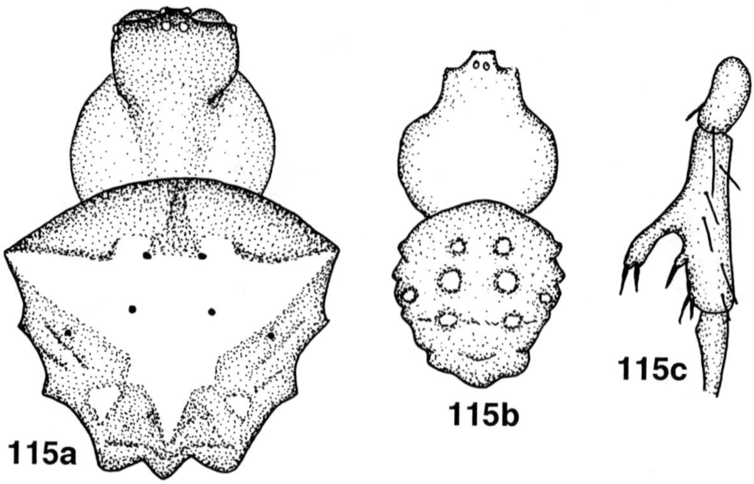

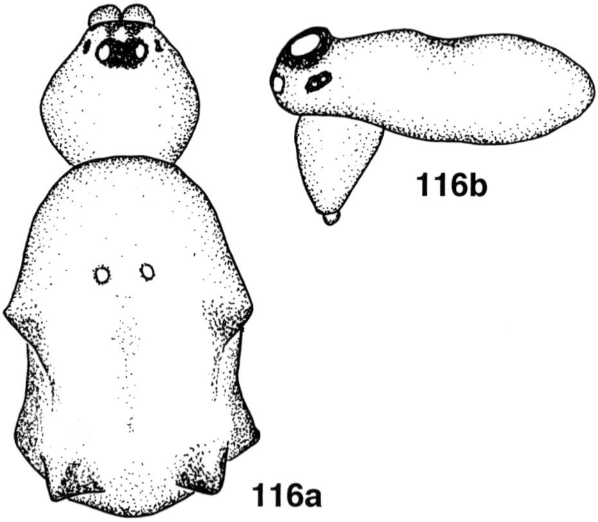

Figs. 115-116. Fig. 115: *Verrucosa*. a. female, dorsal view, b. male, dorsal view, c. male tibia II with spur, lateral view. Fig. 116: *Pronous*, female. a. dorsal view, b. prosoma, lateral view.

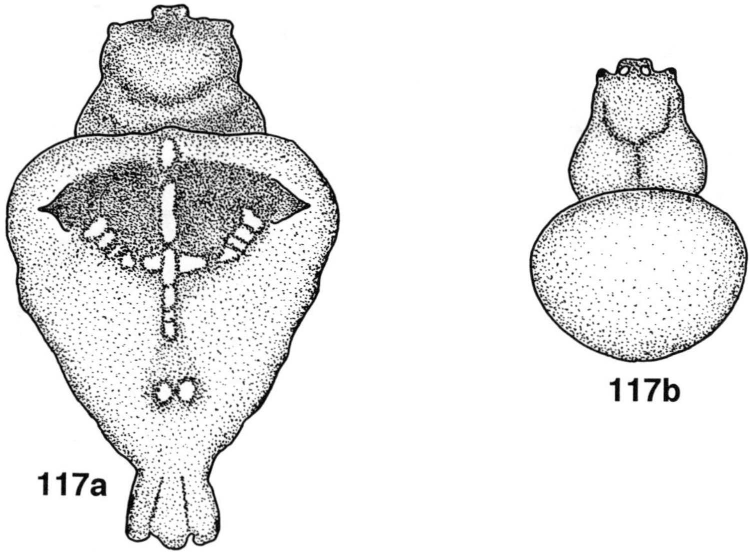

117a

117b

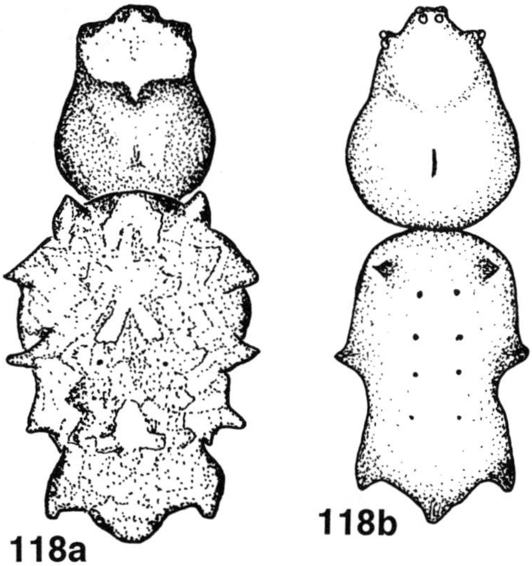

118a

118b

Figs. 117-118. Fig. 117: *Witica*, dorsal view. a. female, b. male. Fig. 118: *Wagneriana*, dorsal view. a. female, b. male.

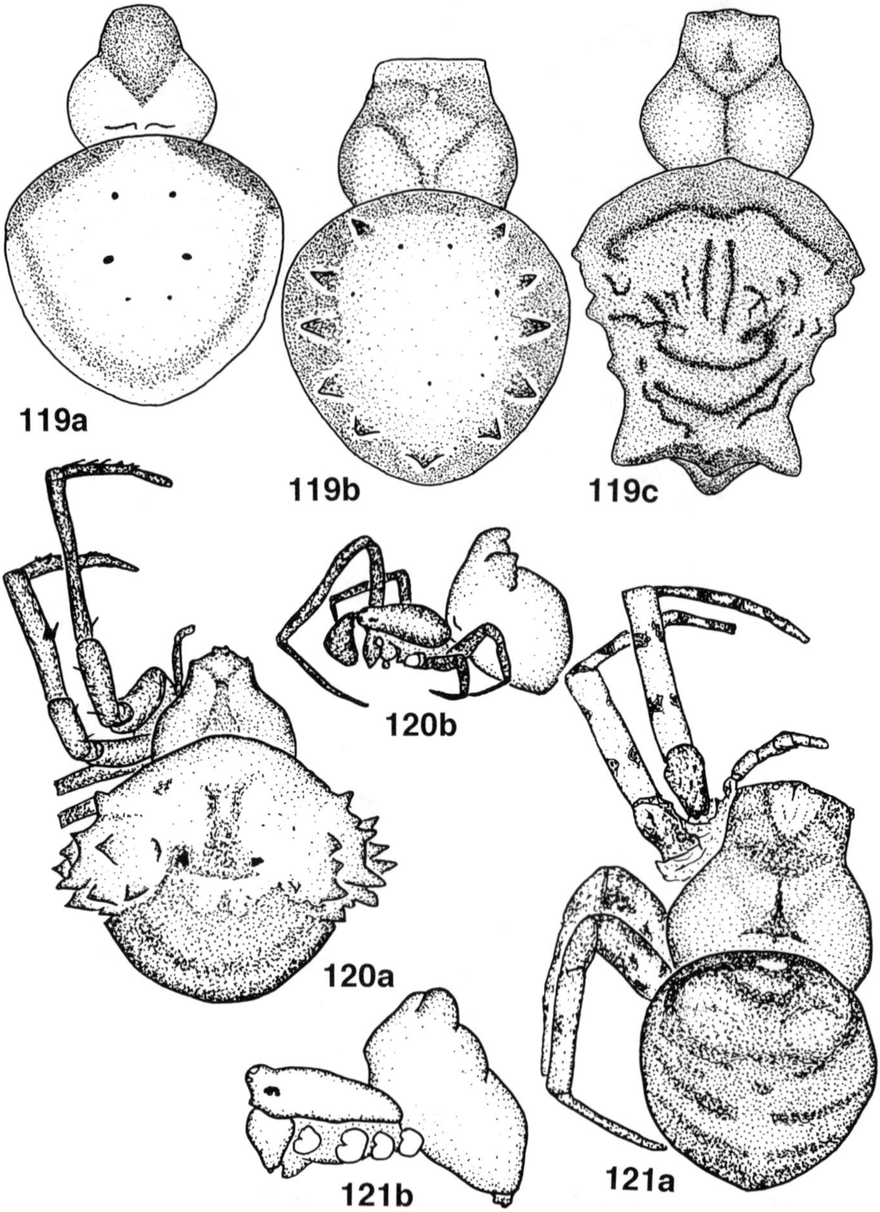

Figs. 119-121. Fig. 119a. - c.: *Parawixia*, females, dorsal view. Fig. 120: *Kaira*. a. female, dorsal view, b. male, lateral view. Fig. 121: *Wixia*. a. female, dorsal view, b. male, lateral view.

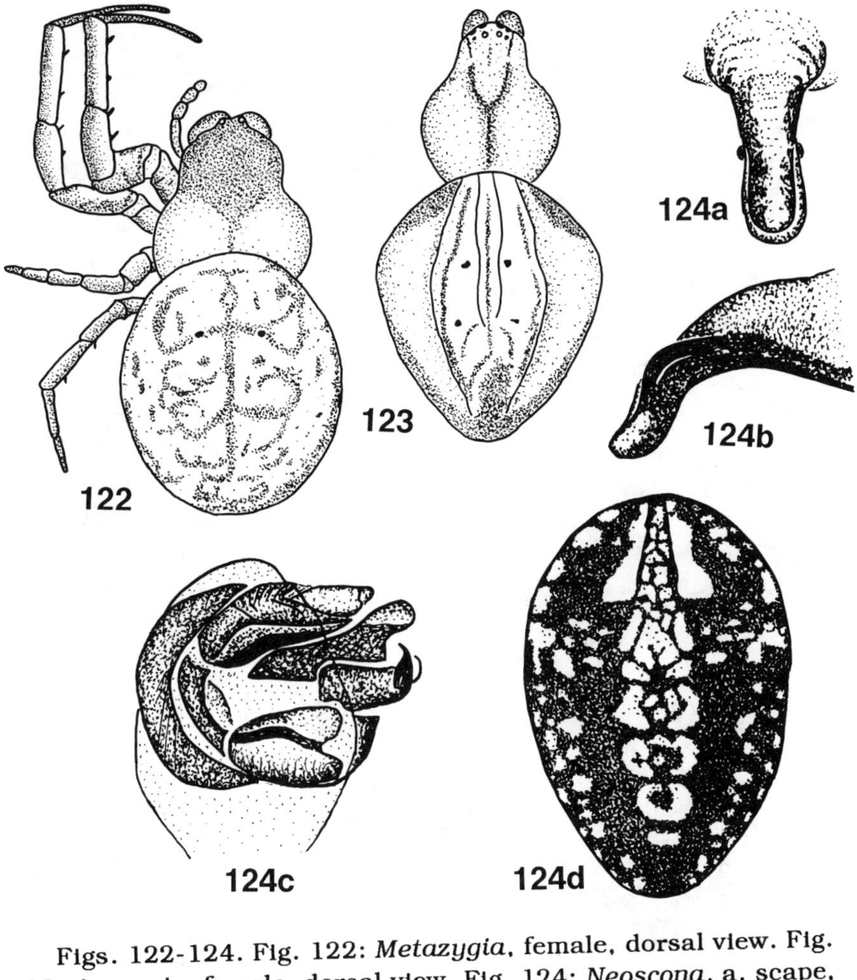

Figs. 122-124. Fig. 122: *Metazygia*, female, dorsal view. Fig. 123: *Acacesia*, female, dorsal view. Fig. 124: *Neoscona*. a. scape, ventral view, b. scape, lateral view, c. male pedipalpus with transparent cymbium, d. female opisthosoma, dorsal view.

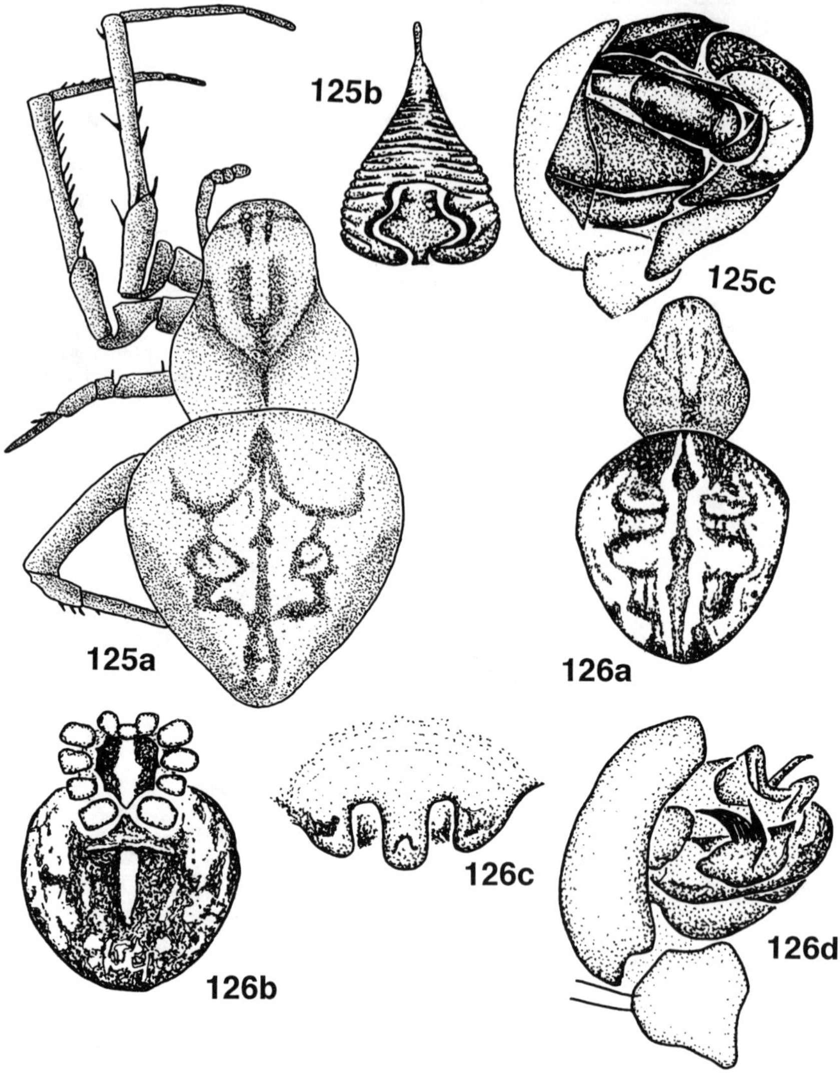

Figs. 125-126. Fig. 125: *Eustala*. a. female, dorsal view, b. epigyne, ventral view, c. male pedipalpus. Fig. 126: *Metepeira*. a. female, dorsal view, b. female, ventral view, c. epigyne, d. male pedipalpus.

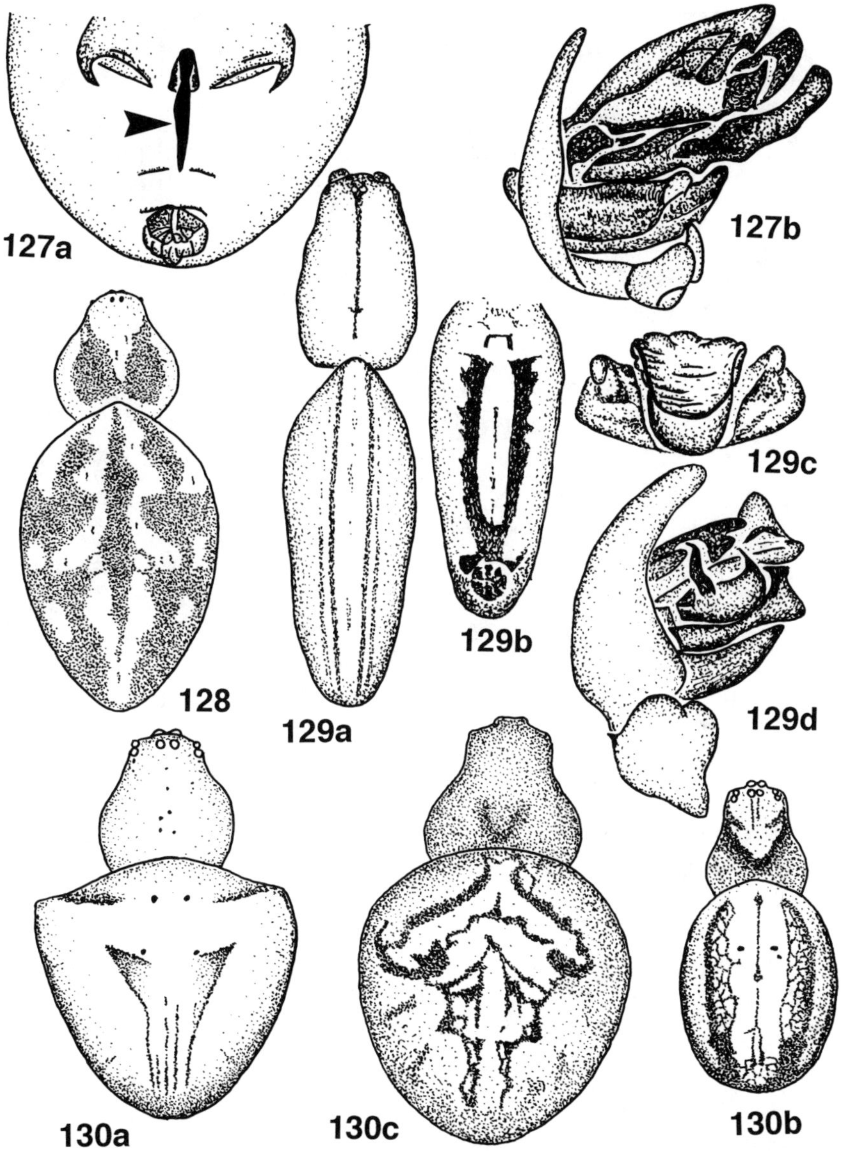

Figs. 127-130. Fig. 127: *Eriophora*. a. opisthosoma, ventral view, showing long scape (arrow), b. male pedipalpus. Fig. 128: *Aculepeira*, female, dorsal view. Fig. 129: *Larinia*. a. female, dorsal view, b.opisthosoma, ventral view, c. epigyne, ventral view, d. male pedipalpus. Fig. 130 a. - c: *Araneus*, females, dorsal view.

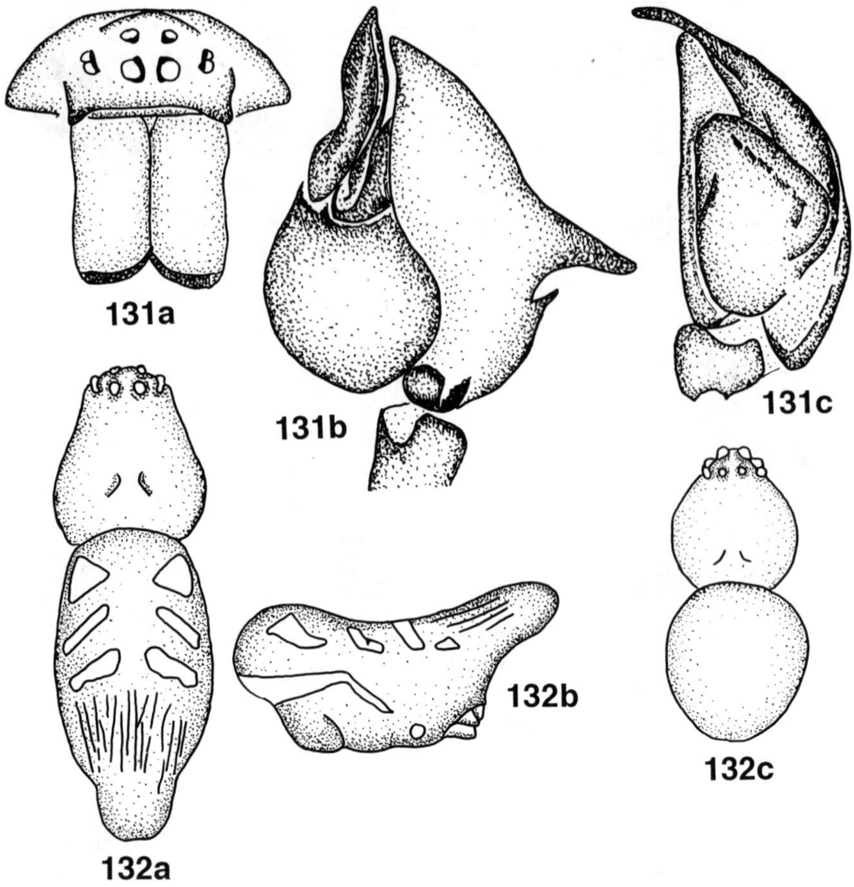

Figs. 131-132. Fig. 131: *Leucauge argyra*. a. female face and chelicerae, from in front, b. c. male pedipalpus. Fig. 132: *Mecynometa*. a. female, dorsal view, b. female, lateral view, c. male, dorsal view.

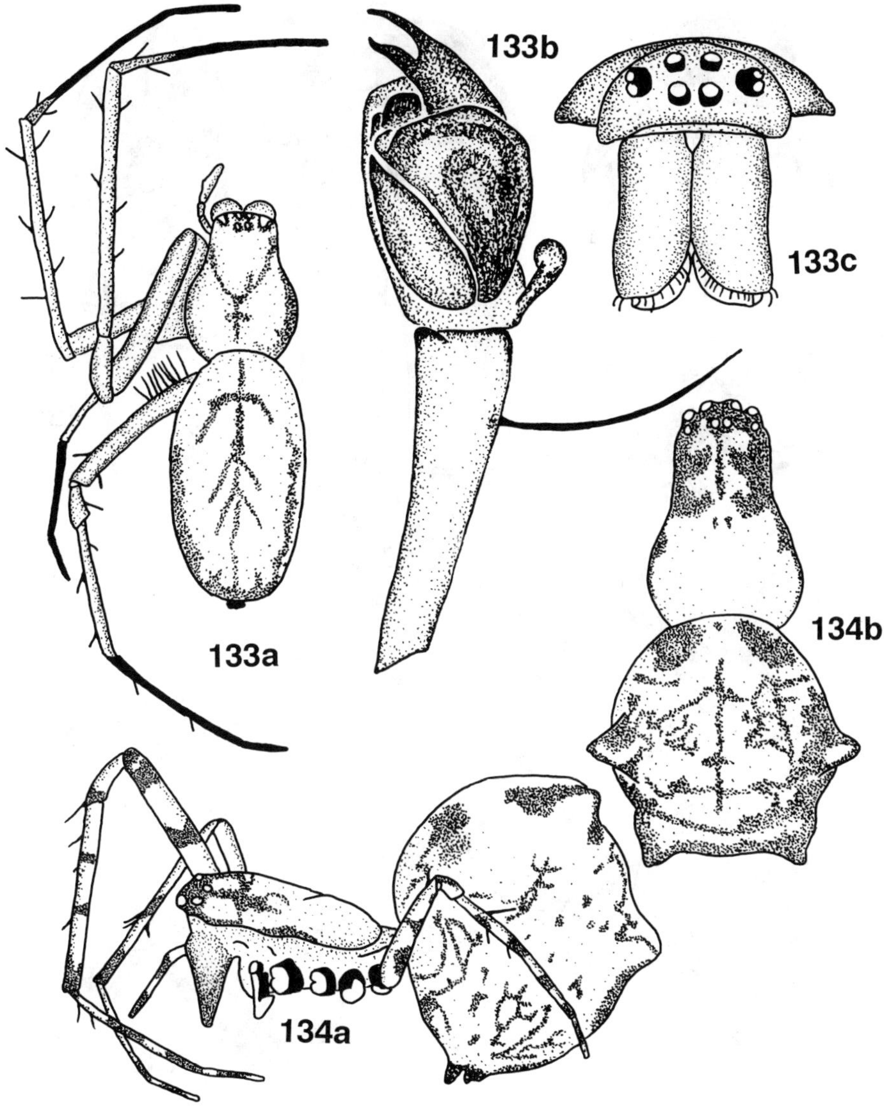

Figs. 133-134. Fig. 133: *Leucauge*. a. female, dorsal view, b. male pedipalpus, c. female face and chelicerae, from in front. Fig. 134: *Dolichognatha*. a. female lateral view, b. female, dorsal view. (*See p. 266 for c-e.*)

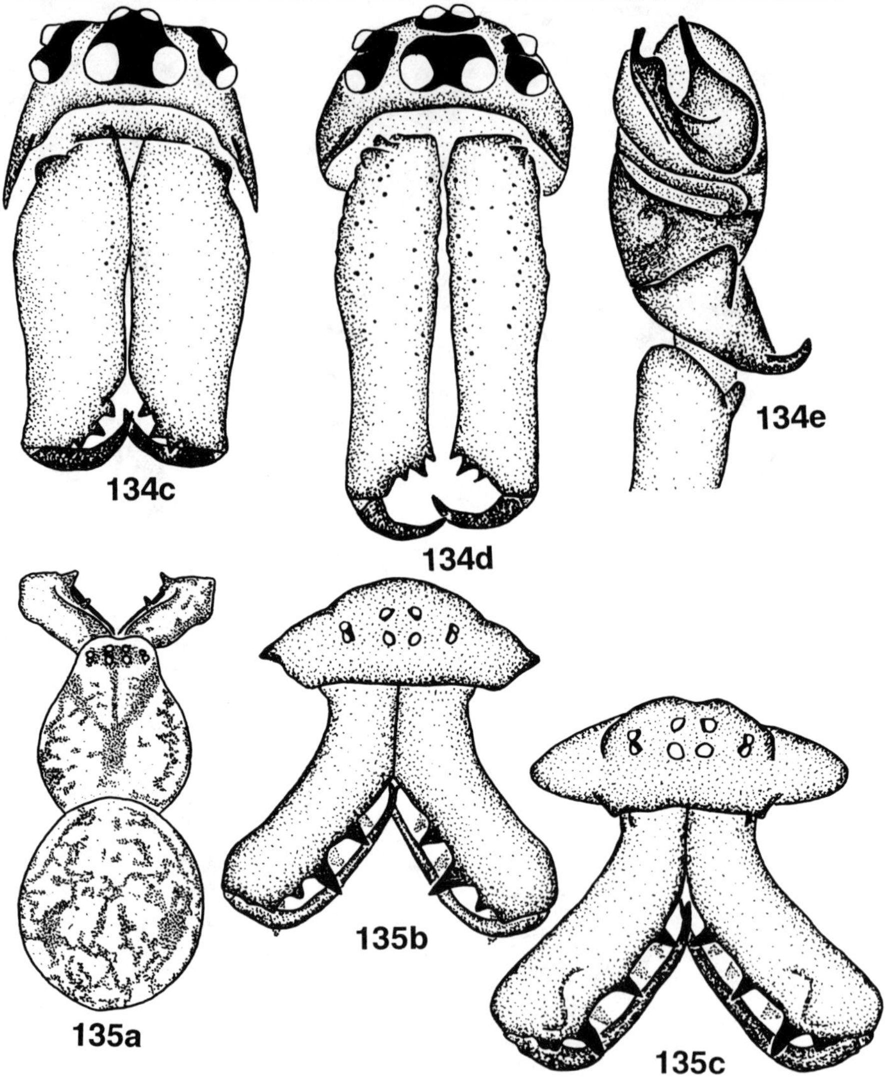

Figs. 134-135. Fig. 134 c. female face and chelicerae, from in front, d. male face and chelicerae, from in front, e. male pedipalpus. Fig. 135: *Glenognatha*. a. female, dorsal view, b. - c. face and chelicerae, from in front. (*See p. 267 for d.*)

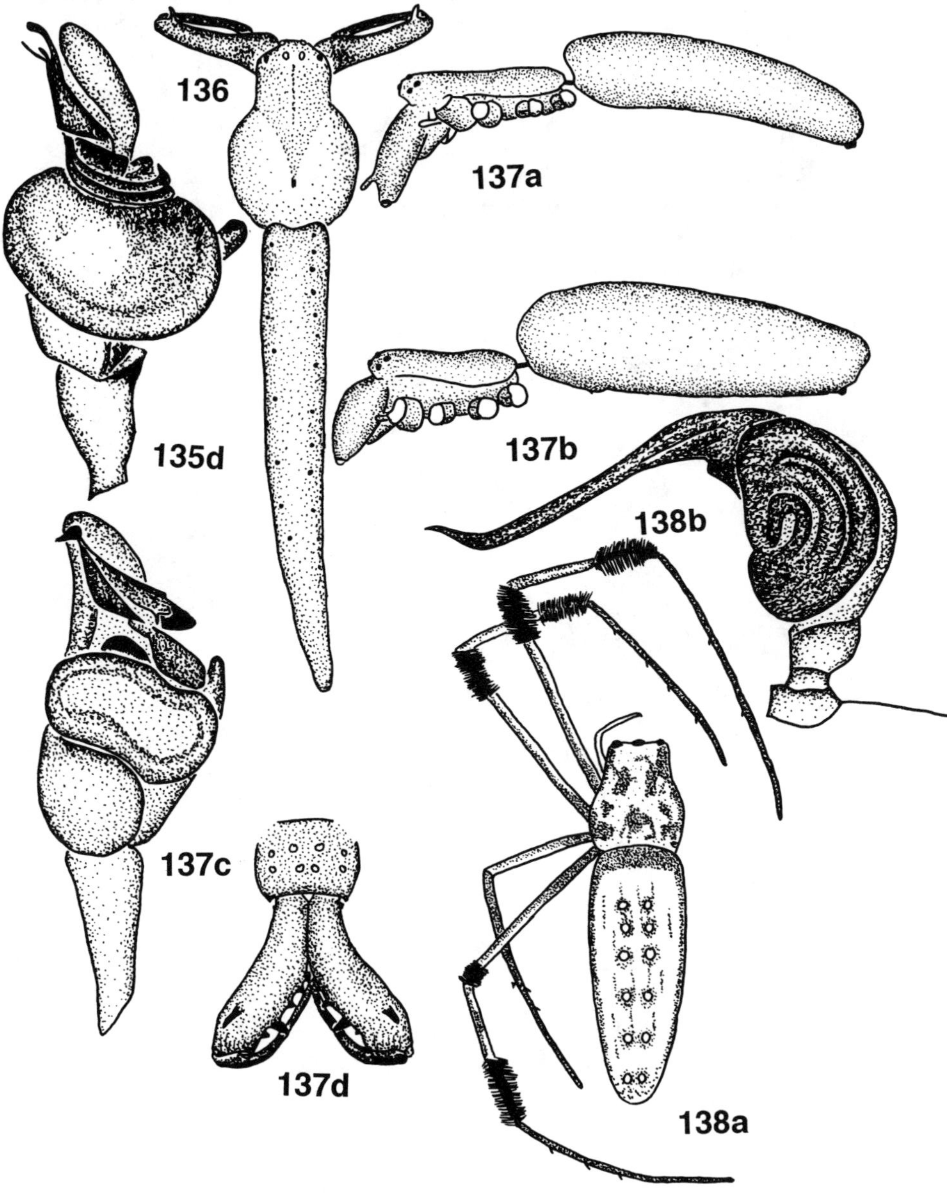

Figs. 135-138. Fig. 135 d. male pedipalpus. Fig. 136: *Cyrtognatha*, dorsal view. Fig. 137: *Tetragnatha*. a. male, lateral view, b. female, lateral view, c. male pedipalpus, d. face and chelicerae, from in front. Fig. 138: *Nephila clavipes*. a. female, dorsal view, b. male pedipalpus.

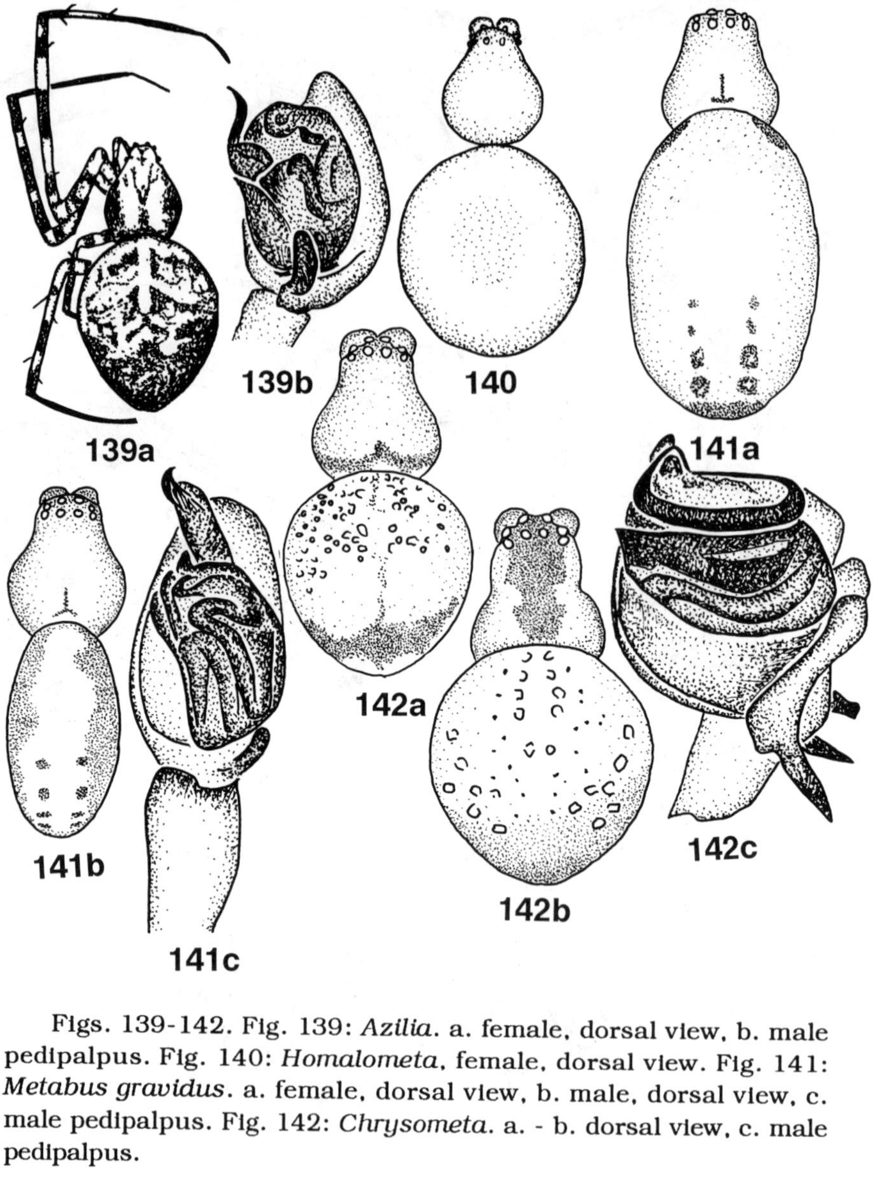

Figs. 139-142. Fig. 139: *Azilia*. a. female, dorsal view, b. male pedipalpus. Fig. 140: *Homalometa*, female, dorsal view. Fig. 141: *Metabus gravidus*. a. female, dorsal view, b. male, dorsal view, c. male pedipalpus. Fig. 142: *Chrysometa*. a. - b. dorsal view, c. male pedipalpus.

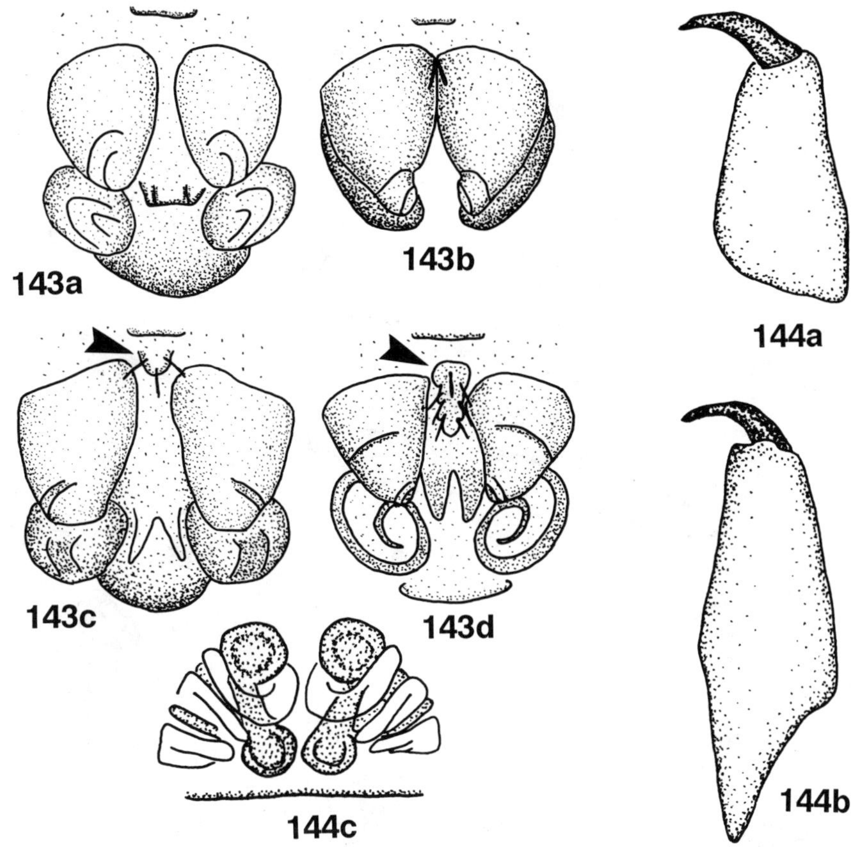

Figs. 143-144. Fig. 143: Spinnerets and colulus (arrow), ventral view. a. no colulus, b. colulus replaced by two setae, c. colulus with three setae, d. colulus with seven setae. Fig. 144: *Latrodectus*. a. left chelicera, posterior view, b. left chelicera, frontal view, c. female genitalia, dorsal view.

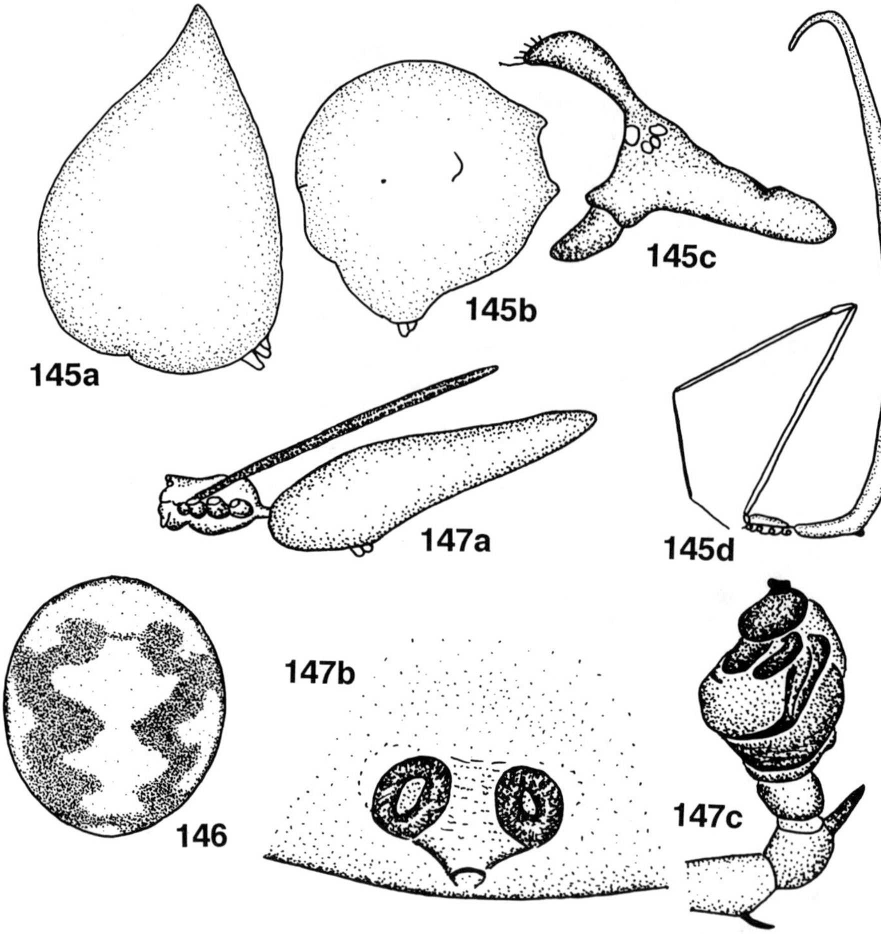

Figs. 145-147. Fig. 145: *Argyrodes*. a. *projiciens*, opisthosoma, lateral view, b. *dracus*, opisthosoma, lateral view, c. *projiciens*, male prosoma, lateral view, d. *attenuatus*, female. Fig. 146: *Steatoda*, female opisthosoma, dorsal view. Fig. 147: *Synotaxus*. a. female, lateral view, b. epigyne, ventral view, c. male pedipalpus.

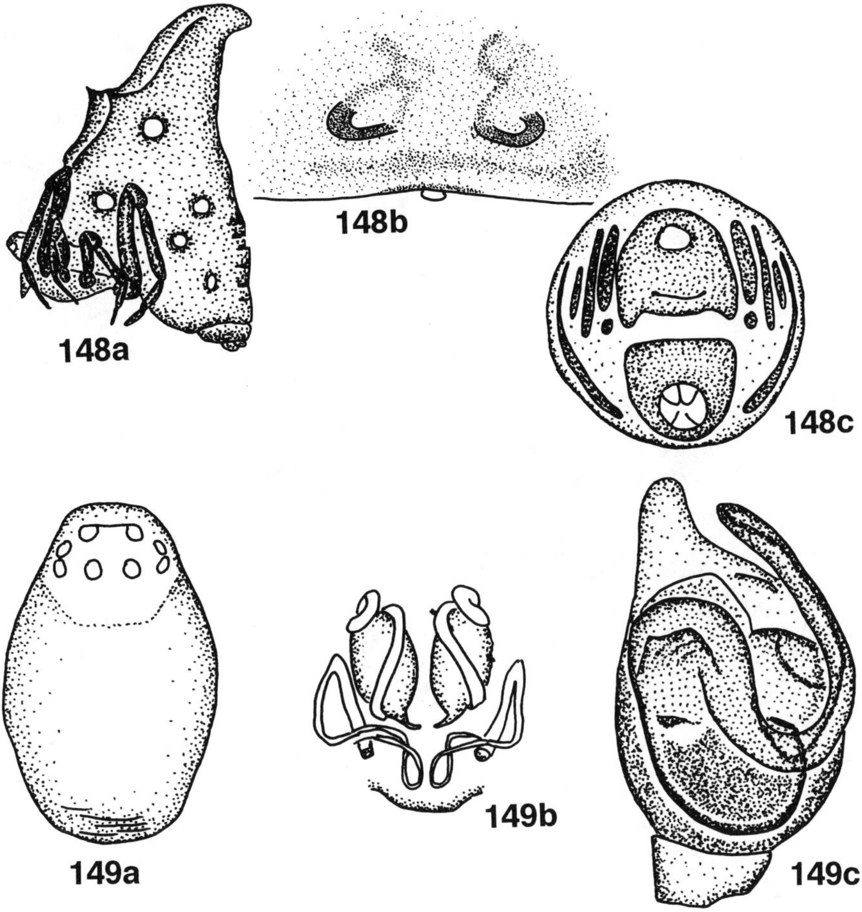

Figs. 148-149. Fig. 148: *Phoroncidia*. a. female, lateral view, b. epigyne, ventral view, c. male opisthosoma, ventral view. Fig. 149: *Helvibis*. a. female prosoma, dorsal view, female genitalia, dorsal view, c. male pedipalpus.

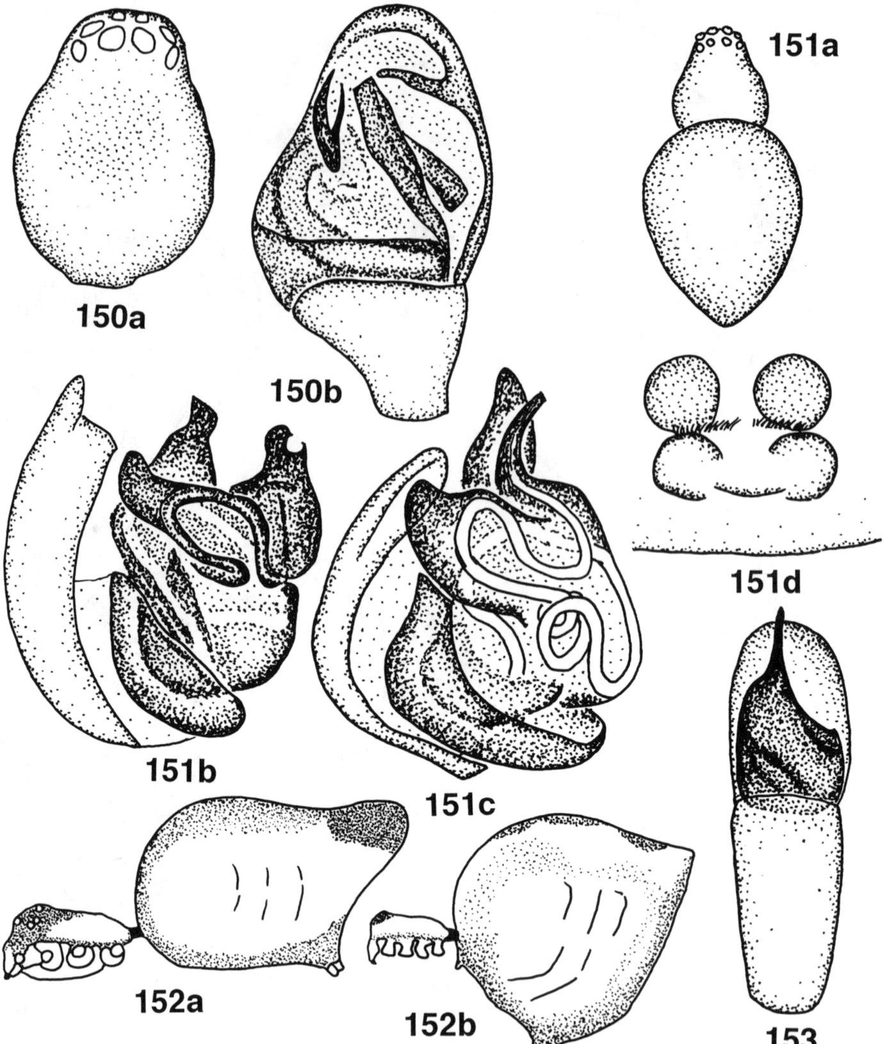

Figs. 150-153. Fig. 150: *Styposis*. a. female prosoma, dorsal view, b. male pedipalpus. Fig. 151: *Euryopis*. a. female, dorsal view, b., c. male pedipalpus, expanded, mesal view, d. epigyne, cleared. Fig. 152 a., b.: *Chrysso*, females, lateral view. Fig. 153: *Paratheridula*, male pedipalpus, ventral view.

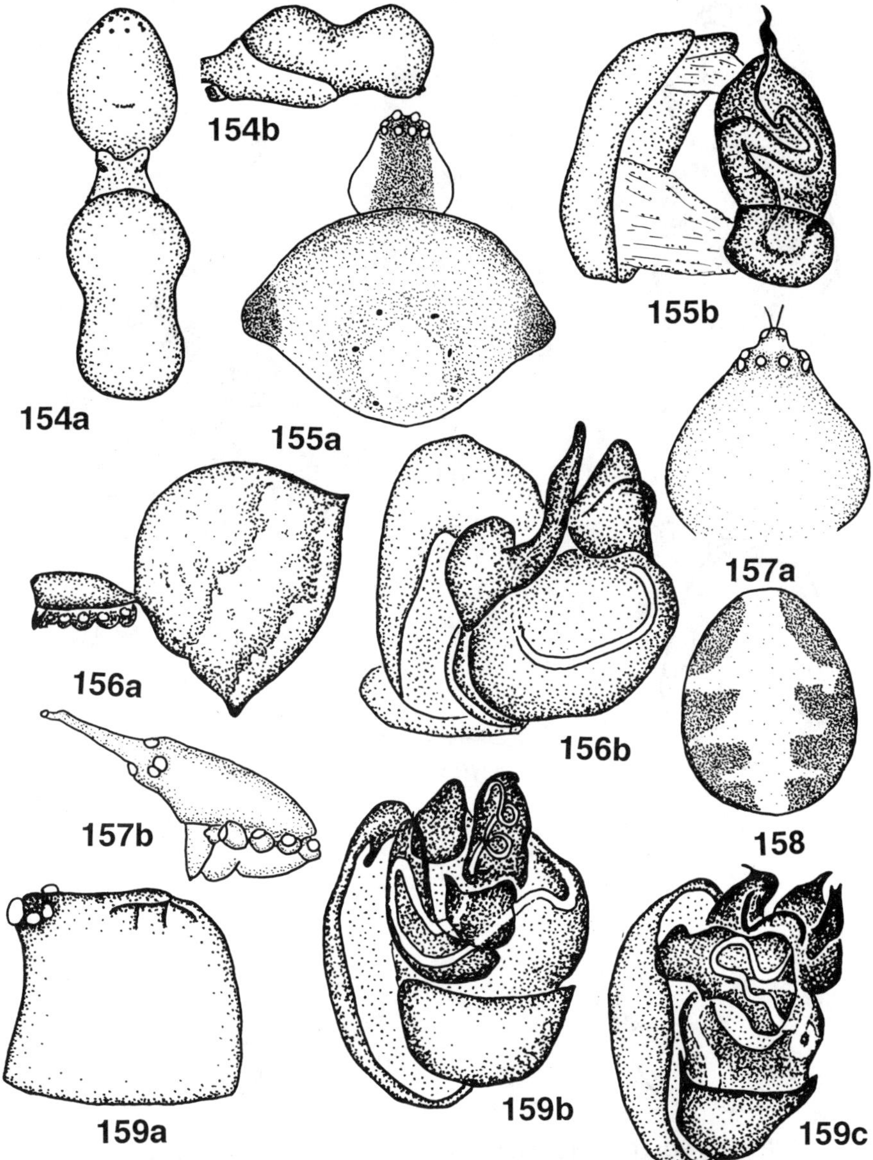

Figs. 154-159. Fig. 154: *Coleosoma*, male. a. dorsal view, b. opisthosoma, lateral view. Fig. 155: *Theridula*. a. female, dorsal view, b. male pedipalpus, expanded, mesal view. Fig. 156: *Achaearanea*. a. female, lateral view, b. male pedipalpus, expanded, subventral view. Fig. 157: *Thymoites*, male prosoma. a. dorsal view, b. lateral view. Fig. 158: *Theridion*, female opisthosoma, dorsal view. Fig. 159: *Dipoena*. a. male prosoma, lateral view, b. - c. male pedipalpus, expanded.

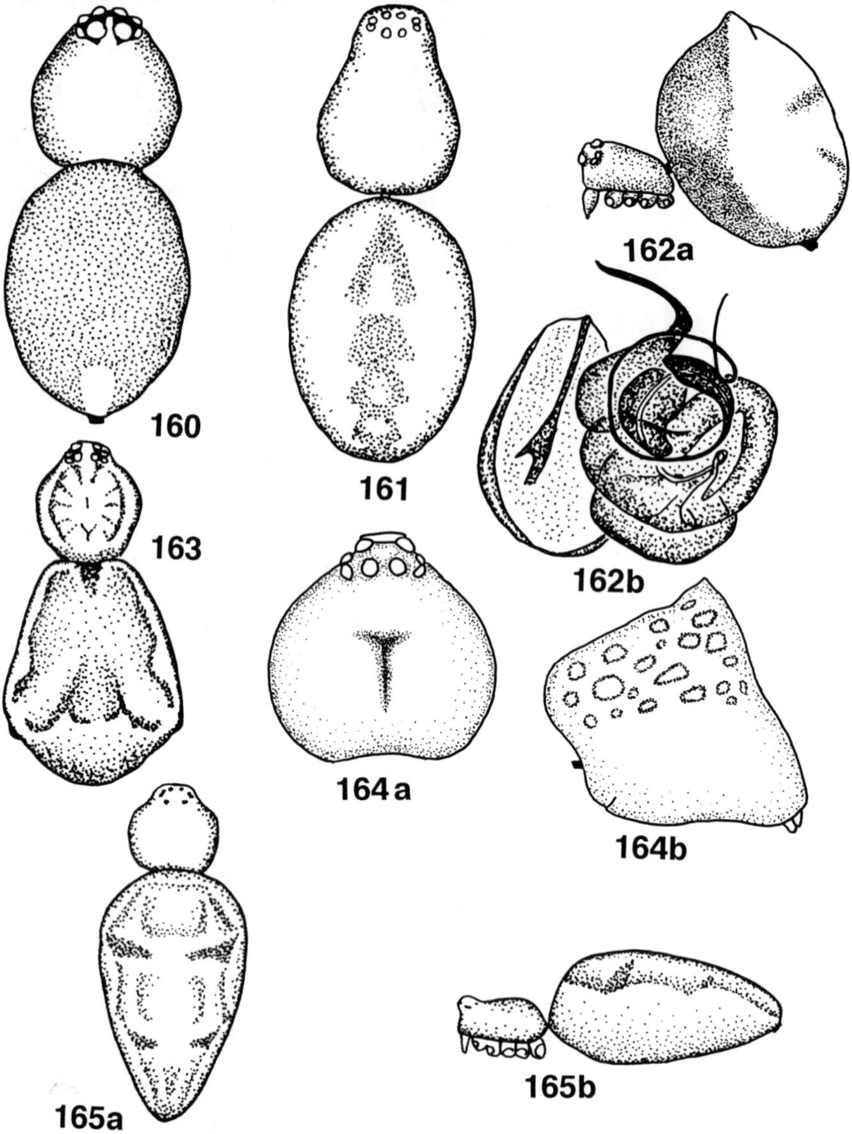

Figs. 160-165. Fig. 160: *Stemmops*, female, dorsal view. Fig. 161: *Anelosimus*, female, dorsal view. Fig. 162: *Chrosiothes*. a. female, lateral view, b. male pedipalpus, expanded, ventral view. Fig. 163: *Episinus*, female, dorsal view. Fig. 164: *Thwaitesia*, female. a. prosoma, dorsal view, b. opisthosoma, lateral view. Fig. 165: *Spintharus*, female. a. dorsal view, b. lateral view.